TOWARD A MORE
EXACT ECOLOGY

TOWARD A MORE EXACT ECOLOGY

The Second Jubilee Symposium to Celebrate
the 75th Anniversary of the British Ecological Society,
St Catherine's College, Oxford 13–15 September 1988
Published as the 30th Symposium of the Society

EDITED BY

PETER J. GRUBB
Botany School,
University of Cambridge

JOHN B. WHITTAKER
Department of Biological Sciences,
University of Lancaster

CAMBRIDGE UNIVERSITY PRESS

CAMBRIDGE UNIVERSITY PRESS
Cambridge, New York, Melbourne, Madrid, Cape Town, Singapore, São Paulo, Delhi

Cambridge University Press
The Edinburgh Building, Cambridge CB2 8RU, UK

Published in the United States of America by Cambridge University Press, New York

www.cambridge.org
Information on this title: www.cambridge.org/9780521839976

First published on behalf of the British Ecological Society by Blackwell Scientific Publications 1989
First published on behalf of the British Ecological Society by Cambridge University Press 2008
This digitally printed version 2008

A catalogue record for this publication is available from the British Library

ISBN 978-0-521-83997-6 hardback
ISBN 978-0-521-10063-2 paperback

CONTENTS

IV. *Evolutionary and Behavioural Ecology*

V. *Interrelationships between Organisms*

VI. *Ecosystem Ecology*

VII. *Applied Ecology*

FOREWORD

The British Ecological Society was founded in 1913, the first national ecological society in the world. For our jubilee 75 years later, we decided that the time was ripe for taking stock of the achievements of our subject, and looking forward toward new developments. Accordingly, the Society held two complementary symposia in 1988. The first was planned as an assessment of the contribution of ecology to understanding the natural world, under the title *Ecological Concepts;* it was held at University College London, the venue of the inaugural meeting of the society, on its anniversary date, 12–13 April. The second was focused on the future, and called *Toward a More Exact Ecology*, aimed at considering the most fruitful current approaches and technologies, and determining the major obstacles and likely profitable lines of advance; it was held at St Catherine's College, Oxford, on 13–15 September. The two symposia volumes thus represent a vision of the ecological community's understanding of the past, present and future of the subject; it is hoped that both will be of value in the planning of future research.

R. J. BERRY
President 1988–89

PREFACE

The last major celebration of the British Ecological Society was in 1963, when it reached its 50th year. It was decided then to mark the occasion by holding a jubilee symposium at University College London, and in due course the proceedings of that meeting were published in the form of a special supplement to the Society's journals. At that time the organizers considered that there were five main areas of development in ecology: ecology and conservation, quaternary ecology, production ecology, experimental and autecological studies, and finally the community concept.

Now, 25 years on, the 75th anniversary of the Society has been marked by two jubilee symposia. As explained in the foreword, the first reviewed the history and current state of ecological ideas, and the second looked forward to exciting new developments.

In its 75 years the Society has grown from a small vegetation committee to a 'broad church' of nearly 5000 members. As it has grown, it has embraced a vast range of expertise and interests. Selecting a limited number of high-profile topics, representative of advances in the subject as a whole, is necessarily a somewhat idiosyncratic task. In the last 25 years, however, ecology has emerged from being a predominantly descriptive subject to one with a more substantial theoretical framework underpinned by evidence from experiments. In this advance, improved technology has undoubtedly played a most significant role; few ecologists could operate without a computer to hand, while some branches of the subject require very sophisticated instrumentation. But we also owe to this period much greater precision in the formation of hypotheses, a more general search for mechanisms, increasing integration of ecology with other disciplines, and increasing demands for application of the lessons learned. These trends have, in part, determined the choice of our six main topics.

We have resisted the tendency for symposia to become more and more specialized, and have brought together botanists, microbiologists and zoologists, evolutionists and behaviourists, physiologists and system modellers to identify the ways in which their particular ecological fields are becoming more exact through application of new techniques, refinement of theoretical concepts, and bridge-building over the artificial divides between cognate disciplines. In doing so, we are constantly reminded of the absolute necessity for concepts, models and experiments to be based on sound biological assumptions founded on good natural history observations.

Emerging again and again in these essays is the question of scale in investigations, ranging from those concerned with dynamics of a few square metres of vegetation to those on whole ecosystems and global phenomena. Linked with this question of scale is the inevitable divergence resulting on the one hand from the reductionism appropriate to some experimental techniques and the need to synthesize information on a vast scale, and on the other hand from the gap between ecological theory and practical management of resources. It is paradoxical that the move toward greater exactness in ecology can involve simultaneous deployment of these widely different approaches.

In addressing the whole question of exactness in ecological science we hope to show how each part of the subject can benefit from being placed in a wider framework, and perhaps to give the lie to Edmund Burke's view that 'it is the nature of all greatness not to be exact.'

We were much helped in the planning of the symposium by Malcolm Cherrett, Charles Godfray, John Grace and John Lawton, and in the editing of this volume by numerous referees. We thank Mick Crawley, Charles Godfray, John Grace and Bill Heal for leading the evening discussions. The Society is especially indebted to Robin McCleery for acting as local organizer; together with the domestic bursar of St Catherine's College and her staff, he ensured that the meeting was both very enjoyable and efficiently arranged. We are also grateful to Susan Sternberg, Rowena Millar and other staff at Blackwell Scientific Publications for their sympathetic handling of the production side.

Finally we thank our wives, Anne and Helen, for their tolerance and support while we were engrossed in our endeavours to prepare the text of this book on time.

PETER J. GRUBB
JOHN B. WHITTAKER

I. INTRODUCTION

Holling (1966) considered the attributes of a good ecological study to be realism, precision, completeness and generality. In the introductory paper, exactness in ecology is defined as accuracy or precision and perfection or completeness of understanding, leading to an increased ability to predict. Exactness may be qualitative or quantitative, but it will be underpinned by sound mechanistic understanding, which may be possible at an ecological level, but will often be supported by physiological or chemical and physical observations.

It is proposed that an ecological principle is a proposition that is always true and has a mechanistic basis, whereas a generalization is a proposition that has a high probability of being true but need not have a theoretical basis. Definition of as complete an array as possible of principles and generalizations will undoubtedly lead to a more exact ecology, but so will the art of asking new or neglected questions. Sometimes this will be by new approaches to old questions, for example by ecological physiologists exploring what sets the limits to the distribution of an organism rather than over-emphasizing adaptation, or it may be by the coming together of disciplines previously pursued separately, as in the recent exciting developments in evolutionary and behavioural ecology. If there is any tendency for these to be somewhat reductionist in nature, this is surely balanced by ecologists working on ecosystem and large-scale applied problems requiring adjustments of scale with which we are only just beginning to learn to cope.

The ways forward are diverse. Exactness need not be the prerogative of the mathematical ecologist, and valid generalizations may be made by anyone with mechanistic insight into patterns of nature. Ecology will advance most effectively when its wealth of intellectual diversity is combined in a community of effort. That should be better understood by ecologists themselves than by most other scientists.

REFERENCE

Holling, C.S. (1966). The functional response of invertebrate predators to prey density. *Memoirs of the Entomological Society of Canada,* **48**, 1–8.

1. TOWARD A MORE EXACT ECOLOGY: A PERSONAL VIEW OF THE ISSUES

P. J. GRUBB
Botany School, University of Cambridge, Cambridge CB2 3EA, UK

INTRODUCTION

The objectives for the meeting, as set out by the Organizing Committee, were: 'to review the most exact studies in a selection of fields within ecology, to consider the approaches and technologies which are being most fruitful at the present time, to determine the major obstacles to progress, and to make suggestions for the most profitable lines of work in the future.' It is certainly not my intention to 'set an agenda', as some would say, because the fields covered are too diverse for any one person to deal with. Rather, I have the three following major objectives: to put forward some very general ideas about the theme of the symposium, to place the topics chosen for analysis in some kind of perspective, and to mention some of the important topics not allocated a place in the programme. Inevitably the perspective provided is an essentially personal one, but I hope it will serve as a basis for useful reflection by others.

The first part of this paper deals with the meanings of 'exact ecology', and the issue of prediction in ecology. The second part deals with exactness in various approaches to ecology: in studies on the distribution and abundance of organisms, in evolutionary ecology, in community ecology, and in ecosystem and applied ecology. The third and final part offers a viewpoint on the issue of empiricism and the naturalist tradition.

THE MEANING OF EXACT ECOLOGY

According to the *Oxford English Dictionary*, the word 'exact' is used to express two kinds of idea. The first is 'accurate' or 'precise', and the second is 'consummate, finished, perfect'. I believe that it is helpful to consider this symposium as concerned with both of these senses of 'exact'. Certainly we are concerned with improving the accuracy and the precision of ecological studies, and some contributors will emphasize this aspect, but I believe that we are also concerned with the issue of perfection, which might perhaps be restated as 'completeness of understanding'. That, in turn, might be taken to be reflected in an increased ability to predict.

Advances in accuracy and/or precision

There are several areas of ecological study in which major advances in accuracy and/or precision are being made at the present time through development of new techniques. The one area chosen for discussion at this meeting is that of measuring the rates of physiological processes in free-living plants and animals. Here the greatest general problem has been the 'inaccessibility' of the organism. Even for plants, which do not run, swim or fly away when you start to study them, the issue of inaccessibility is real, because as soon as you enclose some plant part to measure, say, the rate of output of water or uptake of carbon dioxide, you change the conditions and you fail to measure the rates that would occur in the absence of your apparatus. A great deal of effort has been expended by ecologist physiologists to overcome this problem. One approach has been to determine the rate of uptake of CO_2 into whole stands of plants by aerodynamic methods which do not involve enclosure of plant parts, but this approach is useful only in certain kinds of vegetation, and in any case cannot yield information about individual plants or plant parts. To obtain that, enclosure is necessary, and then it is essential to control the environment of the plant part being studied so as to keep it like that of the unenclosed parts. Pioneering work of this type, done in the Negev Desert of Israel in the 1960s (Koch, Lange & Schulze 1971), ultimately provided information of great value in testing ideas on the optimization of changes in stomatal aperture which control the ratio of water lost to CO_2 taken up (Cowan & Farquhar 1977), so illustrating our second meaning of a more exact ecology — a more complete understanding. Long's paper in this volume on measurement of gas exchange in plants brings up to date the story concerning both meanings of exactness.

For those working with animals the problems are plainly more severe. However, the use of radio- and acoustic-telemetry has hugely increased the quantity of information available about rates of physiological processes in free-living vertebrates, and has incidentally transformed the quality of our understanding of the ways in which the animals concerned are suited to their environment and way of life (Butler, this volume). The use of an indirect technique — 'double labelling' of the water supplied to animals — has yielded independent information of great interest on the particular question of respiratory rates, which are vital for our understanding of many different issues in evolutionary and behavioural ecology (Bryant, this volume).

Other kinds of measurement that might reasonably have been chosen for review at the meeting are listed in Table 1.1. In several cases it might

TABLE 1.1. Areas of current advance in accuracy and/or precision through development of new techniques, not discussed at length in this symposium

Area of advance	Selected references
Determination of genetic identity by 'finger-printing' of DNA or RNA	Schaal (1988)
Determination of animal movements by radio-tracking	Kenward (1987)
Determination of changes in cover and other properties of particular vegetation-types on land, and determination of ever-changing patterns of surface temperature, productivity, etc. of the oceans through remote sensing by satellite	Curran (1985); Platt & Sathyendranath (1988)
Determination of sources of metabolites and rates of metabolism, and determination of past climatic conditions by use of stable isotopes	Peterson & Fry (1987); Bryant (this volume); Gray (1981)

be argued that the techniques which have recently become available do not merely increase attainable accuracy, but make possible the answering of questions that simply could not be tackled before.

Outstanding barriers to the attainment of greater accuracy and precision

Just one field within ecology will be considered here, by way of illustration. The results of pollen analysis have changed fundamentally our understanding of the status of present-day plant communities (West 1964; Watts 1973), and our knowledge of the pace and extent of change in them (Davis 1976, 1986; Walker 1989), but there are innumerable cases where our knowledge of what happened in the past is limited by our inability to identify plants from their pollen further than to genus or even family.

Also striking is the impossibility of reaching any very accurate quantitative reconstruction of past vegetation because of the variation between species in their output and dispersal of pollen, and because of variation in the degree of preservation. In the investigation of many deposits the feasible resolution in the time dimension is also severely limited, although in favourable cases changes over a few decades can be followed in detail (Bennett 1986). Our recognition of limits to both qualitative and quantitative exactness in this particular context leads me to make some general remarks on these two types of exactness.

Qualitative and quantitative exactness

As shown in Table 1.2, the two kinds of exactness are concerned in both the description of the status quo and any forecast about the effects of change in one or more conditions. Qualitative exactness is what we are concerned with in establishing the genetic identity of the individual, the pathway of energy or nutrients through an ecosystem (including not just the animals but also the micro-organisms in such intractable material as soil), the biochemical nature of the mechanisms in, say, plant–herbivore interactions, and the identity of the currency in evolutionary trade-offs or in mutualisms such as those described by Pierce in this volume.

TABLE 1.2. A simple classification of variables studied by ecologists according to the kind of exactness involved

Involving qualitative exactness
(a) Descriptive of the status quo
Genetic identity of individual
Pathway of energy, molecule or atom
Mechanism of interaction between parties
Currencies of trade-offs and mutualisms
Setting for any detailed analytical or experimental study
(b) Forecasting effects of changed conditions
Abundance of x: decrease or increase?
Involving quantitative exactness
(a) Descriptive of status quo
Extent, number, rate
(b) Forecasting effects of changed conditions
Extent, number, rate

I would add to this list the more controversial issue of defining the setting for ecological studies. I once asked Professor Heinz Ellenberg, Honorary Member of the British Ecological Society, what he thought was the most important contribution of European phytosociology to the world's ecological tradition, and he said 'qualitative exactness'. In answering in this way, he had no illusions about achieving absolute exactness in phytosociological work, but he considered as inexcusably inexact the specifications of ecological setting all too often found in British and North American papers reporting otherwise valuable work, for example 'dry oakwood' or 'wet grassland'. For those who doubt the value of Ellenberg's point, I recommend that they read his account of investigative ecological studies placed in a phytosociological context in his *magnum opus*, *The Vegetation Ecology of Central Europe* published at last in English (Ellenberg 1988).

An approach to qualitative exactness is also vital in the production of any inventory of community-types which is to serve as a basis for the choice of sites for 'nature conservation', and this is recognized in the Nature Conservancy Council's support for the current National Vegetation Classification scheme in Britain.

I have put 'effects of change' under the qualitative heading because there are many cases where we do not know whether a certain change in conditions will give a positive or negative result, let alone a large or a small one. More generally, we are, of course, concerned with quantitative exactness, and there is no need to give examples in this case.

PREDICTION IN ECOLOGY

There is, I believe, much loose talk about an ability to predict being a measure of the maturity of ecology, or indeed any science. Consider the predictions an ecologist might make, as set out in Table 1.3. An experienced ecologist might reasonably be expected to be able to predict for an area with a specified climate, geology and topography the kinds of plants, animals and micro-organisms present, even the numbers of species of plants and animals, the kinds of temporal change and the primary productivity. This could be done wholly on the basis of empirical knowledge — that is to say 'based on the results of observation and experiment only' (*Oxford English Dictionary*). In the past 'quacks' were able to earn a living because they could make sufficient correct forecasts on an empirical basis, and even now much successful medical practice is strictly empirical.

TABLE 1.3. Examples of predictions expected of ecologists

(a) What is likely to be living in such and such a place?
  Kinds of plants, animals and micro-organisms?
  Number of species?
  Patterns of relative abundance?
  Primary and secondary productivity?
  Kinds of temporal change likely to occur?

(b) How are we to conserve the predators and prey?

(c) How can the system be exploited on a sustainable basis?

(d) How can the newly arrived pests or pathogens in this system be controlled?

(e) What will happen if an entirely new combination of conditions arises, or if we apply this or that management which has never been tried before?

Many of the questions considered by ecologists involve relatively simple kinds of prediction — this is true not only of 'ivory tower' ecologists in academia but also those working in applied fields such as conservation or restoration. However, both theoreticians and practitioners often have to answer questions of types (c)–(e) in Table 1.3, and then it is found that answers based on empirical knowledge are less and less likely to be reliable. I suggest that it *is* reasonable to say that a measure of the maturity of ecology is the *ability to predict the result of new kinds of manipulation or changing conditions that have not been experienced before.* A sound mechanistic understanding is necessary for success by this criterion. I recognize that there must be an arbitrariness about what is a specifically ecological level of mechanistic understanding. Behind the ecological level will lie successively physiological, chemical and physical levels of expressing mechanism.

Principles and generalizations in ecology

Some will say that we need to establish 'principles' of ecology upon which to base predictions. But what are 'principles'? The term is commonly used but rarely defined. Allee and co-authors in the introduction to their text *Principles of Animal Ecology*, published in 1949, wrote 'A word is in order about "principles". We do not wish, nor are we competent, to enter into a philosophical evaluation and definition of "laws", "concepts", and "principles".' I have to say Amen to that. They go on to write 'Thus the "principles" we shall attempt to formulate and interrelate are simply those generalizations inductively derived from the data of ecology.' And 'In this view, a principle is a means of description of nature in succinct and compressed form.'

My impression in talking to various scientists is that they would wish to use 'principle' as something nearer to 'law', and rather hesitantly I put forward the following distinction between 'principle' and 'generalization'. A principle may be defined as a proposition that is always true and has a mechanistic basis. A useful generalization may be defined as a proposition that has a high probability of being true, and it may or may not have a mechanistic theoretical basis.

Six ecological principles, in my sense, are set out in Table 1.4. Items 1, 3 and 6 have been proposed as 'laws of ecology' by Murray (1979, 1986). I was tempted to add a seventh concerning the idea of a trade-off: 'a unit of resource cannot be used in two ways at one time'. However, while it is true that an atom cannot be in two molecules simultaneously, and thus a unit of nitrogen cannot be used in two enzymes (say) at once, the

TABLE 1.4. Six principles of ecology

1 A population must meet limitation of resources if not limited by predation, disease or other deleterious environmental factors (Malthus 1798)
2 One biotype will replace another under a given set of conditions if it has a greater fecundity or lesser mortality (Darwin 1859)
3 Following from **2**, two biotypes which are not permanently subject to rarefaction by some third party or other environmental factor can coexist indefinitely only if they have different niches or are subject to greater intraspecific competition than interspecific (Gause 1934; but see Krebs 1972, p. 231)
4 A prey species and a predator species cannot coexist indefinitely unless the environment is heterogeneous (Huffaker 1958)
5 Density-dependent effects may either stabilize population size or produce cycles or produce 'chaos', depending on the degree of non-linearity of the relation between N_{t-1} and Nt (May 1974, 1986a)
6 Available energy decreases at each stage of predation (Lindemann 1942)

ecologically more relevant point is that a single substance *can* perform two or even several functions at once. For example, the substances which make up the thick cell walls of tough leaves may serve during their time of residence in those walls (i) to prevent the collapse of the mesophyll and the occlusion of the intercellular spaces when the leaf is partially dehydrated, (ii) to reduce the chance of the cuticle being fractured by wind, raindrops or animal activity, (iii) to reduce the nutritive value and palatability of the leaf to animals, and (iv) — a point I omitted in a recent review (Grubb 1986a) — to maintain leaf inclination, once turgidity is lost, at such an angle that damage by overheating or further desiccation is minimized. I have dealt elsewhere with the more general issue of the limited value of thinking in terms of trade-offs in the evolution of plants or animals (Grubb 1985, p. 600).

Some readers may be surprised not to see in Table 1.4 the '3/2 thinning rule' of Yoda *et al.* (1963) which describes the relationship between density and mean plant mass in a self-thinning stand of plants; in fact that 'rule' is best seen as a statement of the limiting case, and in spontaneous populations of trees, for example, the slope of the relationship between log mean mass and log density may be substantially less than 1.5 (Mohler, Marks & Sprugel 1978).

We may contrast the few principles set out in Table 1.4, which go very little way generally to answering specific practical problems, with the wide array of useful generalizations that we encounter in the ecological literature, and which all of us find so helpful in our daily practice of ecology. A two-by-two classification of ecological generalization is set out in Table 1.5: they may be qualitative or quantitative, and with or without a mechanistic basis.

TABLE 1.5. A two-by-two classification of generalizations in ecology, with one example for each of the four types

	Qualitative	Quantitative
With mechanistic basis	Early successional vascular plants are short-lived (cf. Wells 1976; Grubb 1986b, 1987)	The number of species in one taxonomic group is related to island area by $S = c \cdot A^{0.27}$ MacArthur & Wilson 1967, based on Preston 1962)
Without mechanistic basis	Gramineae are associated with soils of higher nutrient status than Cyperaceae (Grubb 1986b)	Basal metabolic rate is related to body mass by $R = a \cdot W^{0.75}$ (Kleiber 1961)

None of the generalizations in Table 1.5 is absolutely true, but all are sufficiently nearly true to be useful. A fair rationale can be put up for the propositions about the longevity of early-successional vascular plants, or the numbers of species in some animal group on different islands as a function of island size. On the other hand, as far as I know, we have no mechanistic explanation for the difference between Gramineae and Cyperaceae in the ability to occupy soils of different fertility, or the particular relation between the basal metabolic rate of an organism and its weight. One impediment to an exact ecology, in the sense of completeness of understanding, is thus a lack of a mechanistic basis for many generalizations, even for many where a precise mathematical relationship has been established (cf. Peters 1983). The impediment is not, of course, unique to ecology; many of the generalizations concerned might be considered equally within the realm of physiology (cf. Prosser 1973).

If we think for a moment about the generalizations that we do find useful in our ecological work, they are all conditional. Take one example at the level of description. We cannot make a single simple sound generalization about the nature of pioneer plants. Their nature depends on the substratum. The pioneer plants on smooth rocks are generally lichens, and those on rougher rocks are generally mosses, while those on gravel and silt are generally herbaceous dicotyledons and grasses (Grubb 1986b, 1987). And to take one example at the level of manipulation, we may consider the effects of so-called 'intermediate levels of disturbance' on species-richness in animal and plant communities. Again we cannot make a useful single generalization. In communities with what Yodzis (1986) calls 'niche control' of community function, intermediate disturbance decreases species-richness (as long as it is non-selective), but in communities with what he calls 'dominance control' it increases species-richness. And so on. We have to learn to cope with what May (1986b) has called 'the ineluctably contingent nature of such rules and patterns as are to be found governing the organization of communities'. Without doubt one of the means of advancing toward a more exact ecology will be the definition of a more complete array of generalizations that apply to specific sets of circumstances, which are each sufficiently widespread to be of interest.

While this is true, I have to say that I myself believe that many of the most interesting and satisfying moves toward a more exact ecology will come about in a different way — from asking new or neglected questions. I will try to show what I mean as I go through a number of different approaches to ecology.

ECOLOGY AS THE EXPERIMENTAL ANALYSIS OF DISTRIBUTION AND ABUNDANCE

My point can be made forcefully in the case of this approach to ecology. There is no part of this symposium dedicated to the first half of the field of study defined by Krebs (1972), and by Andrewartha & Birch (1954) before him, and for that reason more space in this introduction is devoted to the topic that would otherwise have been appropriate.

It seems to me that a major impediment in the movement toward a more exact ecology has been an excessive concern for the nature of so-called 'adaptations' of animals and plants to particular circumstances. The huge amount of effort put into studying 'adaptations' — in recent years on an increasingly sophisticated physiological, biochemical and biophysical basis — has not been balanced by an equivalent amount of thoughtful work attempting to find out how the limits to distribution of an organism are set. This is a clear example of the need to ask questions of a type not sufficiently often considered.

Distribution-limits of organisms

Part of the problem is, I believe, that physiologically minded people seem disinclined to get involved in the difficult issue of 'competition' between organisms, but this is not the whole story since it is commonly accepted that many limits seem to be set by physiological tolerance rather than by competition, for example, the poleward limits of many animals and plants, and (on a different scale) the upper limits of many animals and plants on rocky shores. Descriptive ecologists are always mapping these limits, and quite often correlating them with environmental variables, but experimental ecologists rarely study them. When they do, they all too often stop at correlating the rank order of occurrence on some gradient and the rank order of sensitivity to some relevant environmental factor. This level of satisfaction is seen in a number of studies on the upper limits of gastropod molluscs and seaweeds on rocky shores (Broekhuysen 1941; Biebl 1962). But the exact limits of the various species as such are not truly understood. I therefore want to mention a few encouraging studies, first for plants and then for animals.

In a part of the Sonoran Desert, Jordan & Nobel (1979) determined the conditions needed for establishment of a new generation of one of the major perennial species, *Agave deserti*. Typically the adults are not accompanied by any very young plants despite copious production of seed. In the laboratory, seedlings were grown for different lengths of time on a soil of specified moisture content, and then subjected to droughts of differing length. The results, shown in Fig. 1.1, indicate that the seedlings can tolerate virtually no desiccation until of a certain size; then they can tolerate longer and longer drought spells the longer the moist-soil spell beforehand. Jordan & Nobel used meteorological records, and a model relating weather to soil moisture content, to plot out the soil moisture conditions for the last 17 years (Fig. 1.1). Only one year came within the tolerance limit of the species (1967). Back at the field site the plants were all aged, and it was shown that the youngest plant had indeed established in that year. This is the kind of study all too rarely attempted by ecological physiologists. In theory, the approach could be applied to the distributional limit of the species in drier and drier semi-desert. Does it ever get a wet enough year for establishment there, or one sufficiently often relative to the longevity of the adult? Of course, this is a very simple case with apparently one stage in the life-cycle more sensitive than any other, and one overwhelming effective environmental constraint.

Commonly we will encounter additive effects at various stages of the life-cycle, as found by Woodward & Jones (1984) in their experimental study of the upper altitudinal limits of certain plants in Wales. In experimental plots at two altitudes, seeds were sown at various densities, and the fates of the plants recorded over 18 months. A simple model of what would be expected to happen to the populations of the various species over a longer period was then produced. The results for two species are shown in Fig. 1.2. One species was super-sensitive at the stage of surviving the first winter (*Eupatorium cannabinum*) and the population collapsed very quickly. Another species (*Potentilla reptans*) was

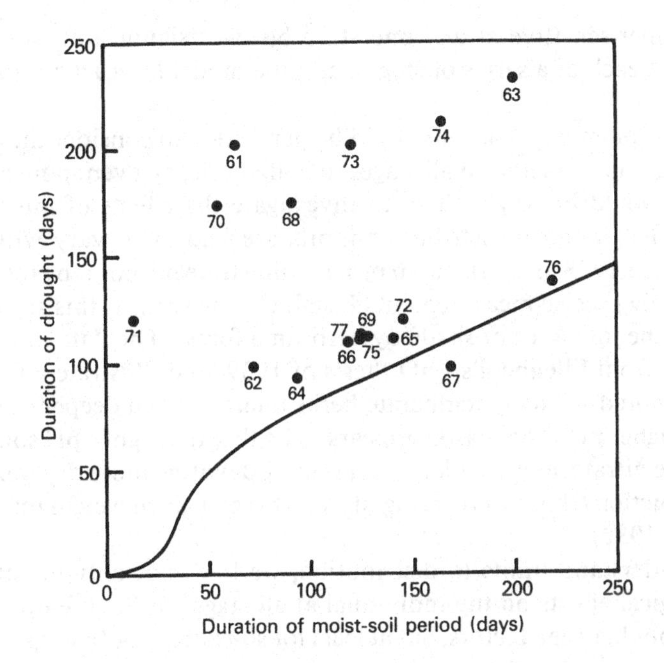

FIG 1.1 The maximum length of dry period tolerated by seedlings of *Agave deserti* as a function of the duration of the moist-soil period beforehand, and the actual combinations of moist and dry conditions believed to have occurred in 17 different years in one study area in the Sonoran Desert; regeneration in fact occurred only in 1967 (from Jordan & Nobel 1979).

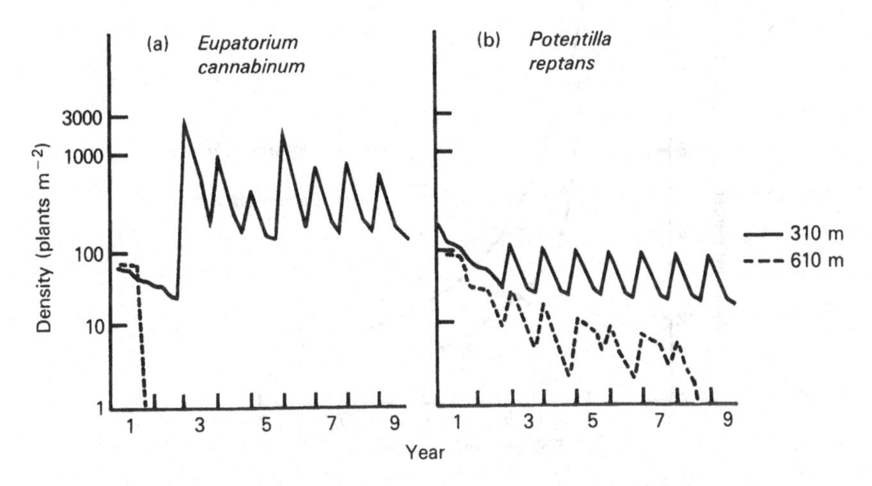

FIG 1.2 The forecast population size of two herbaceous plant species in gaps of a standard size in vegetation at two altitudes in Wales, based on an experiment lasting 18 months and a computer model extending the results to 9 years (from Woodward & Jones 1984).

not so super-sensitive at any one stage but was significantly adversely affected at each of a series of stages, and the model forecast a prolonged decline.

Even if more physiologists could be persuaded to consider the effects of their favourite factor on all stages in a life-cycle, an even more serious problem would be to get them to investigate the effects of interacting factors. Geographical distribution-limits are known to vary with soil-type (Walter 1973, p. 284), but in order to illustrate my point by reference to relatively precise measurements I shall change scale at this stage, and consider the tolerance of shade by herbs in a forest. The data in Fig. 1.3 are taken from Ellenberg's PhD thesis of 1939, and illustrate a fact still widely ignored — many temperate herbs tolerate much deeper shade on soils of higher pH. The reason appears to be that the higher-pH soils provide more nitrate, and the plants respond by devoting more dry weight to leaf production (thus intercepting more light) at a given weight (cf. Peace & Grubb 1982).

To understand limits to distribution, we have not only to consider physiological effects on the individual at all stages of the life-cycle, and interactions between factors, but at least for so-called 'fugitive' species we

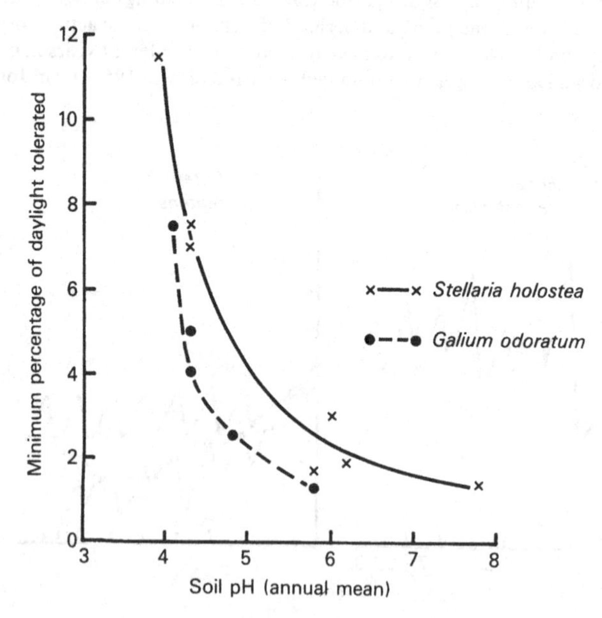

FIG 1.3 The minimum percentage of full daylight tolerated in summer by two herbaceous plant species in various forest-types in northern Germany as a function of soil pH (based on Ellenberg 1939).

have also to weld this approach on to that of studying the properties of marginal populations in terms of epidemiology (Carter & Prince 1981). Tentative preliminary steps in this direction have been taken by Carter & Prince (1988) in their work on the northern limit of the wild annual lettuce, *Lactuca serriola*.

Finally we have to ask, even in those cases where it seems that direct environmental control is dominant, whether in fact competition with other plants of a similar life-form is not involved. For example, in central Europe the upper limit of beech (*Fagus sylvatica*), as documented by Ellenberg (1988), is higher on mountains without natural forest of fir (*Abies alba*) or spruce (*Picea abies*). Also we have to ask about the impact of animals. The small-leaved lime (*Tilia cordata*) rarely makes many seeds at its northern limit, apparently because the growth of the pollen tube down the style is especially sensitive in the relevant temperature range (Pigott & Huntley 1981). However, the tree is very long-lived and periodically experiences hot summers when seeds can be produced abundantly. Why does that not suffice for invasion and regeneration? The answer seems to be that the seedlings are extremely palatable to small rodents, and the regeneration of the tree by seed depends on escaping these herbivores in time or space (cf. Pigott 1985). The chance of doing so decreases to vanishing point when seed production becomes a rare phenomenon. Parker & Root (1981) describe a fascinating case where the ravages of grasshoppers, rather than the physical environment or competition from other plants, is responsible for excluding a common roadside biennial from arid grassland.

Given the complexities, it is perhaps not surprising that so few exact studies have been made on the determination of distribution-limits, but, until more work of the type I have indicated is done, it will not be practicable to formulate anything more than the vaguest generalizations in this field, and certainly we shall not begin to approach an exact ecology.

My one example from the animal side does represent an attempt to generalize. It comes from the very recently published work of Root (1988) on the northern distribution-limits of birds in North America. She used winter censuses to determine effective northern limits, as shown in Fig. 1.4 (a) for the eastern phoebe (*Sayornis phoebe*). She then found the January mean isotherm (air temperature) nearest to the limit, in this case −4°C. Using the data of others, she next found the expected metabolic rate at this temperature — of course the metabolic rate under standard conditions goes up at lower temperatures in a bird, as shown in Fig. 1.4 (b). Putting together the results for a number of species of

(a)

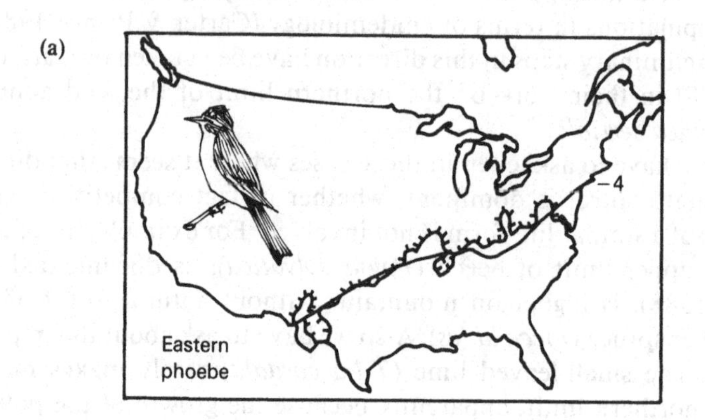
−4
Eastern
phoebe

(b)

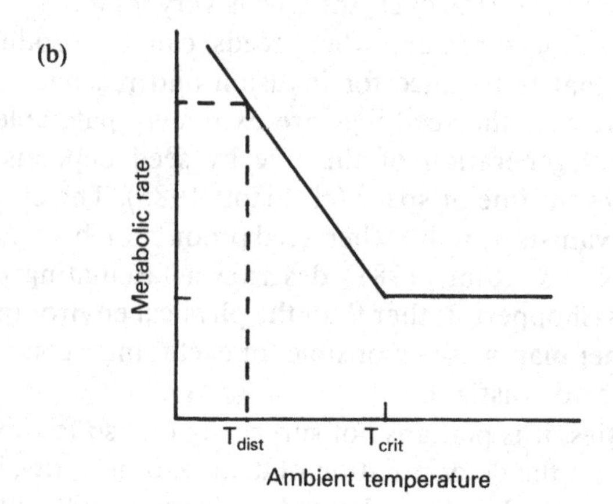
Metabolic rate
T dist
T crit
Ambient temperature

(c)

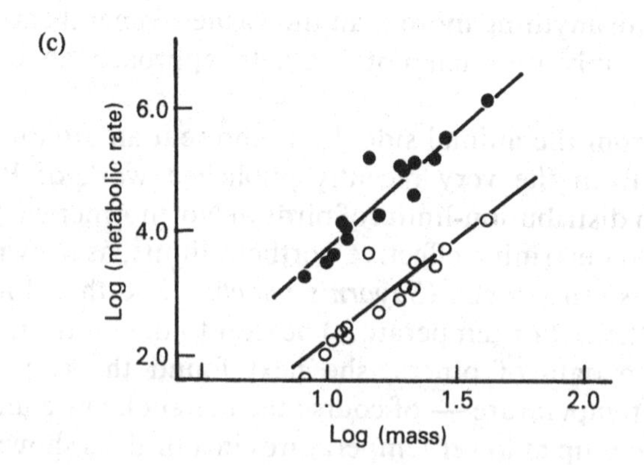
Log (metabolic rate)
6.0
4.0
2.0
1.0
1.5
2.0
Log (mass)

different body-weight, as shown in Fig. 1.4 (c), she concluded that the different northern limits coincide with metabolic rates about 2.5 times the relevant basal metabolic rate. It is surprising that such a relationship should hold; if death were to be related simply to lack of energy, one might expect all sorts of deviations to arise from the birds seeking protected micro-habitats and not having body temperatures closely related to air temperature. And there would certainly have to be complications in the case of birds such as boreal titmice (chickadees) which can undergo hypothermia (Chaplin 1976; Reinertsen & Haftorn 1986). Even more serious grounds for surprise at the regular relationship in Fig. 1.4 (c) arise from the finding that the mean rate of respiration over the day in free-living birds is not greater at lower temperatures, presumably because they are less active in the pursuit of food (Bryant, this volume) and the general supposition traced back to Darwin by Potts & Aebischer (this volume) that birds at their northern limits die from an interaction of low temperatures, reduced food supply and predation. Root's results almost certainly require more complicated interpretation than they have received so far, but her study represents an important provocative development in its attempt to generalize in an area notable for the absence of generalizations. Obviously we would like some mechanistic rationale for the ratio in Fig. 1.4 (c) being 2.5 and not 2.0 or 3.0.

Abundance of organisms

There is, I believe, an analogous problem here in that too few people have attempted to understand the control of absolute numbers of individuals of any given species in a study area, and too many have been content with studies on the patterns of year-to-year variation, the shapes of survival curves, the incidence of density-dependent effects or other fragmentary studies. The problem is particularly severe among plant ecologists, and fortunately some animal ecologists have done much better — a fact that

FIG 1.4 Stages in the analysis by Root (1988) of the relationship between the northern limits of various bird species in North America in winter and the rates of metabolism she expected them to show at those limits: (a) the northern limit in midwinter of one species (the eastern phoebe, line with markers) and the isotherm for average minimum January temperatures of −4 °C (air temperature); (b) the general form of the relationship between metabolic rate and air temperature under standardized conditions (T_{crit} = 'lower critical temperature', i.e. that at which metabolic rate begins to rise; T_{dist} = average minimum temperature at northern distribution limit); (c) the forecast metabolic rate at the northern limit (●) and the basal metabolic rate (○), showing that the former is generally about 2.5 times the latter.

is reflected in the contributions to this volume by Murdoch & Walde on insects, and by Potts & Aebischer on birds.

In the case of plants it is useful to consider them as forming a spectrum from those in which recruitment is tightly coupled to disturbance, to those in which it is not at all coupled to disturbance (Grubb 1988). Plants with tightly coupled recruitment are exemplified by species that come up instantly in new gaps created by windthrow in forest, or by animal activity in grassland. Plants with recruitment which is not coupled are exemplified by annuals of extreme sites with no perennials present, and by the annuals and biennials living in the interstices of closely grazed grassland.

As far as I am aware, there is only one species of plant with 'uncoupled recruitment' for which we have anything like a quantitative understanding of the control of absolute abundance: the ragwort, *Senecio jacobaea*. Dempster & Lakhani (1979) published data for a substantial number of years — a *sine qua non* in this area — and a simple model relating number of plants per unit area in one year to that in the previous year as a function of the abundance of the major herbivore (the cinnabar moth, *Tyria jacobaeae*) and the rainfall in successive periods in the summer. As shown in Fig. 1.5, the two measured factors of herbivore load and rainfall very largely explained the changes in abundance.

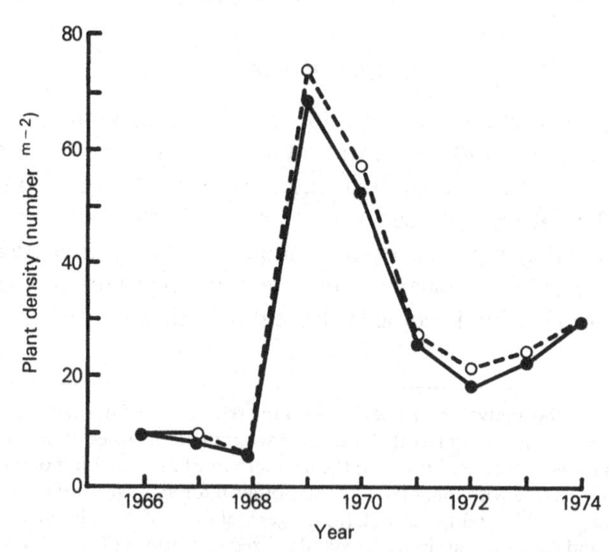

FIG 1.5 The population density of the pauciennial herbaceous plant *Senecio jacobaea* at a site in the Breckland of eastern England: filled symbols and continuous line show actual values; open symbols and pecked line show values forecast by a simple model based on the effects of the major predator (the moth *Tyria jacobaeae*) and the amount of rainfall in summer (from Dempster & Lakhani 1979).

What we want to know for a range of species is how many factors need to be specified for us to be able to forecast plant density. Is *Senecio jacobaea* on sandy soil exceptional in this respect? It *is* a plant with a single major herbivore, and the soil *is* particularly drought-prone. We simply do not know whether the populations of many species with uncoupled recruitment can be forecast similarly easily. It is not the sort of question plant population biologists have been asking. In his contribution to this symposium, van der Meijden discusses other important aspects of population control in plants, and incidentally shows that just occasionally a third factor can have a massive effect on populations of *Senecio jacobaea* on a sandy soil.

As for plants whose populations are tightly coupled to disturbance of the matrix-forming plants, a great deal of modelling work has been done, especially for forest trees, but there seems to be no species for which a quantitative relationship with the major environmental factors has been worked out like that for *Senecio jacobaea* (cf. Shugart & Urban, this volume).

In their different ways, all three contributions in the section of this volume on control of population size advocate a three-pronged strategy of investigation: thoughtfully planned long-term monitoring, modelling and experimentation. To date this combination has been all too rare.

EVOLUTIONARY AND BEHAVIOURAL ECOLOGY

These fields have expanded in a most impressive way in the last two decades. Once more a tripartite approach has been in evidence in many of the best studies. In this case theoretical modelling and experimentation have been combined most powerfully with comparative studies rather than long-term monitoring as such, although the results of the latter certainly have been important in some cases. Theory, comparative studies and experiment receive explicit emphasis in the contributions to this volume by Lloyd, Harvey & Pagel, and Partridge respectively. My concern here is with four general points about different dimensions of this approach to ecology.

First, progress in these fields has come about primarily because people have decided to think about new questions. Of course, mathematical expertise has been important, and in some studies the latest refined techniques of measuring the rates of physiological processes have been used. In the immediate future, the latest techniques of determining genetic identity — particularly parentage — are going to be important. But fundamentally it has been a matter of asking new kinds of questions, so that these whole fields illustrate the most general point I wish to make about approaches toward a more exact ecology.

Second, some comment must be made on the almost universal use of optimization models. At first sight, for the ecologist with a long-term view of evolution in mind, the case against thinking about present-day organisms in terms of optimization seems overwhelming. For both animals and plants the fossil record shows that successive groups of organisms have arisen in time, with the later ones commonly able to explore a greater variety of environments and often better able to control their internal environment. Plants are particularly interesting in this respect because the fossil record suggests that, while new groups of animals evolved mostly *after* the extinction of previously dominant types by some unknown environmental factor(s), each successive major new group of plants ousted the previously dominant plants by competition (Knoll 1986). The only significant exception to this generalization for plants occurred at the Carboniferous–Permian boundary when world climates became dramatically drier. Each successive major new plant group has been able to dominate the sites richest in resources, and must surely therefore be regarded as nearer and nearer to some theoretical optimum plant that has yet to evolve. If we believe that evolution is still going on, with some evolutionary lines in the process of expansion and others on the way to extinction in their respective habitats, it must surely be the case that the former are more nearly optimized than the latter for life in the world in the Present period.

Observation of present-day animals and plants shows that many have evolved organs or ways of life despite their inherited tendencies, for example, the panda with its five bear-type fingers and a 'thumb' developed from wrist bones (Gould 1980) or the tree with the C_4 type of photosynthesis (a system ideally suited to bright light) found in the shaded understorey of rain forest in Hawaii (Pearcy 1983). The existence of rudimentary organs also argues against universal optimization; they can involve what appears to be a significant waste of resources. 'Can we suppose that the formation of rudimentary teeth which are subsequently absorbed, can be of any service to the rapidly growing embryonic calf by the excretion of precious phosphate of lime?' (Darwin 1859, p. 453).

Not only the fossil record, improbable organisms and rudimentary organs argue against optimization. A serious problem is raised by the apparent lack of genetic variation in many characters of organisms. Thus, in the 1970s it was common for ecologists to think in terms of close co-evolution between plants with animal-dispersed fruits and their dispersers. A more mature analysis is reflected in Herrera's paper of 1986 entitled 'Vertebrate-dispersed plants: why they don't behave the way they should'. He spells out two reasons: inertia on the part of the plants, and the fact that the interaction between plant and disperser does not take place in a world where they are virtually the only inhabitants. Similarly, in the 1970s many ecologists thought in terms of an intense 'arms race' between plants and herbivorous insects, but Edwards (this volume) argues that the relevant evolutionary breakthroughs by plants have occurred only rarely, and the overall pace of evolution in plants has been slow. An essentially similar point about the evolution of seed size in plants is made by Silvertown (1989) who emphasizes the apparent paucity of heritable variation in seed mass in a number of species.

Most worrying of all, as emphasized by Gould & Lewontin (1979) and by Harper (1982), is the tendency of some biologists to suppose that every individual trait of an organism has evolved as a result of direct natural selection, and is optimized.

No wonder that some distinguished practitioners of optimization theory are keen to emphasize that use of this approach should *not* be taken to mean that any organism is, in some absolute sense, optimized (cf. Krebs & Houston 1989). It seems that increasingly practitioners write explicitly that optimization is constrained by the available genetic material. However, I seriously doubt whether that point is sufficiently emphasized to beginning students.

Provided that the long-term and large-scale vision of evolution is constantly kept in mind, and students do not expect organisms to be optimized in some absolute sense, and provided the most naive approaches to individual traits are avoided, then of course the use of optimization models can be an important step toward a more exact ecology in that they represent one particular way of using mathematical models to sharpen up hypotheses about how animals, plants and micro-organisms are suited to their environments and ways of life. Increasingly plant ecologists are copying animal ecologists in using the approach, as shown by the many authors in the volume edited by Givnish (1986).

My third general point about the fields of evolutionary and behavioural ecology is that the theory to date is almost wholly concerned with the allocation of resources (materials or time) between alternative uses. A different way of thinking is needed to cope with what I feel is the most exciting point to emerge from the work on reproductive costs in *Drosophila*, reported by Partridge in this volume. The costs appear *not* to relate to division of resources but to result from deleterious physiological effects of certain forms of behaviour. The mechanisms are unknown, but in the light of what is known about the hormone systems of advanced mammals, one could readily imagine hormones involved in the sexual behaviour of insects having effects on general metabolism. For an analogy in plants one might contemplate the 'tactic' of a species meeting drought partly by increasing the concentration of solutes of small molecular weight in the cell sap, and having to face the cost of increased susceptibility to invasion by sap-feeding insects and certain kinds of fungi.

Fourthly, there is the issue of absolute properties as opposed to relative. Lloyd, in this volume, points out that it is possible to make sense of plants with large numbers of small seeds and plants with small numbers of large seeds, but that nobody has attempted a quantitative mechanistic rationale of the absolute seed size appropriate to a given type of plant. In 1977, in my contribution to the admirably motivated *Encyclopaedia of Ignorance*, I pointed out that in a similar way we can understand how it is appropriate for plants in some situations to have relatively large leaves and for others to have relatively small, but nobody has yet come up with a quantitative rationale of the absolute leaf size appropriate to a given type of plant. These problems, and analogous ones with animals, await the arrival of a truly original mind.

COMMUNITY ECOLOGY

There is no section of this symposium entitled 'community ecology', which may seem appropriate in that no very precise meaning can be given to the term 'community'. However, one of the greatest proponents of the generality of continua in nature did see fit to entitle his highly successful textbook *Communities and Ecosystems* (Whittaker 1970)! In fact, there has been exhaustive coverage of the topic in four recent symposia — those edited by Price, Slobodchikoff & Gaud (1984), Strong *et al.* (1984), Diamond & Case (1986) and Gee & Giller (1987). Arising from these, there is among some ecologists a sense of frustration at the 'ineluctably contingent nature of the rules' and even a sense of disillusion over the intractability of some of the central questions such as limiting similarity (though for a brave attempt on the latter see Chesson & Huntly 1988).

Fortunately very real and indeed exciting progress is being made by a number of those working on interrelationships between organisms. For example, studies on the control of relative abundance in mechanistic terms have been largely the province of plant ecologists in the past (starting with grass-clover mixtures), but increasingly they are being undertaken by animal ecologists, as in the case of Gray (1987) on sea-bed organisms and O'Connor (1987) on birds. As it happens, plants are discussed in this context in this volume (Shugart & Urban).

New perspectives are emerging also in plant-animal relations, and in this volume Edwards offers a further advance over the recent reorientation of ideas on plant 'defence' by Coley (1983). Increasingly the importance of mutualisms is appreciated, and Pierce illustrates various approaches toward a more exact ecology in her contribution on this topic.

It is particularly unfortunate that it proved impractical to include in the symposium a review of recent work on host-parasite relations, in which, for example, precise and realistic models are replacing the old qualitative arguments about parasites that kill their hosts too quickly being 'ill-designed' (cf. Anderson & May 1982).

ECOSYSTEM ECOLOGY AND APPLIED ECOLOGY

It is, I think, very significant that the ecologists who still use the term ecosystem, and deem the concept valuable despite the fact that they find it impossible to set precise limits to any one ecosystem (Cousins 1988), are mainly those concerned with productivity in real systems and those trying to solve real environmental problems. There is a very worrying gulf between them and the evolutionary camp — referred to by Moss in this volume as 'increasingly . . . left exploring minutiae'. Those in the evolutionary camp will not like that description.

Ecosystem ecology is notoriously inexact in the sense that very many of the published values for productivity and nutrient flows are presented without any measure of accuracy, and the problem is seriously aggravated by the common practice of quoting both primary data and derived values to excessively large numbers of significant figures, apparently without thought for what the errors might be. See, for example, the summary tables in the official International Biological Programme (IBP) volume edited by Reichle (1981), containing values given to five and even seven significant figures without estimates of variance. More important ultimately is the fact that so much of the sampling for measures undertaken during the IBP was done without any sense of the pattern and process of regeneration in plant communities; only belatedly did the message of Watt (1947) get through (cf. Bormann & Likens 1979). Equally important is the fact that the difficulties of measuring even above-ground primary productivity accurately in herbaceous communities — with their continuous and relatively rapid turnover of leaves — are commonly glossed over (cf. Whittaker 1970, pp. 193–4). Woefully inadequate estimates have been published in all seriousness in prestigious journals, such as some of those discussed by Williamson (1976). The problem is not simply that the leaves and other parts of different species grow and die at different rates, but that the individual leaves on one shoot of one plant of one species generally turn over at appreciably different rates (Martinez-Yrizar 1988). Estimates of flow rates for mineral nutrients are inevitably flawed if the data for primary productivity are suspect.

The quantitative generalization most often quoted concerning net primary productivity (NPP) is that suggested by Rosenzweig (1968): that the NPP of terrestrial communities is proportional to actual evapotranspiration. His graphical plot, so often reproduced or quoted, is a log–log one, and even so has appreciable scatter. Moreover, others have found a curvilinear relationship instead (Lieth 1975, p. 256). The background studies in 'pure' aspects of ecosystem ecology do not always give us a secure base for studies of an applied type.

The sections of this symposium devoted to ecosystem ecology and applied ecology virtually merge, and in the light of successive charges against the British Ecological Society over the years that it is not sufficiently concerned with practical issues, it is good to see one-third of this symposium effectively concerned with them.

My general thesis about the need to change our approaches and ask new questions is perhaps best illustrated by Moss's contribution, in which he shows that the older approaches to study of the growth of vascular plants and planktonic algae in the face of eutrophication must be replaced by approaches based on the interrelations between plants and animals as well as plants and chemicals. Equally there is a need for new approaches to problems as different as microbial ecology and the release of genetically 'engineered' micro-organisms (Paul), the productivity of the sea and allowance for the ever-changing patterns of almost any property one might measure in the water (Ulanowicz), the effects of aerial pollutants on forest growth (Krause) and the characterization of carrying capacity for *Homo sapiens* (Slesser).

A major issue running through the papers on ecosystem and applied ecology is that of how to scale up. For example, how on earth is one to forecast the effects of completely new combinations of environmental conditions on all the world's biota, as in the case of the 'greenhouse effect'? Clearly, the right starting point is a plan for working on a rational sample of systems so that useful generalizations are likely to emerge (cf. Strain *et al.* 1983). But then it is important that a large enough sample of each system be subjected to experimentation. Experiments on short-term effects of elevated CO_2 concentration and higher temperature on plants alone may yield quite misleading results. The effects, direct and indirect, on animals and micro-organisms have to be brought in from the start. It is, in fact, so difficult to predict the behaviour of the whole from studies on the parts, that experimenters will be justified in the short term in carrying out empirical studies on as large and as complete samples as they can handle.

EMPIRICISM AND THE NATURALIST TRADITION

Fenchel, in his new book *Ecology — Potentials and Limitations* (1988 — the first in the series 'Excellence in Ecology', and appropriately published in a golden cover), makes a sharp distinction between the empirical results of natural historians, and the scientific principles to be sought by ecologists. To me this distinction is too sharply drawn. Of course, there have been many naturalists over the years who have been empiricists not concerned with generalization, but I believe that many have perceived patterns in nature, and have sought mechanistic explanations of repeated phenomena. I think that sometimes too much emphasis is placed on mathematical relations as uniquely satisfying and uniquely non-empirical, and that too little emphasis is placed on mechanisms which can often be thought out by those not mathematically gifted. Despite his emphasis on mathematical relations rather than mechanisms, I prefer MacArthur's vision — expressed in his posthumous work *Geographical Ecology* — of the experienced naturalist being involved in the development of ecological theory.

Absolutely appropriately we are taken through mathematical models for all kinds of ecological phenomena, at many different scales, in this volume. However, I do not believe that the day has passed when a valuable contribution can be made in the drive toward a more exact ecology by those with limited mathematical skills but considerable insight mechanistically. I end with a quotation from one Nobel Prize-winning physicist about another. In 1977 Paul Dirac wrote of Nils Bohr, recalling the heady days of the Cavendish Laboratory in Cambridge in the late 1920s, 'his arguments were mainly of a qualitative nature... What I wanted was statements which could be expressed in terms of equations, and Bohr's work very seldom provided such statements.' If Bohr could make such a profound contribution to physics, there is perhaps hope yet for non-mathematical mechanists in ecology!

ACKNOWLEDGMENTS

I thank John Whittaker and Ian Woodward for very helpful comments on drafts of this paper, and John Grace for lending me at a critical moment a copy of Fenchel's new book.

REFERENCES

Allee, W. C., Park, O., Emerson, A. E., Park, T. & Schmidt, K. P. (1949). *Principles of Animal Ecology.* Saunders, Philadelphia.

Anderson, R. M. & May, R. M. (1982). Coevolution of hosts and parasites. *Parasitology,* **85,** 411–26.

Andrewartha, H. G. & Birch, L. C. (1954). *The Distribution and Abundance of Animals.* University of Chicago Press.

Bennett, K. D. (1986). The rate of spread and population increase of forest trees during the postglacial. *Philosophical Transactions of the Royal Society,* **B, 314,** 523–31.

Bormann, F. H. & Likens, G. E. (1979). *Pattern and Process in a Forested Ecosystem.* Springer, New York.

Biebl, R. (1962). Seaweeds. *Physiology and Biochemistry of Algae* (Ed. by R. A. Lewin), pp. 799–815. Academic Press, New York.

Broekhuysen, G. J. (1941). A preliminary investigation of the importance of desiccation, temperature and salinity as factors controlling the vertical distribution of certain intertidal marine gastropods in False Bay, South Africa. *Transactions of the Royal Society of South Africa,* **28,** 255–92.

Carter, R. N. & Prince, S. D. (1981). Epidemic models used to explain biogeographical distribution limits. *Nature, London,* **293,** 644–5.

Carter, R. N. & Prince, S. D. (1988). Distribution limits from a demographic viewpoint. *Plant Population Ecology* (Ed. by A. J. Davy, M. J. Hutchings & A. R. Watkinson), pp. 165–84. Symposia of the British Ecological Society, 28. Blackwell Scientific Publications, Oxford.

Chaplin, S. B. (1976). The physiology of hypothermia in the black-capped chickadee, *Parus atricapillus. Journal of Comparative Physiology,* **112,** 335–44.

Chesson, P. L. & Huntly, N. (1988). Community consequences of life-history traits in a variable environment. *Annales Zoologici Fennici,* **25,** 5–16.

Coley, P. D. (1983). Herbivory and defensive characteristics of tree species in a lowland tropical forest. *Ecological Monographs,* **53,** 209–33.

Cousins, S. H. (1988). Fundamental components in ecology and evolution: hierarchy, concepts and descriptions. *Ecodynamics* (Ed. by W. Wolff, C.-J. Soeder & F. R. Drepper), pp. 60–8. Springer, Berlin.

Cowan, I. R. & Farquhar, G. D. (1977). Stomatal function in relation to leaf metabolism and environment. *Integration of Activity in the Higher Plant* (Ed. by D. H. Jennings), pp. 471–505. Symposia of the Society for Experimental Biology, 31. Cambridge University Press.

Curran, P. J. (1985). *Principles of Remote Sensing.* Longman, London.

Darwin, C. (1959). *The Origin of Species by Means of Natural Selection.* Murray, London.

Davis, M. B. (1976). Pleistocene biogeography of temperate deciduous forests. *Geoscience and Man,* **13,** 13–26.

Davies, M. B. (1986). Climatic instability, time lags, and community disequilibrium. *Community Ecology* (Ed. by J. Diamond & T. J. Case), pp. 269–84. Harper & Row, New York.

Dempster, J. P. & Lakhani, K. H. (1979). A population model for cinnabar moth and its food plant, ragwort. *Journal of Animal Ecology,* **48,** 143–63.

Diamond, J. & Case, T. J. (Eds) **(1986).** *Community Ecology.* Harper & Row, New York.

Dirac, P. A. M. (1977). Recollections of an exciting era. *History of Twentieth Century Physics* (Ed. by C. Weiner), pp. 109–46. Proceedings of the International School of Physics 'Enrico Fermi', Course 57. Academic Press, New York.

Ellenberg, H. (1939). Über Zusammensetzung, Standort und Stoffproduktion bodenfeuchter Eichen-und Buchen-Mischwaldgesellschaften Nordwestdeutschlands. *Mittei-*

lungen der floristisch-soziologischen Arbeitsgemeinschaft in Niedersachsen, **5**, 3–155.

Ellenberg, H. (1988). *The Vegetation Ecology of Central Europe.* Cambridge University Press.

Fenchel, T. (1988). *Ecology — Potential and Limitations.* Excellence in Ecology, 1. Ecology Institute, Oldendorf/Luhe.

Gause, G. F. (1934). *The Struggle for Existence.* Williams & Wilkins, Baltimore.

Gee, J. H. R. & Giller, P. S. (Eds) **(1987).** *Organization of Communities Past and Present.* Symposia of the British Ecological Society, 27. Blackwell Scientific Publications, Oxford.

Givnish, T. J. (Ed.) **(1986).** *On the Economy of Plant Form and Function.* Cambridge University Press.

Gould, S. J. (1980). *The Panda's Thumb.* Norton, New York.

Gould, S. J. & Lewontin, R. C. (1979). The spandrels of San Marco and the Panglossian paradigm: a critique of the adaptationist programme. *Proceedings of the Royal Society,* **B205,** 581–98.

Gray, J. (1981). The use of stable-isotope data in climate reconstruction. *Climate and History: Studies in Past Climate and their Impact on Man* (Ed. by T. M. C. Wigley, M. J. Ingram & G. Farmer), pp. 53–61. Cambridge University Press.

Gray, J. S. (1987). Species-abundance patterns. *Organization of Communities Past and Present* (Ed. by J. H. R. Gee & P. S. Giller), pp. 53–61. Symposia of the British Ecological Society, 27. Blackwell Scientific Publications, Oxford.

Grubb, P. J. (1977). Leaf structure and function. *Encyclopaedia of Ignorance* (Ed. by R. Duncan & M. Weston-Smith), pp. 317–29. Pergamon, Oxford.

Grubb, P. J. (1985). Plant populations and vegetation in relation to habitat, disturbance and competition: problems of generalization. *The Population Structure of Vegetation* (Ed. by J. White), pp. 595–621. Junk, Dordrecht.

Grubb, P. J. (1986a). Sclerophylls, pachyphylls and pycnophylls: the nature and significance of hard leaf surfaces. *Insects and the Plant Surface* (Ed. by B. E. Juniper & Sir Richard Southwood), pp. 137–50. Arnold, London.

Grubb P. J. (1986b). The ecology of establishment. *Ecology and Design in Landscape* (Ed. by A. D. Bradshaw, D. A. Goode & E. Thorp), pp. 83–97. Symposia of the British Ecological Society, 24. Blackwell Scientific Publications, Oxford.

Grubb, P. J. (1987). Some generalizing ideas about colonization and succession in green plants and fungi. *Colonization, Succession and Stability* (Ed. by A. J. Gray, M. J. Crawley & P. J. Edwards), pp. 81–102. Symposia of the British Ecological Society, 26. Blackwell Scientific Publications, Oxford.

Grubb, P. J. (1988). The uncoupling of disturbance and recruitment, two kinds of seed bank, and persistence of plant populations at the regional and local scales. *Annales Zoologicali Fennici,* **25,** 23–36.

Harper, J. L. (1982). After description. *The Plant Community as a Working Mechanism* (Ed. by E. I. Newman), pp. 11–26. British Ecological Society Special Publication, 1. Blackwell Scientific Publications, Oxford.

Herrera, C. M. (1986). Vertebrate-dispersed plants: why they don't behave the way they should. *Frugivores and Seed Dispersal* (Ed. by A. Estrada & T. H. Fleming), pp. 5–18. Junk, Dordrecht.

Huffaker, C. B. (1958). Experimental studies on predation: dispersion factors and predator-prey oscillations. *Hilgardia,* **27,** 343–83.

Jordan, P. W. & Nobel, P. S. (1979). Infrequent establishment of seedlings of *Agave deserti* (Agavaceae) in the northwestern Sonoran desert. *American Journal of Botany,* **66,** 1079–84.

Kenward, R. E. (1987). *Wildlife Radio Tracking: Equipment, Field Techniques and Data Analysis.* Academic Press, London.

Kleiber, M. (1961). *The Fire of Life.* Wiley, New York.

Knoll, A. H. (1986). Patterns of change in plant communities through geological time. *Community Ecology* (Ed. by J. Diamond & T. J. Case), pp. 126–41. Harper & Row, New York.

Koch, W., Lange, O. L. & Schulze, E.-D. (1971). Ecophysiological investigations on wild and cultivated plants in the Negev Desert. I. Methods: a mobile laboratory for measuring carbon dioxide and water vapor exchange. *Oecologia,* **8,** 296–309.

Krebs, C. J. (1972). *Ecology: the Experimental Analysis of Distribution and Abundance.* Harper & Row, New York.

Krebs, J. R. & Houston, A. I. (1989). Optimization. *Ecological Concepts* (Ed. by J. M. Cherrett), pp. 309–38. Symposia of the British Ecological Society, 29. Blackwell Scientific Publications, Oxford.

Lieth, H. (1975). Modeling the primary productivity of the world. *Primary Productivity of the Biosphere* (Ed. by H. Lieth & R. H. Whittaker), pp. 237–63. Springer, Berlin.

Lindemann, R. L. (1942). The trophic-dynamic aspect of ecology. *Ecology,* **23,** 399–418.

MacArthur, R. H. (1972). *Geographical Ecology.* Harper & Row, New York.

MacArthur, R. H. & Wilson, E. O. (1967). *The Theory of Island Biogeography.* Princeton University Press.

Malthus, T. R. (1798). *An Essay on the Principle of Population.* Johnson, London.

Martinez-Yrizar, A. (1988). *Above-ground productivity, nutrient dynamics and leaf characteristics in a chalk grassland.* Ph.D. dissertation, University of Cambridge.

May, R. M. (1974). Biological populations with nonoverlapping generations: stable points, stable cycles and chaos. *Science,* **186,** 645–7.

May, R. M. (1986a). The search for patterns in the balance of nature: advances and retreats. *Ecology,* **67,** 1115–26.

May R. M. (1986b). Species interactions in ecology. *Science,* **231,** 1451–2.

Mohler, C. L., Marks, P. L. & Sprugel, D. G. (1978). Stand structure and allometry of trees during self-thinning of pure stands. *Journal of Ecology,* **66,** 599–614.

Murray, B. G. (1979). *Population Dynamics: Alternative Models.* Academic Press, New York.

Murray, B. G. (1986). The structure of theory, and the role of competition in community dynamics. *Oikos,* **46,** 145–58.

O'Connor, R. J. (1987). Organization of avian assemblages — the influence of intraspecific habitat dynamics. *Organization of Communities Past and Present* (Ed. by J. H. R. Gee & P. S. Giller), pp. 163–83. Symposia of the British Ecological Society, 27. Blackwell Scientific Publications, Oxford.

Parker, M. A. & Root, R. B. (1981). Insect herbivores limit habitat distribution of a native composite, *Machaeranthera canescens. Ecology,* **62,** 1390–2.

Peace, W. J. H. & Grubb, P. J. (1982). Interaction of light and mineral nutrient supply in the growth of *Impatiens parviflora. New Phytologist,* **90,** 127–50.

Pearcy, R. W. (1983). The light environment and growth of C3 and C4 tree species in the understorey of a Hawaiian forest. *Oecologia,* **58,** 19–25.

Peters, R. H. (1983). *The Ecological Implications of Body Size.* Cambridge University Press.

Peterson, B. J. & Fry, B. (1987). Stable isotopes in ecosystem studies. *Annual Review of Ecology and Systematics,* **18,** 293–320.

Pigott, C. D. (1985). Selective damage to tree-seedlings by bank voles *(Clethrionomys glareolus). Oecologia,* **67,** 367–71.

Pigott, C. D. & Huntley, J. P. (1981). Factors controlling the distribution of *Tilia cordata* at the northern limit of its geographical range. III. Nature and cause of seed sterility. *New Phytologist,* **87,** 817–39.

Platt, T. & Sathyendranath, S. (1988). Oceanic primary production: estimation by remote sensing at local and regional scales. *Science,* **241,** 1613–20.

Preston, F. W. (1962). The canonical distribution of commonness and rarity. Parts I & II. *Ecology,* **43,** 185–215 & 410–32.

Price, P. W., Slobodchikoff, C. N. & Gaud, W. S. (Eds) **(1984).** *A New Ecology: Novel Approaches to Interactive Systems.* Wiley, New York.

Prosser, C. L. (Ed.) **(1973).** *Comparative Animal Physiology,* 3rd edn. Saunders, Philadelphia.

Reichle, D. E. (Ed.) **(1981).** *Dynamic Properties of Forest Ecosystems.* Cambridge University Press.

Reinertsen R. E. & Haftorn, S. (1986). Different metabolic strategies of northern birds for nocturnal survival. *Journal of Comparative Physiology,* **B, 156,** 655–63.

Root, T. (1988). Energy constraints on avian distributions and abundances. *Ecology,* **69,** 330–9.

Rosenzweig, M. L. (1968). Net primary production for terrestrial communities: prediction from climatological data. *American Naturalist,* **102,** 67–74.

Schaal, B. (1988). Somatic variation and genetic structure in plant populations. *Plant Population Ecology* (Ed. by A. J. Davy, M. J. Hutchings & A. R. Watkinson), pp. 47–58. Symposia of the British Ecological Society, 28. Blackwell Scientific Publications, Oxford.

Silvertown, J. (1989). The paradox of seed size and adaptation. *Trends in Ecology and Evolution,* **4,** 24–6.

Strain, B. R., Bazzaz, F. A. and 15 others (1983). Terrestrial plant communities. *CO_2 and Plants* (Ed. by E. R. Lemon), pp. 177–222. Westview Press, Boulder, Colorado.

Strong, D. R., Simberloff, D., Abele, L. G. & Thistle, A. B. (Eds) **(1984).** *Ecological Communities: Conceptual Issues and the Evidence.* Princeton University Press.

Walker, D. (1989). Diversity and stability. *Ecological Concepts* (Ed. by J. M. Cherrett), pp. 115–45. Symposia of the British Ecological Society, 29. Blackwell Scientific Publications, Oxford.

Walter, H. (1973). *Ecology of Tropical and Subtropical Vegetation.* Oliver & Boyd, Edinburgh.

Watt, A. S. (1947). Pattern and process in the plant community. *Journal of Ecology,* **35,** 1–22.

Watts, W. A. (1973). Rates of change and stability in vegetation in the perspective of long periods of time. *Quaternary Plant Ecology* (Ed. by H. J. B. Birks & R. G. West), pp. 195–206. Symposia of the British Ecological Society, 14. Blackwell Scientific Publications, Oxford.

Wells, P. V. (1976). A climax index for broadleaf forest: an *n*-dimensional ecomorphological model of succession. *Central Hardwood Conference* (Ed. by J. S. Fralish, G. J. Weaver & R. C. Schlesinger), pp. 131–76. Department of Forestry, Southern Illinois University, Carbondale.

West, R. G. (1964). Inter-relations of ecology and Quaternary palaeobotany. *Journal of Ecology,* **52** (Supplement), 47–57.

Whittaker, R. H. (1970). *Communities and Ecosystems.* Macmillan, New York.

Williamson, P. (1976). Above ground primary production of chalk grassland allowing for leaf death. *Journal of Ecology,* **64,** 1059–75.

Woodward, F. I. & Jones, N. (1984). Growth studies on selected plant species with well-defined European distributions. I. Field observations and computer simulations of plant life cycles at two altitudes. *Journal of Ecology,* **72,** 1019–30.

Yoda, K., Kira, T., Ogawa, H. & Hozumi, K. (1963). Self thinning in overcrowded pure stands under cultivated and natural conditions. *Journal of Biology, Osaka City University,* **14,** 107–29.

Yodzis, P. (1986). Competition, mortality, and community structure. *Community Ecology* (Ed. by J. Diamond & T. J. Case), pp. 480–91. Harper & Row, New York.

II. PHYSIOLOGICAL PROCESSES IN FREE-LIVING ORGANISMS

Physiologists working on ecological problems have had special difficulties arising from the fact that often they have had available to them rather exact methodologies for use in the laboratory, but it has been technically impossible to apply them in conditions at all natural for the organism. So the gulf between the laboratory physiologist and the field ecologist has often been uncomfortably large, and extrapolation from the laboratory to the field a matter of conjecture. The 1980s have seen the greatest progress in the development of techniques to solve these problems, so much so that it is not an exaggeration to suggest that the more significant advances in ecology in the remaining part of the century may well be in this area.

Long shows that although there is a tradition of measuring gas exchange of plants in the field that stretches back to the 1950s and even earlier, the technologies that have recently become available offer exciting prospects for significantly more exact measurements; there is also a shift in the kinds of question being asked.

Highly mobile animals have always presented a challenge to the ecologists and physiologists trying to bring their disciplines together. Where these have met, for example, in ecological energetics, discrepancies have been all too obvious. Butler and Bryant describe elegant techniques which have, in different ways, advanced enormously the study of free-living animals, able to behave in as natural a manner as possible. The complementary techniques of telemetry (able to convey all manner of messages) and double isotopic labelling (of specific use in measurement of respiration rates) have both made possible substantially changed views of the energy budgets of animals in the wild.

2. GAS EXCHANGE OF PLANTS IN THE FIELD

S. P. LONG
Department of Biology, University of Essex,
Colchester CO4 3SQ, UK

INTRODUCTION

Twenty-five years ago a consideration of gas exchange of plants in the field would have been limited largely to what might be possible in the future. At that time measurement of gas exchange in the field required (i) a field or mobile laboratory, capable of housing bulky equipment, (ii) a large team capable of maintaining the apparatus and reducing the data gained, and (iii) a high level of financial support for the heavy capital outlay and frequent maintenance. Today off-the-shelf portable gas-exchange systems, which can be carried easily and operated by a single person are available and typically include a microprocessor for data reduction and storage. The complexity of the early systems meant that these were limited to a very few groups of researchers. However, at that time the stage was set for rapid developments.

There had been important advances in understanding the physiology of photosynthesis and transpiration, and the physical and micrometeorological principles governing gas transfer were being applied to exchange between leaves and the atmosphere. At the same time infra-red gas analysers for laboratory use were well developed, although too large to be readily portable. Simultaneously, the potential of $^{14}CO_2$, already utilized to elucidate the pathways of photosynthetic carbon metabolism, was recognized as a means of developing portable apparatus for the measurement of photosynthesis in the field (Sestak, Catsky & Jarvis 1971). Indeed these remained the only means of measuring CO_2 uptake of plants in sites inaccessible to vehicles until the late 1970s (Long & Incoll 1979). These developments, coupled with the stimulus provided by the IBP, made possible the rapid realization of field measurements of gas exchange by plants and stands of plants in a wide range of communities during the late 1960s and early 1970s (cf. Monteith 1976). The 1970s saw considerable reappraisal of the approaches to field measurements, and the continued development of mathematical models relating leaf measurements of photosynthesis to canopy photosynthesis, so extending the value of individual leaf measurements (Hesketh & Jones 1980). At the same time developments in laboratory studies of leaf gas exchange led to the development of models relating leaf gas exchange to biochemical limitations, thus allowing quantitative separation of limitations assignable to stomata, carboxylation and acceptor regeneration (Farquhar, von Caemmerer & Berry 1980; Farquhar & Sharkey 1982). Finally, the early 1980s saw the commercial launch of two fully portable gas exchange systems, both now widely used.

Measurements of gaseous exchange for plants in the field may involve a wide range of molecules: CO_2 and O_2 in photosynthesis and respiration, water vapour in transpiration, and SO_2, NO_x, O_3, CO and many other gases as pollutants. This review will be concerned primarily with uptake of CO_2 in photosynthesis.

THE ROLE OF CO_2 EXCHANGE MEASUREMENTS

The growth of plants and stands of plants, in terms of dry-weight or carbon gain, has traditionally been measured by sampling, drying, weighing and chemical analysis of the dried material. Direct measurement of CO_2 uptake provides a complementary approach with important advantages over productivity information gained from analysis of dry weight changes alone.

1 It is instantaneous, measuring production *in vivo* on the time-scale of both *in vitro* studies of subcellular photosynthetic processes and of *in vivo* slow chlorophyll fluorescence and leaf absorptance transients.

2 It is non-destructive so that the same leaf, plant or stand may be measured repeatedly throughout its life.

3 The immediate effects of sudden changes in microclimate or experimental treatments on photosynthetic productivity may be determined, where change in dry-weight gain would require days before an effect might be detectable. Transient effects of, for example, environmental change or herbicide application on production may be revealed by measurement of gas exchange, yet be of too short duration to be apparent in dry-weight changes.

4 It accounts for all photosynthetic carbon-gain, including the large fraction which may be lost to the rhizosphere.

5 It allows separate investigation of individual organs, plants or groups of plants.

6 It allows separation of photosynthetic gain from respiratory losses of carbon for a system as a whole.

However, it cannot be used simply to estimate net primary productivity. Here the major limitation is the difficulty in separating root respiration from other forms of soil respiration. The technique may be used to estimate gross carbon uptake.

Figure 2.1 provides examples of how measurements of CO_2 exchange can provide temporal and spatial resolution of carbon-gain by stands, plants and individual organs of plants in the field.

Approaches to estimating CO_2 uptake by vegetation fall into two categories (i) micro-meteorological methods, and (ii) enclosure methods. These techniques and the associated equipment have been described in detail by Sestak, Catsky & Jarvis (1971), with more recent advances described in the volumes edited by Marshall & Woodward (1985), Coombs *et al.* (1985) and Gensler (1986).

MICRO-METEOROLOGICAL TECHNIQUES

The theory and methods for estimating transfer of momentum, heat and mass (including CO_2) between the atmosphere and vegetation have been reviewed in detail previously (Thom 1975; Denmead & Bradley 1985). Basically, these techniques estimate CO_2 fluxes from the gradient of gas concentration above the canopy and the vertical flux of air in that gradient. The major advantage of this approach is that it is non-intrusive, and thus neither alters the environment of the area under investigation nor damages the vegetation. Transfer of CO_2, and of other gases, between the atmosphere and the vegetation takes place in turbulent eddies generated in two ways: (i) by friction at the surface creating a differential in speed between air moving at the surface and at different heights above the surface, and (ii) by buoyancy of the gas near the surface, commonly created by surface heating leading to instability. Two micro-meteorological methods of measuring vertical fluxes have been widely employed: flux gradient analysis and eddy correlation.

Flux gradient analysis

In this procedure air samples are taken continuously at a range of heights through and above the canopy and their CO_2 content is determined. By analogy with molecular diffusion the flux of CO_2 into the canopy (A_c) is assumed to be proportional to the vertical gradient of mean CO_2 concentration ($\mathrm{d}C/\mathrm{d}z$) averaged over a period of minutes:

$$A_c = -\mathrm{K}\mathrm{d}C/\mathrm{d}z. \qquad (2.1)$$

The constant K relating the gradient to the flux is the eddy diffusivity for mass. The validity of this approach is doubtful under unstable conditions and where the surface is not uniform, e.g. when the stand consists of large plants with large spaces or where marked surface heating is likely. For short or uniform herbaceous vegetation, mass fluxes calculated by the

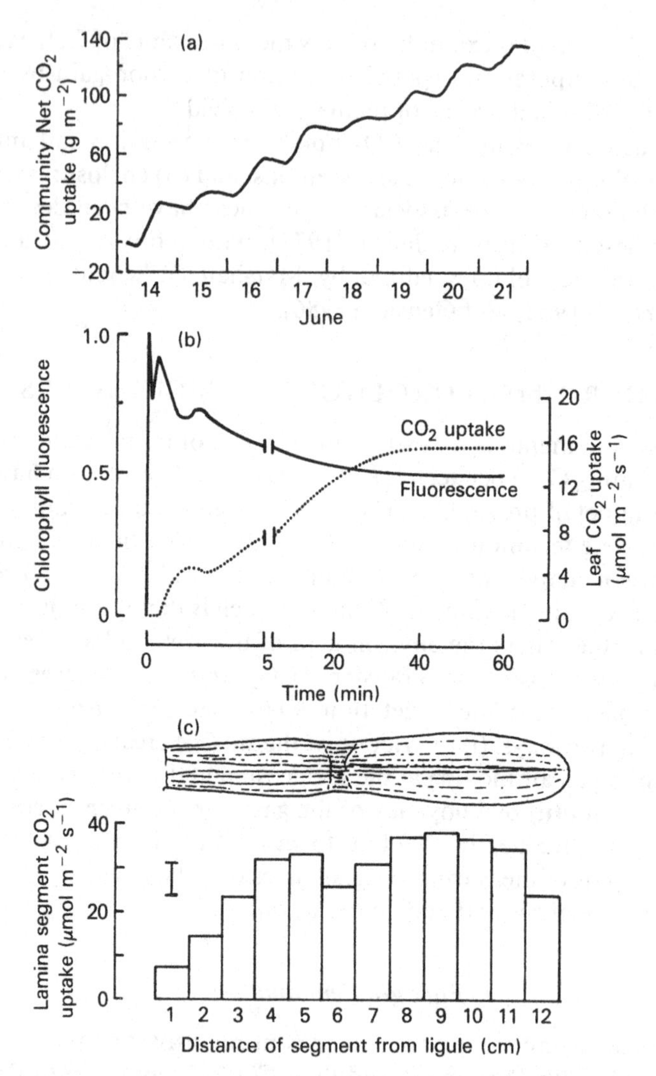

FIG 2.1 Three examples of temporal and spatial variation in CO_2 uptake: (a) cumulative net CO_2 exchange determined by flux gradient analysis for a field of barley (*Hordeum sativum* cv. Proctor) over 8 consecutive days (redrawn from Biscoe, Scott & Monteith 1975), (b) CO_2 uptake measured simultaneously with chlorophyll fluorescence at 680 nm, expressed in relative units, during light induction of photosynthesis in a leaf of maize, *Zea mays* cv. LG11 (redrawn from Ireland, Long & Baker 1984), and (c) the distribution of CO_2 uptake within the lamina of the second leaf of maize, which was chilled during development, inducing a band at about 6 cm from the ligule (redrawn from Long *et al.* 1989).

aerodynamic method agree well with those estimated by other procedures (Biscoe, Scott & Monteith 1975). However, over natural mixed grassland in Canada errors were estimated at +20%. In Australia, where intense local heating can rapidly move to instability, the error even for a uniform wheat crop was as high as 50%. Over forests the aerodynamic method may underestimate fluxes by a factor of 2–3 (Thom *et al.* 1975).

Determination of K is central to this method. Two procedures have been used. The energy balance method assumes that K is the same for heat and mass transfer, temperature gradients and latent heat transfer being measured simultaneously. This method appears valid over a wide range of stability conditions, but is inaccurate when available energy is small, e.g. at night. The alternative aerodynamic method derives K for momentum from wind profiles and assumes empirical relationships between K for momentum and that for mass, from measurements made over bare soil or short grass. It is the failure of these empirical relationships to apply to some vegetated surfaces which underlies much of the error noted for this technique. Despite these shortcomings, flux gradient analysis has provided us with some of the most detailed pictures available for the exchange of CO_2 and water vapour between plant communities and the atmosphere. Some examples include a barley crop in England (Biscoe, Scott & Monteith 1975), short-grass prairie (Saugier & Ripley 1978), coniferous plantations (Jarvis, James & Landsberg 1976) and tropical rain forest (Allen & Lemon 1976).

A major limitation is that a large area, several hectares of the vegetation under consideration, is required to ensure that the observed CO_2 concentration profile results from the photosynthetic and respiratory activity of the canopy under consideration rather than the activity of adjacent vegetation of different composition. Smaller areas are adequate for the alternative method of eddy correlation.

Eddy correlation

At the end of the 1960s and in the early 1970s flux gradient analysis was seen as the key development for assessing gaseous fluxes between plant communities and the atmosphere in the field (Lemon & Wright 1969). However, by the mid 1970s empirical measurements had shown the serious errors that could develop in estimates of fluxes (Raupach & Legg 1984). Several objections to the approach have been noted, but of particular importance is the point that gas concentration may vary in a canopy on a scale similar in size to the turbulent eddies, so that K (eqn 2.1) is not a unique function of the rate of turbulent mixing, but differs for each entity (Legg & Monteith 1975; Raupach & Legg 1984). An alternative micro-climatological procedure which avoids the assumptions underlying the estimation of K is the eddy correlation method. Here anemometers and gas analysers with fast response times (<0.1 s) are used to determine the rapid fluctuations in vertical wind speed (w') and gas concentration (c') which occur at a point due to the passage of individual eddies (Grace, Ford & Jarvis 1981). The rate of exchange of CO_2 between the plant canopy and atmosphere is given by the time average of the product of the two measurements:

$$A_c = \overline{w'c'} \ . \tag{2.2}$$

Realization of the eddy correlation technique required two practical developments. First, the development of open-path analysers, i.e. analysers which could measure the infra-red absorption of CO_2 over a short path of atmosphere, rather than within a gas cell. Drawing a parcel of air through the cell of a gas analyser would destroy the structure of the air parcel and thus CO_2 must be measured *in situ.* A number of instruments have been developed specifically for this purpose, and are capable of measuring concentrations with a resolution of <0.5 μmol mol^{-1} of CO_2 in air with a response time of <0.1 s (Bingham *et al.* 1981; Brach, Desjardins & StAmour 1981; Ohtaki & Matsui 1982). Secondly, sensitive and rapid measurement of the vertical wind profile became practicable with the development of miniature sonic transducer anemometers, so avoiding the stall problems associated with propeller anemometers and allowing direct measurement at the exact point of CO_2 concentration measurement (Campbell & Unsworth 1979). In addition, the continued development of microprocessors/microcomputers which are cheap, relative to the costs of the rest of the system, have increased the feasibility of the approach. The method requires measurement of w' and c' at frequent intervals (<0.1 s) so that large quantities of data are generated, requiring storage and analysis.

Although the eddy correlation technique appears to hold great promise, most studies to date appear to be technological exercises testing the validity and potential of equipment rather than attempts to answer serious ecological questions. Nevertheless the validity of the technique has been demonstrated with towers over vegetation and even in measuring gaseous exchanges between the sea and atmosphere using aircraft-mounted instrumentation (Desjardins *et al.* 1982: Jarvis & Sandford 1985). It has been successfully employed for the longer-term monitoring of CO_2 exchange by rice and wheat crops (Ohtaki 1980; Ohtaki & Matsui 1982).

All micro-meteorological techniques suffer from two potential weaknesses. First, the techniques concern exchange from an area of vegetation. The exact area from which exchange is measured is also difficult to define and may change with prevailing weather. It therefore follows that the technique cannot be used easily to separate the contributions of different species within a community or different parts of the plant. For example, in savanna the technique could not be used to separate the contributions of tree and grass species. Secondly, in calculating photosynthetic uptake correction must be made for soil respiration. This can be done by enclosing a portion of the surface (Biscoe, Scott & Monteith 1975) since simple measurement of the night efflux will include the shoots which are likely to make a lower or zero daytime contribution. Enclosure of an area of the surface also presents problems, since stems cannot be enclosed, and thus placement of the enclosure will be non-random and will probably avoid the areas of highest rooting density. A further problem is that measurement of the efflux requires some alteration of the surface CO_2 concentration. Indeed many techniques are based on chemical removal of all CO_2, e.g. that used by Biscoe, Scott & Monteith (1975). Thus the natural diffusion gradient is altered and some error in the estimated flux is inevitable.

ENCLOSURE METHODS

Most studies of CO_2 exchange have involved enclosure of leaves, plants or stands of plants in transparent chambers. The rate of CO_2 assimilation by the material enclosed is then determined by measuring the drop in the CO_2 concentration of the air flowing through the chamber. Thus, if the vegetation is assimilating CO_2, a decrease in concentration will occur in the chamber in proportion to the net rate of CO_2 uptake. The technique allows separation of the contributions of different organs of the plant, and separation of respiration from photosynthesis without any dependence on the theoretical assumptions that have dogged micro-meteorological approaches. Its main disadvantage is that unlike the micro-meteorological approach it is intrusive and some modification of the environment of the vegetation being measured is unavoidable. Two approaches to chamber design are apparent.

The first and most widely used is to design the chamber so that the environment of the plant part is altered as little as possible by enclosure, and thus rates of gas exchange by the enclosed plant parts may be assumed to approach those of similar unenclosed portions of the vegetation. To achieve this, the light, temperature and gas concentrations must be altered as little as possible by the enclosure. This requirement can be approached by careful choice of chamber materials, for example walls with a high thermal conductivity and windows with a high transmissivity in both the visible and infra-red wavelengths. In addition, materials with a low adsorptivity and a low permeability to the gases being measured are essential. The large errors resulting from use of an inappropriate material, such as acrylic plastic (Perspex), were well illustrated by Bloom *et al.* (1980). Materials appropriate for chamber construction and for the connecting tubing were reviewed by Bloom *et al.* (1980), Long & Hällgren (1985) and Parkinson (1985). The chamber gases must be well stirred to prevent the development of thermal and gaseous gradients within the chamber. The larger the chamber the more difficult it becomes to meet these requirements. Two widely used commercial systems, the LI-COR 6000/6200 (LI-COR Inc., Lincoln, Nebraska, USA) and the LCA/ASU/PLC system (Analytical Development Co., Hoddesdon, Herts., UK) are both based on small chambers designed to enclose a few square centimetres of leaf or shoot surface, with minimal modification of the ambient environment (Long 1986; Welles 1986).

The second approach is to control the environment of the chamber, such that the response of a leaf or plant in the field to variation in temperature, light, CO_2 concentration or humidity can be established for a given point in time. This approach is used in the Minicuvette commercial system (H. Waltz, Effeltreich, West Germany). Such systems are particularly useful for testing hypotheses developed from simple observations of CO_2 uptake in the field. Variation in many microclimatic parameters, e.g. leaf temperature, photon flux density and leaf–air water vapour deficit are highly correlated, and thus it is difficult to establish which may be causal to a change in CO_2 uptake and which simply correlated. Field systems in which the chamber environment may be controlled allow this separation. Given that some modification of the natural environment is inevitable in an enclosure, a further adaptation of controlled environment chambers is to track the natural environment. Thus, external sensors measure the temperature, humidity and CO_2 concentration of the environment of the vegetation outside the chamber and this information is then used to adjust the internal environment of the chamber to track that of the outside. Although the advent of portable multichannel microprocessors have made this approach more practical, such systems are still inevitably complex.

System configurations

Configurations of equipment for the measurement of CO_2 assimilation by the enclosed vegetation or plant parts may be classified into three groups: closed, semi-closed and open systems. All three have been used in field gas exchange measurements. The technical advantages and limitations of each type have been reviewed in detail previously (Sestak, Catsky & Jarvis 1971; Long & Hällgren 1985). In a closed system air is pumped from the chamber enclosing the leaf or plant into an IRGA which continuously records the CO_2 concentration of the system. The air is then recycled back to the chamber, and thus no air leaves or enters the system. If the vegetation is photosynthesizing, then the concentration of CO_2 in the system will fall in proportion to the net rate of CO_2 uptake. This configuration is the simplest, since no accurate measure of flow is required, and since recirculation effectively amplifies the depletion of CO_2, a relatively insensitive 1–5 μmol mol^{-1} infra-red CO_2 analyser (IRGA) is adequate. However, this configuration also has important practical and theoretical limitations. Since the gas is recirculated, errors resulting from leakage of CO_2 into the system or adsorption/desorption from chamber and tubing walls will be amplifed by the continuous recirculation. Since the CO_2 concentration will be continuously changing through a measurement, steady-state CO_2 uptake cannot exist.

The semi-closed system is a variation on the closed system which avoids this latter limitation. Here the CO_2 sensor is used as a null-point device, which senses any decline in the CO_2 concentration of the system. If this falls, then a flow of CO_2 into the system is triggered until the system concentration has returned to its pre-set value. The rate of CO_2 uptake will equal the rate of CO_2 addition to the system. Such systems therefore require both an accurate flowmeter and IRGA and a feedback control system. In addition, a supply of compressed CO_2 is needed, limiting the portability of these systems. However, the advantage is that response to a range of CO_2 concentrations can readily be obtained. Field semi-closed systems have been described by Bingham & Coyne (1977).

The third type of system configuration is the open system, characterized by a net flow of air through the system (Fig. 2.2). A differential IRGA is required to compare the CO_2 concentration of the air before and after the leaf chamber. In theory, the rate of CO_2 uptake per unit surface area (A) may be determined as follows:

$$A = \frac{f(c_e - c_o)}{s} \qquad (2.3)$$

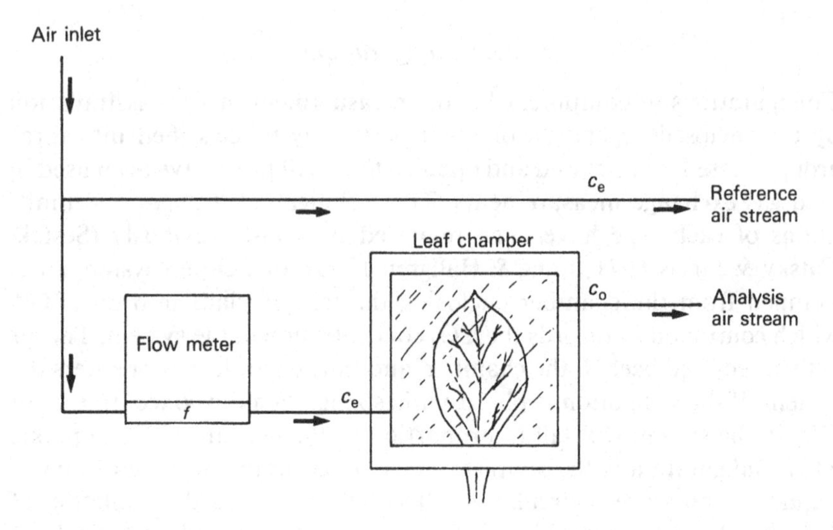

FIG 2.2 Typical arrangement of a simple open gas-exchange system, where f is the mole flow of air through the leaf chamber and $c_e - c_o$ is the difference in the mole fraction CO_2 in air across the chamber. Subscripts e and o refer to the chamber entrance and outlet, respectively.

where f is the flow of air through the leaf chamber (mol s^{-1}), c_e is the CO_2 concentration at the chamber entrance (μmol mol^{-1}), c_o is the CO_2 concentration at the chamber outlet (μmol mol^{-1}), s is the leaf surface area enclosed (m^2).

Although forms of this equation were widely used in the earlier work involving open systems, serious errors arising from interference from water vapour have now been recognized. These may all be removed by drying the air before it enters the infra-red gas analyser. However, this can introduce further problems since many driers can themselves absorb CO_2, although some, for example calcium chloride and magnesium perchlorate, have very little effect (Long & Hällgren 1985). Even when there is no chemical interference, addition of drying columns will retard the time response of the system and increase the bulk and complexity of a portable system. Humidity produces three problems in gas-exchange systems. First, most infra-red CO_2 analysers have a cross-sensitivity to water vapour, so that if the humidity of the air increases across the chamber an increase in CO_2 concentration may be indicated. This erroneous signal may be eliminated by addition of optical filters which remove the water vapour-absorbing bands from the radiation beam before it reaches the detector. Secondly, increase in the water vapour concentration in the chamber by transpiration will effectively dilute the concentration of all other gases, including CO_2. Thus, if in reality there is no net CO_2 uptake but the leaf is transpiring, an uptake of CO_2 would be indicated by eqn 2.3 simply because the addition of water vapour has caused a reduction in c_o. If the change in humidity across the chamber is known, then this dilution of CO_2 may be corrected for as follows:

$$A = \frac{f}{s} \frac{(c_e - c_o)(1 - e_o)}{(1 - e_e)} \tag{2.4}$$

where e_o is the concentration of water vapour at the outlet (mol mol^{-1}), e_e is the concentration of water vapour at the inlet (mol mol^{-1}).

Thirdly, the presence of water vapour alters the absorption spectrum of CO_2, causing band-broadening. This will therefore alter the response of the analyser to a change in CO_2 concentration. The resulting error, for a given humidity, will depend on the properties of the source, filters and detector used in the analyser. Since the properties of sources and detectors can change with age this error may even vary between analysers of the same make and thus empirical correction for this error is essential. These considerations are of equal importance to closed and semi-closed systems.

In all three systems the accuracy with which CO_2 uptake is determined will depend on the accuracy of the CO_2 analyser, whilst in semi-closed and open systems it will depend also on the accuracy of the flowmeter used. Eqn 2.3 shows that error in A will be equally dependent on the accuracy of both the gas analyser ($c_e - c_o$) and the flowmeter (f). These key components of systems for measuring gas exchange in the field are considered in the subsequent sections.

INFRA-RED GAS ANALYSERS

The most widespread method of determining CO_2 concentration in gas exchange systems and in micro-meteorological approaches is by infra-red gas analysis. The basic structure and operation of these instruments have been described in detail by Long & Hällgren (1985), and recent developments in these instruments reviewed by Jarvis & Sandford (1985). This section will therefore be limited to an outline of these instruments and developments of particular importance for measurement of CO_2 uptake in the field.

Principles

Heteratomic gas molecules absorb radiation at specific sub-millimetre infra-red wavebands, each gas having a characteristic absorption spectrum. The major absorption band of CO_2 is at 4.25 μm. The IRGA is a spectrophotometer specifically designed to measure the concentration of CO_2 in a gas cuvette. Table 2.1 summarizes the capabilities of some of the instruments currently available. An IRGA consists of three key parts: infra-red source, gas cell and detector arranged in series along an optical bench. The loss of infra-red radiation reaching the detector will be directly proportional to the concentration of CO_2 in the gas cell. Most IRGAs used in biological applications are of the non-dispersive type, that is, broad-band radiation is emitted by the source and passed through the gas cell. Selectivity is achieved at the detector, either by using a detector which can sense only the CO_2-absorbing wavebands or by use of a 4.25 μm band-pass filter in front of the detector. For true differential measurements of CO_2 uptake a dual-beam instrument is required. Here the beam is split between reference and analysis gas cells, and the decrease in radiation intensity between the two cells compared. The most widely used detector for infra-red gas analysers of sufficient sensitivity for measurement of CO_2 uptake in photosynthesis has been the Luft type, which operates on the principle of positive filtration, i.e. it absorbs infra-red in the CO_2 absorption bands. This is achieved simply by filling the detector with CO_2. The detector is divided into two chambers separated by a thin diaphragm which forms one electrode of a diaphragm condenser. In parallel configuration, radiation passing through the reference gas cell enters one detector chamber and radiation passing through the analysis cell enters the other. Both detector chambers will absorb radiation in the CO_2-absorbing bands, the amount available being inversely proportional to the amounts absorbed in the cells. Any difference in CO_2 concentration between the two cells will cause a pressure difference across the detector membrane in proportion to this difference. As the radiation is chopped the membrane will vibrate, the amplitude being proportional to the pressure difference and measured as a change in condenser capacity.

Detector developments

Whilst providing an extremely sensitive method of detecting small differences in CO_2 concentration (e.g. 0.1 μmol mol^{-1}), Luft detectors have two serious limitations for use in field instruments. First, the principle of detection used limits the minimum size of these detectors, preventing miniaturization without substantial loss of precision. Secondly, the presence of a vibrating membrane makes the instrument inherently sensitive to external vibration. Two solutions are illustrated in the current generation of portable gas analysers.

In the Binos-II (Leybold-Heraeus GMBH, Hanau, West Germany) the Luft detector has been modified by the addition of a connecting tube, incorporating a miniature mass-flow detector, between the two detector chambers. A difference in pressure between the two cells therefore creates a flow between the two cells in proportion to the pressure and thus CO_2 difference. This allows the production of a considerably smaller and less bulky detector which is insensitive to vibration. However, because the pressure difference is generated by a small temperature difference, the detector must be thermostatically controlled. This increases the power consumption of the instrument and hence reduces its portability.

A further alternative to the Luft detector is provided by the solid-state detector. Generally, these are broad-band pyroelectric detectors. They are internally polarized and produce a voltage proportional to temperature change, i.e. when they receive a pulse of infra-red radiation. These detectors are therefore very sensitive to ambient temperature change. In the portable dual-beam LI-6251 (Li-Cor Inc.) the temperature of the solid-state detector is cooled and maintained at a constant temperature by a peltier unit. Not only does this isolate the detector from ambient temperature fluctuations, but because of the effects of temperature on these detectors also increases the sensitivity. By using cells of 15 cm length, this instrument, weighing only 5 kg, is capable of a resolution of up to 0.1 $\mu mol\ mol^{-1}$, and thus whilst fully portable it is comparable in sensitivity to much heavier and larger mains-operated laboratory IRGAs (Table 2.1).

In the LCA II (ADC Ltd) a different solution to the temperature sensitivity of solid-state analysers is seen. This instrument uses a single gas cell. Here, radiation from the source passes through the sample gas and into the detector via a narrow band-pass thin-film filter (to isolate the 4.25 μm absorption band). For measurement of the absolute concentration the instrument pumps the sample gas through the cell for 2 s, and then diverts it by the switching of a solenoid valve through a column of soda-lime. This removes all CO_2 from the gas stream prior to entry into the cell. After 2 s the valve is switched back, allowing the sample gas, complete with its CO_2, to enter the cell again and so starting another cycle. The difference in energy reaching the detector in the two half-cycles is thus proportional to the CO_2 concentration of the sample. Although this is a single-celled instrument, a pseudo-differential measurement may be made. A further solenoid valve allows the cell to analyse first the reference gas stream and then the analysis gas stream. The two absolute concentrations are stored and then subtracted at the end of two measurement cycles. This technique of gas-alternation or gas-chopping counteracts the inherent instability of the solid-state detectors

TABLE 2.1. Infra-red gas analysers: manufacturers' specifications

Instrument	1	2	3	4	5	6	7	8	9	10	11	12
Differential capability	+		+		(+)	+	+	+	+	+		+
Maximum range (%)	0.15	100	0.25	50	0.11	100	100	100	100	100	0.3	100
Minimum range (μmol mol^{-1})	1500	1000	25(5)	500	1100	50	20	20	5	400	3000	30
No. of ranges	1	2	6	2	1	1	4	4		3	1	4
Precision (μmol mol^{-1})	0.2	20	0.2	3	1	0.2	0.5	0.1	0.1	2		1
Detector:												
Type[1]	S	S	M	S	S	F	M	M	S	M	F	M
Configuration[2]			P			S/P	S	S		P		S
Sample tube:												
Volume (cm^3)	11.9	0.5–100	(0.2)–107	10–15	0.8		8.5					0.1–50
Length (cm)[3]	15.2	0.1–20	(0.5)–25	0.1–20	6.0	V	6.0	21.0		V		0.04–20
Analogue output:												
mA		(+)		+	+	+	+	+		(+)		+
mV/V	+	+	+	+		+	(+)	+	+	+	+	+
Digital output		(+)	(+)		(+)	(+)	(+)	(+)	(+)	(+)		+
Integral printer									+			+
Total weight (kg)	5.0	8.0	27.0	2.5	5.0	6.5	38.0	27.0	23.6	20.0	4.0	15.0
Dimensions:												
Width (cm)	33.0	16.0	50.0	23.0	13.0	21.0	57.0	42.0	83.0	22.0	17.0	43.0
Depth (cm)	20.0	45.0	38.0	13.0	25.0	21.0	45.0	29.0	34.0	28.0	15.0	40.0
Height (cm)	11.0	20.0	28.0	35.0	18.0	13.0	36.0	30.0	19.0	57.0	29.0	17.0
Casing:												
Metal	+	+	+	+		+	+	+	+	+	+	+
Weatherproof								+				
Power supply[4]	1/2/D	(1)/2	1/2	2	(1)/2/D	1/2	1/2	1/2	1/2	1/2	1/2	2
Operating environment:												
Maximum temp. (°C)	50	40	45	40	45	45	40	45	50	50	50	50
Minimum temp. (°C)	0	0	0	10	0	0	0	5	0	0	0	0
Humidity (%RH)	80	90	90		80		75	75	90	95		

Instruments:

1. LI-6251 Infrared CO_2 Analyzer: Li-Cor inc., Lincoln, Nebraska, USA.
2. SB200 Portable IRGA: Analytical Development Co. Ltd., Hoddesdon, UK.
3. Type 225 Mk. 3 IRGA: Analytical Development Co. Ltd., Hoddesdon, UK.
4. Model QGD-07 Portable IRGA: Beijing Analytical Instrument Factory, Beijing, China.
5. L.C.A.2: Analytical Development Co. Ltd., Hoddesdon, UK.
6. BINOS (1 channel): Leybold-Heraeus, Hanau, West Germany.
7. UNOR 4N: H. Maihak A.G., Hamburg, West Germany.
8. Uras 3G: Hartman & Braun AG., Frankfurt, West Germany.
9. Miran 80 Computing Gas Analyser: Foxboro Co., Foxboro, Mass., USA.
10. LIRA Model 3000: M.S.A., Coatbridge, UK.
11. CO_2 Analyser: Siemens, Congleton, UK.
12. Diamant 6000: COSMA, Igny, France.

[1] Detector type: M = microphone; F = microflow: S = solid state.
[2] Detector configuration: S = series; P = parallel.
[3] Length: V = various.
[4] Power supply: 1 = 110/120 Vac 50–60 Hz; 2 = 220/240 Vac 50–60 Hz; D = d.c.

without the need to thermostat the detector. The LCA instrument has a sensitivity of 1(−0.5) μmol mol^{-1} using a path-length of just 5 cm. The instrument complete with power supply weighs just 2.8 kg. A new version of the LCA, introduced in 1988, promises an even higher resolution of 0·2 μmol mol^{-1} (R. Coombs, personal communication). The new LCA may therefore promise to provide the sensitivity of a laboratory instrument in a highly portable mode. The potential disadvantages of the instrument are also a result of the 'gas-chopping'. First, the flow rate must be sufficient to flush the cell within the 2 s cycle. Given the sensitivity of the analyser this means that the instrument is limited in applications using small leaf areas, e.g. fine-leaved grasses, or for 'stressed' material where rates of leaf gas exchange may be markedly reduced. Secondly, since CO_2 is removed from the gas stream during the course of measurement, the LCA cannot be used in systems in which all or part of the air-stream is recirculated; thus it is inappropriate for semi-closed and closed systems. Thirdly, since 8 s are required to complete a single pseudo-differential measurement, the instrument cannot be used for resolving transient responses. However, for a number of field applications these may not be limitations, but will merely restrict the versatility of the instrument.

Previously, developments in the instrumentation available for the study of CO_2 exchange by plants have largely reflected the needs of the chemical and defence industries, and more recently the pollution control industry. The fact that some of these instruments could be used in plant ecology was perhaps purely fortuitous. The development of both the LI-6251 and the LCA shows a clear departure from this pattern, since both instruments are designed specifically to meet the needs of the field ecologist. This recognition by manufacturers may mean that physiological ecologists will be able to consider increasingly what equipment is required to answer their ecological questions rather than what ecological question might be answerable with the available equipment. These instruments (Table 2.1) form the basis of two portable gas exchange systems, whilst the Binos-II IRGA (Leybold-Heraeus) is used in the 'Minicuvette' and 'CO_2/H_2O porometer' portable systems from H. Walz Mess und Regeltechnik (Effeltrich, West Germany).

FLOW MEASUREMENT

In both open and semi-closed systems the accuracy of the estimates of CO_2 uptake is directly proportional to the accuracy with which flow is measured (eqn 2.3). It is surprising, then, that measurement of flow, in contrast to the measurement of CO_2 concentration, has received little attention in texts on measurement of plant gas exchange. Sestak, Catsky & Jarvis (1971) devoted thirty-five pages to the measurement of CO_2 concentration, but just sixteen lines to measurement of flow rate. The topic of flow rate measurement in gas exchange systems has been reviewed recently (Long & Ireland 1985), whilst more general surveys of pipeline flowmeters are provided by Ower & Pankhurst (1977) and Brain & Scott (1982). The purpose of this section is to consider recent developments of particular relevance to field measurements.

Until the 1980s most gas-exchange systems for the field measurement of CO_2 uptake used variable-area flowmeters. These consist of a transparent graduated tube with a slightly tapered bore, in which the diameter decreases towards the base of the tube while the gas flows upwards. A float of diameter slightly smaller than the minimum bore of the tube is forced by the flow of the gas to the point where its weight is balanced by the force of gas flowing past. With long tubes an accuracy of $\pm$ 2% may be achieved. However, calibration values are altered in a complex manner as a result of fluctuations in temperature and pressure, by small traces of dirt in the tube, and by any deviation from a vertical attitude (Ower & Pankhurst 1977). Thus, these flowmeters are poorly suited to field systems and inappropriate for any electrical recording of flow.

A solution to these problems is provided by the thermal mass flowmeter. Although first described over 70 years ago (reviewed by Long & Ireland 1985), these instruments have been used in gas-exchange systems only within the past decade. Thermal mass flowmeters utilize the thermodynamic principle that the heat carried by a gas flow is related to the heat capacity of the gas and the mass of gas flowing. They consist of a sensor tube which carries the flow, and which is precisely heated such that in the absence of any net flow the temperature distribution is symmetrical about the mid-point. Two temperature sensors, typically platinum resistance thermometers or thermocouples, are situated one each side of and equidistant from the heated mid-point. With no gas flow the temperature at both sensors will be equal. Gas flow will transfer heat downstream, causing an increase in temperature downstream which will be a function of the rate of mass flow. Since these instruments measure the mass of gas flowing they are in theory insensitive to fluctuations in temperature and pressure, and by design produce an electrical signal. Several commercial mass flowmeters suitable for incorporation into portable gas-exchange systems are now available, weighing in the order of 200 g, with a power requirement of under 3–4 W and providing an accuracy of $\pm$1% to $\pm$0.1% (Bingham & Long 1988). These instruments are thus nearly ideal for the requirements of field gas-exchange systems. It is surprising that the potential of this technique, well known in the engineering and physics literature, was not realized earlier by plant biologists.

ADVANCES IN UNDERSTANDING OF PHOTOSYNTHESIS IN THE FIELD AND NEW DIRECTIONS

So far this review has considered developments in the technical procedures for measuring CO_2 exchange in the field. Some of the contributions of field measurements to understanding photosynthesis will now be considered. These may be divided between two main categories: (i) description of diurnal, weekly and seasonal patterns of gas exchange and productivity, as illustrated under the section on micro-meteorological techniques; and (ii) the responses of leaf CO_2 uptake to environmental variables. The latter category encompasses a wide range of variables from atmospheric pollutants and mineral nutrients to humidity and light (see review volumes of Grace, Ford & Jarvis 1981; Mitsui & Black 1982; Beadle *et al.* 1985; Baker & Long 1986). Of fundamental importance among these is the response of photosynthesis to light.

Response to light

Understanding of the response of CO_2 uptake to light is important not simply because of the fundamental role of light in photosynthesis, but also because the character of this response is of importance to interpretation of responses to other variables. For example, the character of the response of CO_2 uptake to temperature changes markedly with light (Fig. 2.3). Analysis of the response of CO_2 uptake by leaves to light has a long history. Blackman (1905) provided the first clear evidence of the biphasic nature of this response, with a light-limited and a light-saturated phase, whilst Bjorkman (1968) demonstrated genetically fixed differences in this response between sun and shade ecotypes.

The light response may be described by three basic parameters. A_{sat}, the light-saturated rate of CO_2 uptake, ϕ, the maximum quantum efficiency, and θ, a measure of the convexity of the portion of the curve connecting the two linear phases (Jarvis & Sandford 1986). At high light levels, A_{sat} will dominate, whilst at low light levels ϕ will dominate (Fig. 2.4). Several different formulae have been developed to describe this

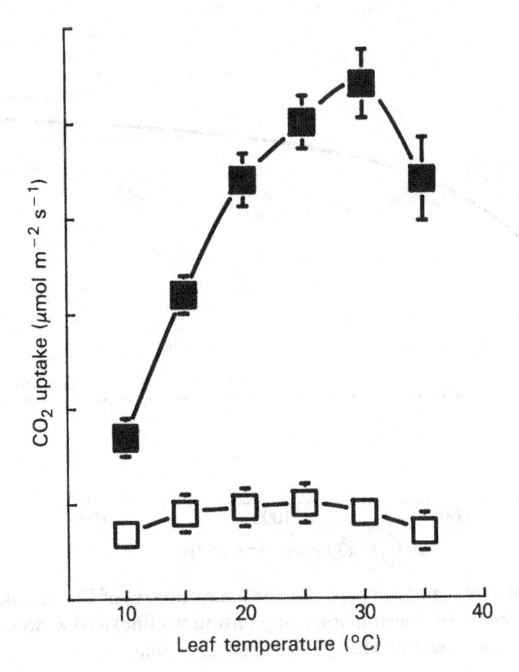

FIG 2.3 The response of CO_2 uptake (A) in leaves of *Spartina anglica* to variation in leaf temperature in a photon flux density (Q) of 130 μmol m^{-2} s^{-1} (□) or 2500 μmol m^{-2} s^{-1} (■); unpublished data of S. P. Long.

response, but a simple form of this response is given by the non-rectangular hyperbola which is the positive solution to the simultaneous equation given by Thornley (1977):

$$A = \frac{\phi Q + A_{sat} - \sqrt{[(\phi Q + A_{sat})^2 - 4\theta(\phi Q A_{sat})]}}{2\theta} \tag{2.5}$$

where Q is the photon flux density incident on the leaf and ϕ, θ and A_{sat} are the quantum efficiency, convexity coefficient and light-saturated rate of CO_2 uptake.

Most ecological studies have concentrated on CO_2 uptake under light-saturating conditions, yielding a large body of data on the effects of a wide range of environmental and physiological variables on A_{sat}. These data show that A_{sat} is highly variable, differing by orders of magnitude between species, varying with age, with position within the leaf and canopy, and with a wide range of environmental variables. Several factors now bring into question this predilection for A_{sat}. First, agricultural studies have shown that there is little or no correlation between A_{sat} and productivity within many crops. Secondly, in the field situation and

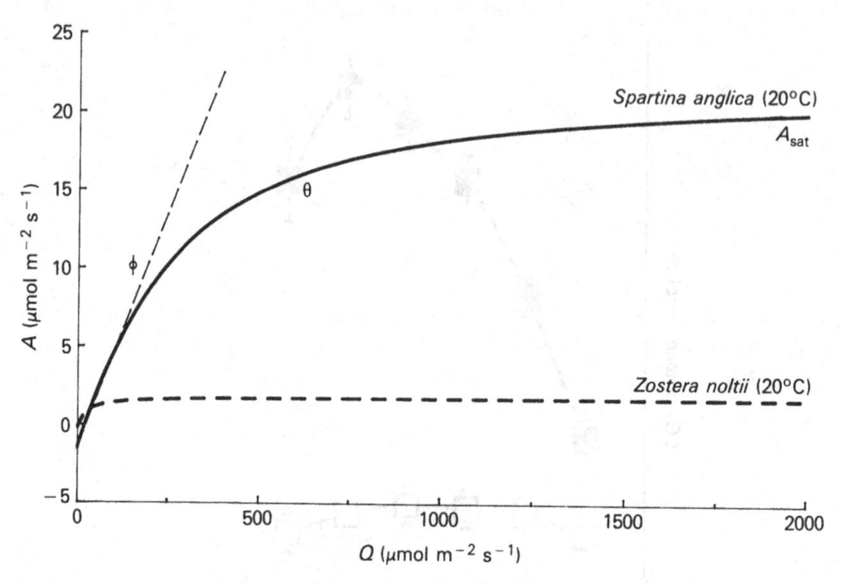

FIG 2.4 Fitted non-rectangular hyperbolae for the responses of CO_2 uptake (A) to photon flux density (Q) for leaves of *Spartina anglica* (continuous line) and *Zostera noltii* (thick pecked line); unpublished data of M. J. Roberts & S. P. Long.

particularly in canopies with a high leaf area index (LAI), a large proportion of the leaves will experience light-limiting conditions, even in full sunlight, since many leaves will be shaded by others. Only those leaves at the top of the canopy are likely to receive sufficient light to become saturated for any significant period and then even these leaves will be saturated only for a part of the day, assuming sunny conditions. Under dull conditions, or outside the summer at high latitudes, there may be periods when no leaves approach light saturation. The significance of this may be seen in measurements of CO_2 assimilation by whole stands. Unlike single leaves, stands of plants will show a much longer phase in which CO_2 assimilation responds linearly to increase in light. For example, the rate of CO_2 uptake by a canopy (LAI = 5) of *Cynodon dactylon* showed a linear increase with increase in photon flux density from 0 to 2500 μmol $m^{-2}s^{-1}$, i.e. full sunlight (Brown 1982). This suggests that CO_2 uptake by canopies is determined more by ϕ than by A_{sat}.

Indirectly, studies of dry matter production also suggest the same importance of ϕ. Monteith (1977) demonstrated that dry matter production for a number of temperate C_3 crops showed a common linear relationship of productivity on intercepted light. This could be explained if ϕ, which is relatively constant between species, rather than A_{sat}, which is highly variable, dominated the total canopy rate of CO_2 uptake.

Although earlier studies suggested that ϕ varied with temperature and between species (Ludlow & Wilson 1971), more recent studies have suggested that within a photosynthetic type ϕ is remarkably constant (Ehleringer & Bjorkman 1977; Pearcy & Ehleringer 1984; Bjorkman & Demmig 1987). In C_3 plants ϕ decreases from about 0.065 at 10 °C to 0.040 at 38 °C because of changes in the affinity of ribulose bisphosphate carboxylase-oxygenase (Rubisco) for O_2 relative to CO_2 and in the relative solubilities of the two gases. This dependence on temperature is removed if ϕ is measured in 1% O_2 so that oxygenation of RubP is effectively eliminated. Under these conditions a range of C_3 species showed a ϕ of 0.073 ± 0.001 (Ehleringer & Björkman 1977) at all temperatures. Similarly, C_4 species which inherently lack photorespiration, show a ϕ of 0.060–0.065 (Long 1983; Pearcy & Ehleringer 1984). This value is virtually insensitive to temperature above the ranges which will induce chilling injury, and below those which will cause heat damage.

Analysis of the energy requirements of photosynthesis suggest that these empirically observed values are very close to the theoretical maxima, the lower value in C_4 species resulting from the increased energetic cost of assimilation of CO_2 via the various photosynthetic C_4 pathways (Pearcy & Ehleringer 1984; Long 1985).

The conserved nature of ϕ might be of considerable ecological significance. When competing for light, any plant with a reduced ϕ would be at an immediate disadvantage, and it might therefore be expected that there is very strong selective pressure for the maintenance of a high ϕ. A test of the significance of ϕ in determining the success of species is provided by the distribution of C_4 and C_3 grasses. In a normal atmosphere C_3 grasses have a higher ϕ than C_4 grasses at low temperatures, and a lower ϕ at high temperatures. Thus we might expect that, along a thermal gradient, C_4 grasses will replace C_3 as the mean temperature increases. Ehleringer showed that the abundance of C_3 and C_4 species in the Great Plains could be closely predicted by this theory (Ehleringer 1979). In cool climates such as that of the British Isles, the few native C_4 plants are largely species which grow in habitats where few other species can grow, e.g. *Spartina anglica* in the pioneer zone of salt marshes and *Salsola kali* on coastal foredunes. In both situations the level of interspecific competition for light will be low or non-existent (Long 1983). One exception is *Cyperus longus*, which occurs on pond and lake margins in south-western England (Jones, Hannon & Coffey 1981). The main photosynthetic organs of this species are the bracts of the inflorescence, which are borne on stems about 1 m tall. One possibility is that this species is able to overcome the disadvantage of a relatively lower ϕ by bearing its photosynthetic organs high in the canopy in a way which minimizes shading from other species and thus reduces the disadvantage of a lower ϕ.

When grown under optimal conditions leaves appear to attain a ϕ which is close to the theoretical maximum for a given photosynthetic pathway. The key question now from an ecological viewpoint is the extent to which sub-optimal conditions can reduce ϕ, since, given the importance of ϕ in determining canopy photosynthesis, any environmental factor reducing ϕ must be expected to lower the fitness of the individual if the ϕ of other individuals shows a greater resistance. The variability in ϕ in the field is poorly understood. Technically, measurement of ϕ in the field is more difficult than measuring A_{sat}. To determine ϕ the initial slope of the light response curve must be established. This requires measurement of A at a number of low photon flux densities to determine both the slope and its duration. Thus, the light level around the leaf must be controlled very closely, and this increases the complexity of the measurement. In conventional gas-exchange chambers the light response curve is based on the incident light level. However, ϕ may vary because the absorptivity of the leaf has changed or because of a change in the efficiency of energy transduction inside the leaf. In a dense canopy a decrease in absorptivity may be of little significance to canopy photosynthesis, whereas a decrease in the efficiency of energy transduction within the leaf is likely to have a marked effect. Separation of these two factors is therefore important. In the laboratory, the absorptivity of the leaf may be determined after measurement in a conventional leaf chamber by use of a Taylor integrating sphere (Bongi & Long 1987). This, however, is not practical in the field. An alternative is to use an Ulbricht integrating sphere for the simultaneous measurement of light absorptivity and CO_2 uptake. This approach has been used for the determination of ϕ_{abs} in the laboratory (Idle & Proctor 1983) and has recently been adapted in a portable chamber for determination of ϕ_{abs} under field conditions (Baker *et al.* 1989).

Photoinhibition

One factor which is known to decrease ϕ_{abs} under field conditions is excess light. High light, in excess of full sunlight, produces a reduction in both ϕ_{abs} and A_{sat} through a type of damage to the photosynthetic apparatus termed photoinhibition (Long 1983; Powles 1984). The threshold for this damage is reduced if other environmental factors influencing photosynthesis are sufficiently sub-optimal to cause a partial inhibition of photosynthesis. Thus, under chilling or freezing conditions, under drought or at high temperatures, photoinhibition will occur at photon flux densities well below full sunlight (Powles 1984). By the use of an integrating sphere it has been possible to show that ϕ_{abs} is decreased in crops of maize (*Zea mays*) growing in eastern England during periods of the spring when chilling temperatures coincide with high photon flux densities (Farage & Long 1986) and in crops of winter rape (*Brassica napus*) during the winter when freezing conditions coincide with sunshine (P. K. Farage & S. P. Long, unpublished data). These decreases in ϕ_{abs} are closely correlated with a decrease in light conversion efficiency by the crop. Similar changes have recently been observed during the winter and spring in natural stands of *Ilex aquifolium* in southern England (Q. Groom, S. P. Long & N. R. Baker, unpublished data) and *Pinus sylvestris* in Sweden (G. Öquist, personal communication).

Stomata and limitation to CO_2 uptake by plants in the field

Twenty-five years ago it was widely accepted that stomata represented a major limitation, if not *the* major limitation, to uptake of CO_2 by leaves. The electrical analogue model of CO_2 uptake by leaves was first proposed in the 1920s (cf. Sestak, Catsky & Jarvis 1971). Basically this viewed the diffusion pathway as a series of resistors to CO_2 diffusion, principally the boundary layer, the stomata and the mesophyll. Since these are resistors in series, the limitation imposed by each on the rate of photosynthesis is directly proportional to its magnitude. Stomatal and boundary layer resistances are conveniently determined from simultaneous measurements of transpiration, since water vapour escapes from the leaf via the route of CO_2 uptake. Stomatal resistance is commonly found to increase under circumstances where photosynthesis is inhibited, e.g. drought, exposure to SO_2 and chilling damage (Farquhar & Sharkey 1982). Application of the resistance analogue suggested that this change in stomatal resistance was often the key cause of reduction in photosynthetic CO_2 uptake. The flaw in this approach was that the analogue assumes a linear response of flux to CO_2 concentration and, whilst this is correct for the gaseous diffusion pathway, within the mesophyll the uptake of CO_2 shows a hyperbolic response to intercellular CO_2 concentration (c_i).

Although this flaw was recognized, its full significance was not realized until the late 1970s. Two factors indicated that the conclusions drawn from the resistance analogue analyses overstated the role of stomata. First, it was noted that under many situations where CO_2 uptake is inhibited, c_i remains constant or rises. If the stomata were the major limitation, a decrease in c_i would be expected. Secondly, Farquhar & Sharkey (1982) developed a new method of separating stomatal and mesophyll limitations using the response of A to c_i. After measurement at the normal atmospheric CO_2 concentration (340 μmol mol^{-1}) A is subtracted from A_o, the rate which would occur if there was no stomatal limitation, i.e. the value of A interpolated from the response curve at c_i = 340 μmol mol^{-1}. The relative limitation (l) which the stomata impose may then be calculated as

$$l = (A_o - A)/A_o. \tag{2.6}$$

The method has the important advantage, when calculated graphically, that it makes no assumptions about the shape of the response of A to c_i. By applying this technique in a reappraisal of some data on the effects of water stress on A, Farquhar & Sharkey (1982) have shown that even where marked decreases in stomatal conductance occur, c_i may rise and l diminish, indicating that increased limitation within the mesophyll must in fact dominate the decline in A. However, in estimating the limitation imposed by stomata, both the method of Farquhar & Sharkey (1982) and the earlier resistance analogues assume that all stomata on a leaf or portion of a leaf behave in a similar manner. The recent studies of Terashima *et al.* (1988) show that there may be marked heterogeneity in the stomatal response in sunflower leaves, such that, whilst the stomata serving some parts of the leaf are open, those in other parts of the leaf serving specific regions of the intercellular air-space are so tightly closed that there may be no CO_2 uptake there. In this situation photosynthesis in half of the leaf could be completely inhibited by stomatal closure, whilst in the other half there would be little or no stomatal limitation. An A/c_i curve determined for such a leaf could erroneously suggest little or no stomatal limitation (Terashima *et al.* 1988).

WHERE NEXT?

The technical advances of the past 25 years have made an important contribution toward a more exact ecology. They have allowed detailed resolution of CO_2 uptake in time and space, separation of respiratory losses from photosynthetic gains, and observation of CO_2 exchange concurrent with measurements of other physiological parameters. These in turn have suggested key questions which could be resolved only under controlled environment conditions or by manipulating the field environment. The wider availability of laboratory instrumentation for determining gas exchange of whole plants and individual leaves has led to a profusion of information about responses to individual environmental variables, and development of models of these responses. Field measurements have played a key role in the testing of these hypotheses and validation of models generated from these laboratory studies. There has therefore been a great advance in the exactness of our understanding of what really regulates photosynthesis in the field, and how this varies between species and between communities. There have also been some setbacks, most notably with regard to the role of stomata. What is less clear is just how important differences in rates of photosynthesis are in determining the distribution and abundance of species.

The advances made in understanding the gas exchange of plants in the field have largely reflected technical developments. The next 25 years will almost certainly see increased miniaturization of equipment. Lasers are already widely used in infra-red spectroscopy (Sheehy 1985) and it should be only a short time before these become available for CO_2. Because they produce monochromatic, narrow-beam radiation, many of the problems of conventional instruments, discussed earlier, would be avoided. The development of miniature laser sources, coupled with further developments of infra-red solid-state detectors, would lead to further miniaturization of infra-red CO_2 analysers, such that the analyser could form an integral part of a small leaf chamber, rather than an external instrument. This would mean that a field gas-exchange system could consist simply of a hand-held cuvette with a dramatically improved response time. Such an instrument would be particularly valuable in assessing the significance of transients in the field. Analysis of the responses of CO_2 uptake to environmental variables have concentrated on steady-state conditions. However, the field environment is continually changing, particularly with respect to light. In the determination of productivity in the field environment, the speed with which a leaf can respond to environmental changes could be as important as the final steady-state value obtained, or even more important.

Other probes of photosynthesis have seen considerable development over the past few years. Thus, it is already possible to measure photosystem II fluorescence *in vivo* and the absorptance of the photosystem I reaction centre simultaneously with measurement of CO_2 uptake and transpiration (Ireland, Long & Baker 1984). The potential is already there for field systems in which CO_2 uptake may be measured concurrently with measurement of the rate of electron transport through the two reaction centres, whilst by further manipulation of the CO_2

concentration in the chamber the carboxylation efficiency and stomatal limitation could be assessed. Thus, future systems will not only be capable of detecting a change in the capacity for CO_2 uptake, but also be capable of pin-pointing *in vivo* the step of the photosynthetic process that has effected the change.

REFERENCES

Allen, L. H. & Lemon, E. R. (1976). Carbon dioxide exchange and turbulence in a Costa Rican tropical rain forest. *Vegetation and the Atmosphere,* Vol. 2 (Ed. by J. L. Monteith), pp. 265–308. Academic Press, London.

Baker, N. R., Bradbury, M., Farage, P. K., Ireland, C. R. & Long, S. P. (1989). Measurements of quantum yield of carbon assimilation and chlorophyll fluorescence for assessment of photosynthetic performance of crops in the field. *Philosophical Transactions of the Royal Society,* **B, 323,** 295–308.

Baker, N. R. & Long, S. P. (Eds) **(1986).** *Photosynthesis in Contrasting Environments.* Elsevier, Amsterdam.

Beadle, C. L., Long, S. P., Imbamba, S. K., Olembo R. J. & Hall, D. O. (1985). *Photosynthesis in Relation to Plant Production in Terrestrial Ecosystems.* United Nations Environment Programme, Tycooly International, Oxford.

Bingham, G. E. & Coyne, P. I. (1977). A portable, temperature-controlled, steady-state porometer for field measurements of transpiration and photosynthesis. *Photosynthetica,* **11,** 148–60.

Bingham, G. E., Gillespie, C. H., McQuaid, J. H. & Dooley D. F. (1981). A miniature, battery powered, pyroelectric detector-based differential infra-red absorption sensor for ambient concentrations of carbon dioxide. *Ferroelectrics,* **34,** 15–19.

Bingham, M. J. & Long, S. P. (1988). *Equipment for Crop and Environmental Physiology: Specifications, Sources and Costs.* University of Essex, Colchester.

Biscoe, P. V., Scott, R. K. & Monteith, J. L. (1975). Barley and its environment. III. Carbon budget of the stand. *Journal of Applied Ecology,* **12,** 269–93.

Björkman, O. (1968). Further studies on differentiation of photosynthetic properties in sun and shade ecotypes of *Solidago virgaurea. Physiologia Plantarum,* **21,** 84–99.

Björkman, O. & Demmig, B. (1987). Photon yield of O_2 evolution and chlorophyll fluorescence characteristics at 77K among vascular plants of diverse origin. *Planta,* **170,** 489–504.

Blackman, F. F. (1905). Optima and limiting factors. *Annals of Botany,* **19,** 281–312.

Bloom, A. J., Mooney, H. A., Björkman, O. & Berry, J. (1980). Materials and methods for carbon dioxide and water vapour exchange analysis. *Plant, Cell and Environment,* **3,** 371–6.

Bongi, G. & Long, S. P. (1987). Light-dependent damage to photosynthesis in olive leaves during chilling and high temperature stress. *Plant, Cell and Environment,* **10,** 241–9.

Brach, E. J., Desjardins, R. L. & StAmour, G. T. (1981). Open path carbon dioxide analyser. *Journal of Physics E: Scientific Instruments,* **14,** 1415–19.

Brain, T. J. S. & Scott, R. W. W. (1982). Survey of pipeline flowmeters. *Journal of Physics E: Scientific Instruments,* **15,** 967–80.

Brown, R. H. (1982). Response of terrestrial plants to light quality, light intensity, temperature, CO_2, and O_2. *Handbook of Biosolar Resources,* Vol. 1, Part 2 (Ed. by A. Mitsui & C. C. Black), pp. 185–212. CRC Press, Boca Raton.

Campbell, G. S. & Unsworth, M. H. (1979). An inexpensive anemometer for eddy correlation. *Journal of Applied Meterology,* **18,** 1072–7.

Coombs, J., Hall, D. O., Long, S. P. & Scurlock, J. M. O. (Eds) **(1985).** *Techniques in Bioproductivity and Photosynthesis,* 2nd edn. Pergamon, Oxford.

Denmead, O. T. & Bradley, E. F. (1985). Flux gradient relationships in a forest canopy. *The Forest Atmosphere Interaction* (Ed. by B. A. Hutchinson & B.B. Hicks), pp. 421–42. Reidel, Dordrecht.

Desjardins, R. L., Brach, E. J., Alvo, P. & Schuepp, P. H. (1982). Aircraft monitoring of surface carbon dioxide exchange. *Science,* **216,** 733–5.

Ehleringer, J. R. (1979). Photosynthesis and photorespiration; biochemistry, physiology, and ecological implications. *Hortiscience,* **14,** 219–22.

Ehleringer, J. W. & Björkman, O. (1977). Quantum yields for CO_2 uptake in C_3 and C_4 plants. *Plant Physiology,* **59,** 86–90.

Farage, P. K. & Long, S. P. (1986), Damage to maize photosynthesis in the field during periods when chilling is combined with high photon fluxes. *Proceedings VIIth International Congress on Photosynthesis Research,* Vol. 4 (Ed. by W. J. Biggins), pp. 139–43. Nijhoff, Dordrecht.

Farquhar, G. D. & Sharkey, T. D. (1982). Stomatal conductance and photosynthesis. *Annual Reviews of Plant Physiology,* **33,** 317–76.

Farquhar, G. D., von Caemmerer, S. & Berry, J. A. (1980). A biochemical model of photosynthetic CO_2 assimilation in leaves of C_3 species. *Planta,* **149,** 78–90.

Gensler, W. G. (1986). *Advanced Agricultural Instrumentation.* Nijhoff, Dordrecht.

Grace, J., Ford, E. D. & Jarvis, P. G. (Eds) **(1981).** *Plants and their Atmospheric Environment.* Symposia of the British Ecological Society, 21. Blackwell Scientific Publications, Oxford.

Hesketh, J. D. & Jones, J. W. (Eds) **(1980).** *Predicting Photosynthesis for Ecosystem Models.* CRC Press, Boca Raton.

Idle, D. B. & Proctor, C. W. (1983). An integrating sphere leaf chamber. *Plant, Cell and Environment,* **6,** 437–40.

Ireland, C. R., Baker, N. R. & Long, S. P. (1985). The role of carbon dioxide and oxygen in determining chlorophyll fluorescence quenching during leaf development. *Planta,* **165,** 477–85.

Ireland, C. R., Long, S. P. & Baker, N. R. (1984). The relationship between CO_2 fixation and chlorophyll *a* fluorescence during induction of photosynthesis in maize leaves at different temperatures and CO_2 concentrations. *Planta,* **160,** 550–8.

Jarvis, P. G., James, G. B. & Landsberg, J. J. (1976), Coniferous forest. *Vegetation and the Atmosphere,* Vol. 2 (Ed. by J. L. Monteith), pp. 171–240. Academic Press, London.

Jarvis, P. G. & Sandford, A. P. (1985). The measurement of carbon dioxide in air. *Instrumentation for Environmental Physiology* (Ed. by B. Marshall & F. I. Woodward), pp. 29–57. Cambridge University Press.

Jarvis, P. G. & Sandford, A. P. (1986). Temperate forests. *Photosynthesis in Contrasting Environments* (Ed. by N. R. Baker & S. P. Long), pp. 199–236. Elsevier, Amsterdam.

Jones, M. B., Hannon, G. E. & Coffey, M. D. (1981). C_4 photosynthesis in *Cyperus longus* L., a species occurring in temperate climates. *Plant, Cell and Environment,* **4,** 161–8.

Legg, B. J. & Monteith, J. L. (1975). Heat and mass transfer within plant canopies. *Heat and Mass Transfer in the Biosphere* (Ed. by D. A. DeVries & N. Afgan), pp. 167–86. Scripta Book Co., Washington, D. C.

Lemon, E. R. & Wright, J. L. (1969). Photosynthesis under field conditions. XI. Assessing sources and sinks of carbon dioxide in a corn crop using a momentum balance approach. *Agronomy Journal,* **61,** 408–11.

Long, S. P. (1983). C_4 photosynthesis at low temperatures. *Plant, Cell and Environment,* **6,** 345–63.

Long, S. P. (1985), Leaf gas exchange. *Photosynthetic Mechanisms and the Environment* (Ed. by J. Barber & N. R. Baker), pp. 453–99. Elsevier, Amsterdam.

Long, S. P. (1986). Instrumentation for the measurement of CO_2 assimilation by crop leaves. *Advanced Agricultural Instrumentation* (Ed. by W. G. Gensler), pp. 39–91. Nijhoff, Dordrecht.

Long, S. P., Bolhar-Nordenkampf, H. O., Croft, S. L., Farage, P. K., Lechner, E. & Nugawela, A. (1989). Analysis of spatial variation in CO_2 uptake within the intact leaf and its significance in interpreting the effects of environmental stress on photosynthesis. *Philosophical Transactions of the Royal Society,* **B, 323,** 385–95.

Long, S. P. & Hällgren J.-E. (1985). Measurement of CO_2 assimilation by plants in the field and the laboratory. *Techniques in Bioproductivity and Photosynthesis,* 2nd edn (Ed. by J. Coombs, D. O. Hall, S.P. Long & J. M. O. Scurlock), pp. 62–94. Pergamon, Oxford.

Long, S. P. & Incoll, L. D. (1979) The prediction and measurement of photosynthetic rate of *Spartina townsendii* in the field. *Journal of Applied Ecology,* **16,** 879–91.

Long, S. P. & Ireland, C. R. (1985). The measurement and control of air and gas flows for determination of gaseous exchanges of living organisms. *Instrumentation for Environmental Physiology* (Ed. by B. Marshall & F. I. Woodward), pp. 123–37. Cambridge University Press.

Ludlow, M. M. & Wilson, G. L. (1971). Photosynthesis of tropical pasture plants. I. Illuminance, carbon dioxide concentration, leaf temperature, and leaf-air vapour pressure difference. *Australian Journal of Biological Sciences,* **24,** 449–66.

Marshall, B. & Woodward, F. I. (Eds) **(1985).** *Instrumentation for Environmental Physiology.* Cambridge University Press.

Mitsui, A. & Black, C. C. (Eds) **(1982).** *Handbook of Biosolar Resources,* Vol. 1. CRC Press, Boca Raton.

Monteith, J. L. (Ed.) **(1976).** *Vegetation and the Atmosphere,* Vol. 2. Academic Press, London.

Monteith, J. L. (1977). Climate and the efficiency of crop production in Britain. *Philosophical Transactions of the Royal Society.* **B, 281,** 749–74.

Ohtaki, E. (1980). Turbulent transport of carbon dioxide over a paddy field. *Boundary-Layer Meteorology,* **19,** 315–36.

Ohtaki, E. & Matsui, M. (1982). Infra-red device for simultaneous measurement of atmospheric carbon dioxide and water vapour. *Boundary-Layer Meteorology,* **24,** 109–19.

Ower, E. & Pankhurst, R. C. (1977). *The Measurement of Air Flow,* 5th edn. Pergamon, Oxford.

Parkinson, K. J. (1985). Porometry. *Instrumentation for Environmental Physiology* (Ed. by B. Marshall & F. I. Woodward), pp. 171–91. Cambridge University Press.

Pearcy, R. W. & Ehleringer, J. R. (1984) Comparative ecophysiology of C_3 and C_4 plants. *Plant, Cell and Environment,* **7,** 1–13.

Powles, S. B. (1984). Photoinhibition of photosynthesis induced by visible light. *Annual Reviews of Plant Physiology,* **35,** 15–44.

Raupach, M. R. & Legg, B. J. (1984). The uses and limitations of flux gradient relationships in micrometeorology. *Agricultural Water Management,* **8,** 119–31.

Saugier, B. & Ripley, E. A. (1978). Evaluation of the aerodynamic method of determining fluxes over natural grassland. *Quarterly Journal of the Royal Meteorological Society,* **104,** 257–70.

Sestak, Z., Catsky, J. & Jarvis, P. G. (Eds) **(1971).** *Plant Photosynthetic Production: Manual of Methods.* Junk, The Hague.

Sheehy, J. E. (1985). Radiation. *Instrumentation for Environmental Physiology* (Ed. by B. Marshall & F. I. Woodward), pp. 5–28. Cambridge University Press.

Terashima, I., Wong, S. -C., Osmond, C. B. & Farquhar, G. D. (1988). Characterisation of non-uniform photosynthesis induced by abscisic acid in leaves having different mesophyll anatomies. *Plant Cell Physiology,* **29,** 385–94.

Thom, A. S. (1975). Momentum, mass and heat exchange in plant communities. *Vegetation and the Atmosphere,* Vol. 1 (Ed. by J. L. Monteith), pp. 57–110. Academic Press, London.

Thom, A. S., Stewart, J. B., Oliver, H. R. & Gash, J. H. C. (1975). Comparison of aerodynamic and energy budget estimates of fluxes over a pine forest. *Quarterly Journal of the Royal Meteorological Society,* **101,** 93–105.

Thornley, J. M. M. (1977). *Mathematical Models in Plant Physiology.* Academic Press, London.

Welles, J. (1986). A portable photosynthesis system. *Advanced Agricultural Instrumentation* (Ed. by W. G. Gensler), pp. 21–38. Nijhoff, Dordrecht.

3. TELEMETRIC RECORDING OF PHYSIOLOGICAL DATA FROM FREE-LIVING ANIMALS

P. J. BUTLER
School of Biological Sciences, University of Birmingham, Birmingham B15 2TT, UK

INTRODUCTION

Strictly speaking to telemeter means 'to transmit to a distant receiving set or station' which may be in direct contact, via a wire or cable, with the source of the signal. However, the present review will be primarily concerned with the monitoring of physiological data from animals that are not tethered to any recording equipment, so it will not include any reference to wired or cable telemetry. Weller (1984) included storage techniques, in his review of 'telemetric' methods, and, with the increasing development of solid-state mass storage devices, this new form of telemetry is already proving to be extremely useful, partly because it is not restricted by the relationship between the power output of a transmitter and the distance from a receiver.

The most commonly used method to transmit over relatively long distances is radiotelemetry but even this method is of little or no use in water containing even small quantities of solutes when severe attenuation occurs, so acoustic (or ultrasonic) and electromagnetic systems have been developed, particularly for work in sea water (Pincock & Church 1984; Priede, French & Duthie 1984).

Whatever method of transmission is used, there must be a signal to transmit and the most easily obtained signals related to physiological variables are those that originate in the animal itself, e.g. electrocardiogram (ecg) and electromyogram (emg), and merely require amplification before transmission, or those for which the sensor does not have to be in direct contact with the tissues of the animal and which require very little power to energize them, e.g. temperature. Not surprisingly, therefore, these are the most regularly telemetered signals. More useful physiological variables, such as rates of respiratory air flow and blood flow, blood pressure and partial pressure of gases in respiratory air and blood, have been determined much less frequently by telemetric techniques.

Often the animal carrying the telemetric device is in a confined (even if largely unrestrictive) space and its behaviour can be regularly monitored and related to the physiological data. It is becoming more common, however, to release animals with telemetric equipment into their natural environment, in which case it is desirable to have some means of monitoring their behaviour. Thus, using transmitters to track animals may become an essential part of physiological studies, and the recent use of satellites such as the NIMBUS-6 and ARGOS systems makes this an even more powerful tool, as shown by Jennings & Gandy (1980) with dolphins and Priede (1984) with basking sharks. Systems are being developed which enable the position of relatively small animals such as golden eagles (*Aquila chrysaetos* and Andean condors (*Vultur gryphus*) to be determined (French 1986; Strikwerda *et al.* 1986). It is also intended to incorporate the transmission of physiological signals to the satellite (Howey *et al.* 1984).

Telemetry has been used to study a variety of physiological variables in a number of different animals ranging from crabs (Wolcott 1980; Graham 1981) to polar bears (Oritsland, Stallman & Jonkel 1977) so to attempt to cover all of the literature in such a short review would be absurd. Many important data have been obtained by monitoring body temperature alone from animals such as tuna, lamnid sharks (Carey *et al.* 1971), alligators (Smith 1975), tits (Reinertsen & Haftorn 1986), emperor penguins (Boyd & Sladen 1971), muskrats (MacArthur 1979), ground squirrels (Wang 1978) and northern elephant seals (McGinnis & Southworth 1971) under natural or semi-natural conditions. The present review, however, will attempt to cover the various applications of telemetry during one particular activity state in animals, namely exercise, partly because this is an area in which the author is involved, but also because this is an aspect of animal physiology and ecology where telemetry clearly comes into its own as the animals need to be free and not tethered to recording equipment. Most of the studies on exercise involving telemetric techniques have been confined to fish, birds and mammals so it is these three groups of vertebrates that will be considered.

FISH

Activity

An important aspect of studying exercise in any animal is to be aware of its normal level of activity. An understanding of the migratory behaviour of commercially important marine fish is of particular importance if an optimum balance between exploitation and conservation of stocks is to be achieved, and acoustic telemetry has proved most useful in this respect (Harden Jones & Arnold 1982). Such studies have indicated that many fish make use of *selective tidal stream transport* by vertically migrating into the tidal flow of water and then returning to the bottom at the next slackwater (Greer Walker, Harden Jones & Arnold 1978). Monitoring displacement alone, therefore, does not give an indication of active expenditure of energy in marine fish. In fact the simultaneous recordings of tail beat frequency and depth have indicated that the eel actively swims downwards during its vertical migration and does not, as had been suggested, glide downwards as a result of being negatively buoyant (Westerberg 1984). Most fish, however, may not be very active for most of their life.

Holliday, Tytler & Young (1972–3) using sonic tags and directional hydrophones found that brown trout, *Salmo trutta*, in a loch had a mean hourly swimming speed of 0.2 body lengths per second (bl s^{-1}) which peaked at 0.28 bl s^{-1} in March. When migrating upriver, sea trout (the migratory form of brown trout) swim at between 0.5–1 bl s^{-1} *over the ground*) for periods of 3–4 h at a time (D. Solomon, personal communication) and, as water velocity may be anything from 0.3–1.5 m s^{-1}, these fish appear to be exercising at high sustainable levels even if only to maintain position in midstream. Such data are obtained by inserting acoustic radio tags into the stomach of the fish via the oesophagus (Solomon 1982) and logging the time that they pass recording stations (Fig. 3.1).

Indirect indicators of metabolic rate

As with other species of animals, once their normal levels of activity are known, these can be simulated in water channels, in wind tunnels or on treadmills and a number of physiological variables can be recorded, using either conventional hard-wired or telemetric techniques. The relationship between heart rate and aerobic metabolism has been studied in a number of species in such conditions in an attempt to use heart rate as an indicator of metabolism in wild, free-living animals. The relationship between heart rate and oxygen uptake is given by the Fick equation:

$$\mathring{V}O_2 = HR \times SV (C_aO_2 - C_{\bar{v}}O_2) \tag{3.1}$$

where $\mathring{V}O_2$ = rate of oxygen uptake, HR = heart rate, SV = cardiac stroke volume, C_aO_2 = oxygen content of arterial blood, $C_{\bar{v}}O_2$ = oxygen content of mixed venous blood.

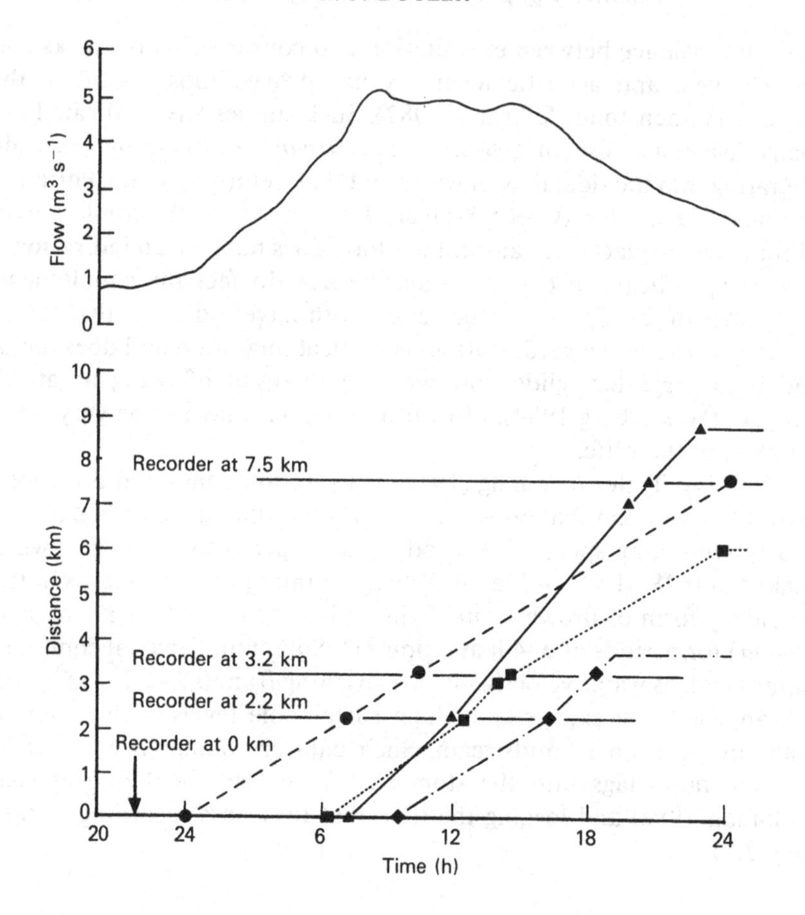

FIG 3.1 The movements of four salmon fitted with acoustic radio tags during a 32h period after leaving a pool at zero distance, where they had been lying, and then moving upstream past four recorders in response to an increase in river flow following rain (from Solomon 1985).

It is clear that variations in SV and ($C_aO_2 - C_{\bar{v}}O_2$) will modify the relationship between $\dot{V}O_2$ and HR, and that in poikilotherms the 'calibration' temperature must be the same as the field temperature. Priede & Tytler (1977) found that there was not a good linear relationship between these two variables in a number of fish species, but they suggested that an oxygen uptake/heart rate 'quadrangle of scope' (Fig. 3.2) is a practical way of obtaining the range of oxygen uptake for a particular heart rate from free-living fish. This technique was used (Priede & Young 1977) to determine maximum oxygen consumption for a given heart rate telemetered from wild brown trout in a loch and to estimate maximum mean oxygen uptake over the recording period (Fig. 3.3). There were no extended periods of high heart rate (or, therefore, of high oxygen consumption) and maximum heart rate occupied less than 0.5% of the recording time. Maximum mean oxygen uptake was estimated to be 1.55 times the basal level and slightly over half of the 'active' level. These data are compatible with the low levels of locomotor activity recorded from the same species under the same conditions (Holliday, Tytler & Young 1972–3). Priede (1977) has proposed that the time that an animal spends with extreme heart rates (high or low) is not only an indicator of aerobic metabolism but can also be used to predict the probability of that fish dying.

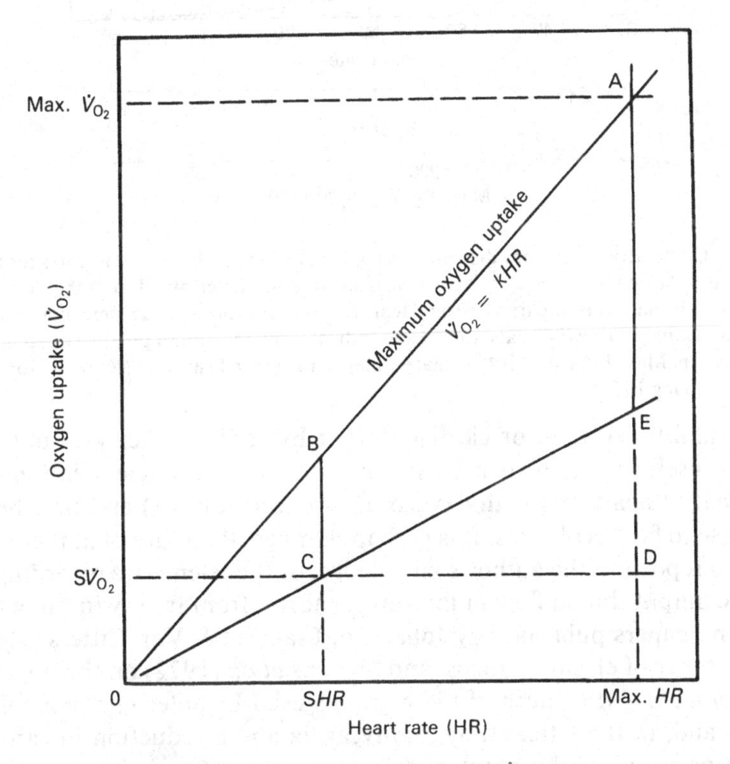

FIG 3.2 The general relationship between oxygen uptake ($\dot{V}O_2$) and heart rate; all data theoretically lie within the 'quadrangle of scope' ABCE. Max. $\dot{V}O_2$, maximum oxygen uptake; $S\dot{V}O_2$, standard oxygen uptake; Max. *HR*, maximum heart rate; S*HR*, standard heart rate; $k = SV \times (C_aO_2 - C_vO_2)$ in the Fick equation (see text for further details). Line OBA, *k* is constant at maximum value, and line OCE, *k* is constant at its value at standard oxygen uptake and standard heart rate (modified from Priede & Tytler 1977).

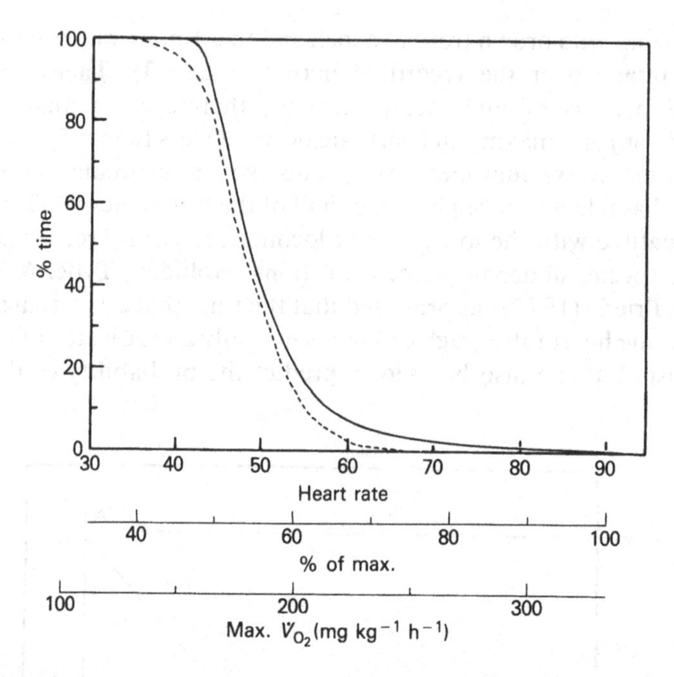

FIG 3.3 Cumulative frequency distribution curves of time at different heart rates for two brown trout (*Salmo trutta*) in a Scottish loch; heart rate was determined from the ECG transmitted by an ultrasonic transmitter. Heart rate is also expressed as a percentage of the assumed maximum rate (90 beats min^{-1}) and indicates the maximum possible oxygen consumption, Max. $\dot{V}O_2$ (see third ordinate scale) at any given heart rate (redrawn from Priede & Young 1977).

An ability to monitor cardiac output by radiotelemetry would not only be useful in itself; it may also provide a more linear relationship than that of heart rate with oxygen uptake (see eqn 3.1) and thus be of more use to field ecologists. It is perhaps strange, therefore, that there has been no report, to the author's knowledge, of the telemetric recording of cardiac output (blood flow in the ventral aorta) from free-swimming fish since the papers published by Johansen, Franklin & Van Citters (1966) on a number of elasmobranchs, and Stevens *et al.* (1972) on the lingcod, *Ophiodon elongatus*. Both of these groups used Doppler ultrasonic flow probes and, in the latter study, bradycardia and a reduction in cardiac output occurred at the onset of spontaneous exercise, whereas, in the former, heart rate increased but cardiac stroke volume remained unchanged during prolonged (>20 min) swimming. This is clearly an area where the application of modern electronic techniques could be profitable, provided that the anatomy of the fish permits the placing of flow probes around the ventral aorta!

A clear linear relationship has been obtained between telemetered electromyogram (EMG) signals obtained from the 'mosaic' (white) epaxial muscle mass of rainbow trout and oxygen uptake (Weatherley *et al.* 1982). However, the slope of the relationship is different between fish swimming in a water channel at constant velocities and those swimming spontaneously (Fig. 3.4). This may be because, in the latter, other muscle masses, associated with turning, changing speed, etc., are involved as well as the epaxial muscles. There appears to be a shift from one slope to the other at a fairly precise value of average EMG (approximately 5 μV) so this variable may be a useful indicator not only of oxygen uptake but also of the type of locomotor behaviour (Fig. 3.4). Rogers & Weatherley (1983) have also demonstrated a linear relationship between average EMG activity from the levator arcus palatini muscle (involved in abduction of the operculum) and oxygen uptake in the rainbow trout during forced swimming in a water channel but not during spontaneous activity. Whatever the reason(s) for this difference (one may be that the fish use ram ventilation more frequently under natural conditions), this study does illustrate that much care is required when

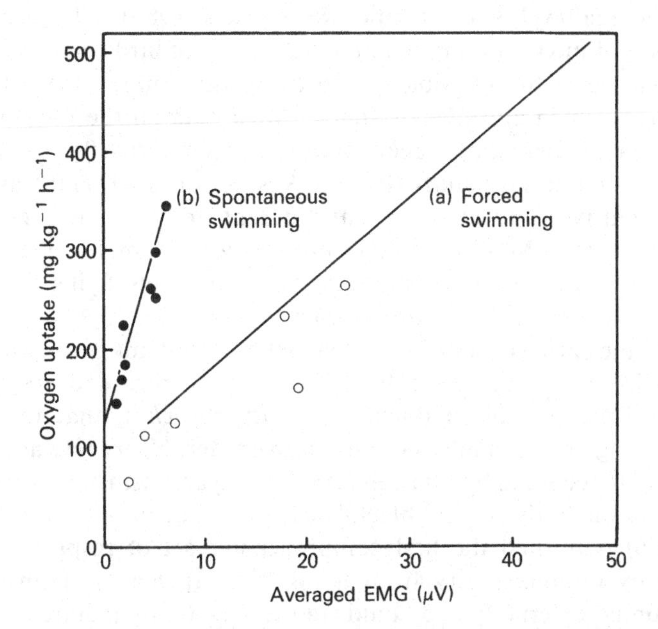

FIG 3.4 The relationships between oxygen uptake and the corresponding telemetered electromyograph (EMG) values (in μV) for a 32.4 cm rainbow trout: (a) swimming in a swim channel at a range of velocities from 20 to 80 cm s^{-1}, and (b) swimming spontaneously (modified from Weatherley *et al.* 1982).

trying to use an indirect indicator of oxygen uptake and that it is not always justified to extrapolate directly from the situation in a water channel to the field. Fairly extensive laboratory studies have been performed on the pike, *Esox lucius*, and the evidence suggests that for this fish, heart rate measured by telemetry would provide a good indication of metabolic rate in the field (Armstrong 1986).

BIRDS

Flight

Flight is the major form of locomotion in birds and is of particular importance during migration. Migration routes and the distances covered have largely been determined by ringing studies and radar but, as already indicated, the use of telemetry, particularly via satellite links, is becoming more feasible (French 1986; Strikwerda *et al.* 1986).

From the more conventional studies it is known that, when migrating, some birds may fly non-stop for distances of 3000–5000 km while others may fly at altitudes up to 9000 m where the partial pressure of oxygen is 1/3 of the sea-level value (Butler & Woakes 1985). The ecological significance of migration in the annual life-cycle of birds is obvious, but our knowledge of the physiology and energetic requirements of flight, particularly at high altitudes, is sparse. Most data on the physiological adjustments to flight have been obtained from birds flying in wind tunnels, but in one such study (Butler, West & Jones 1977) it was noted that the flight pattern was somewhat different from that in freely flying birds from which EMG of the flight muscles was telemetered (Fig. 3.5). Telemetry is an obvious technique to use with freely flying birds, but unfortunately it can have its limitations.

One of the earliest uses of radiotelemetry with flying birds was when Lord, Bellrose & Cochran (1962) accidentally recorded respiratory frequency from a mallard duck, *Anas platyrhynchos*, that they were tracking using an externally mounted transmitter. Heart rate and blood pressure were recorded by Eliassen (1963) using a transmitter attached to the backs of mallards or great black-backed gulls, *Larus marinus*. Hart & Roy (1966) published the first comprehensive set of respiratory data obtained by radiotelemetry from flying birds (pigeons). Transmitters were mounted externally and could transmit ECG, temperature, respiratory frequency and tidal volume. When measuring tidal volume, a mask was also attached to the birds. These attachments must have affected the

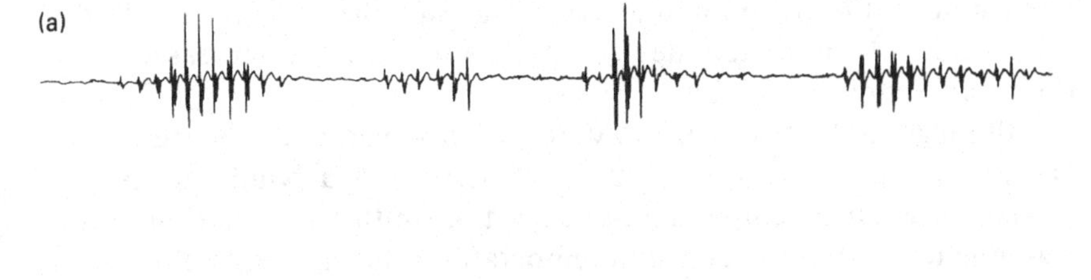

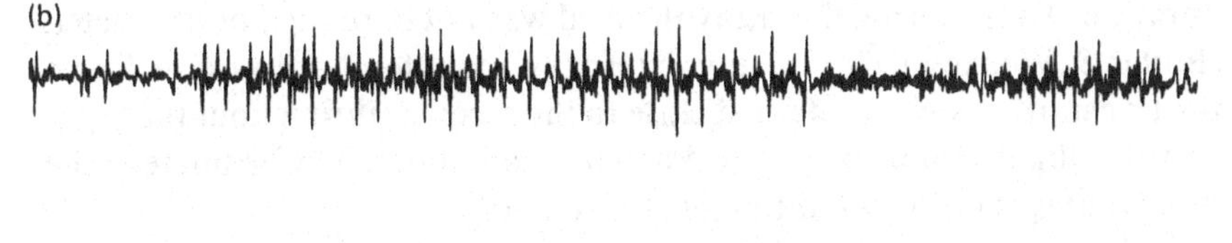

FIG 3.5 Traces of electromyograms (EMG) from the pectoralis muscle of a pigeon flying at 10 m s^{-1}; (a) in a wind tunnel, and (b) during a free spontaneous flight. During the latter the EMG was transmitted by an FM transmitter attached to the back of the bird. The time marker is in seconds (from Butler, West & Jones 1977).

energy requirements of the birds. To prevent the loss of the bird and transmitter and, presumably, to keep the transmitter within range of the receiver, the birds were restrained by a nylon line tied to a harness so that flights were, on average, of only 9 s duration. The birds had barely taken off and were nowhere near a steady state, when they landed again. Thus, all the measured values were considerably different from what they would have been after several minutes of flight (see Butler, West & Jones 1977). Similar criticisms can be levelled at the subsequent papers by these authors (Berger, Hart & Roy 1970; Berger, Roy & Hart 1970), which is a pity because in the latter paper measurements of oxygen uptake were obtained.

Kanwisher *et al.* (1978) attached long-range (80-km) transmitters externally to herring gulls, *Larus argentatus* and used these to monitor heart rate during flights of up to 20 km in distance. This was clearly a great advance over the earlier systems as far as the duration of the flights was concerned, but external mounting and the lack of information on the bird's behaviour during such long flights were still defects of the technique. Indeed, a recent study by Gessaman & Nagy (1988) indicates that external mounting of transmitters increases both flight time and total energy expenditure of homing pigeons. Over a distance of 320 km, total energy expenditure was approximately 85% greater in birds with an externally mounted transmitter of <5% body mass compared with unencumbered birds. As well as the extra mass that they may provide, such externally mounted devices may also produce increased drag (Obrecht, Pennycuick & Fuller 1988).

Although not measuring any respiratory variables, Torre-Bueno (1976) combined the use of radiotelemetry and a wind tunnel. He implanted a small temperature-sensing transmitter into starlings, and was able to record core and skin temperatures during flights of 0.5–2 h duration. Thus, during the study the bird was not burdened or distracted (Butler & Woakes 1976) by any externally mounted hardware or leads, the transmitter was maintained close to the receiver throughout the flight and the flight conditions were known at all times. Unfortunately the disadvantages of the wind tunnel still remained.

It has since been demonstrated that implantable, short-range transmitters can be used with freely flying birds trained to fly behind a moving vehicle (Butler & Woakes 1980). Perhaps the most important feature of this study was the demonstration that it is possible to obtain data on some physiological variables from free-flying birds that are unstressed by the restrictions of a wind tunnel or externally mounted leads or equipment and yet are close enough to obtain accurate measurements of their air speed and behaviour. Data obtained in such a way can then be checked against those obtained from a wind tunnel, thus giving an indication of the 'ecological usefulness' of the latter. For free-flying birds the use of implantable data-storage packages (see section on diving mammals) may be of much greater use.

Swimming and diving

Although flight is the predominant form of exercise in birds, it is by no means the only one, and another aspect of the behaviour of air-breathing vertebrates that has fascinated physiologists for many decades is the ability of the aquatic species, particularly the homeotherms, to remain submerged under water for extended periods while feeding. The current view as to how this is achieved is largely the result of data obtained via telemetry from freely diving birds. Millard, Johansen & Milsom (1973) used radiotransmitters to record heart rate and blood flow in a few major arteries of Adelie and Gentoo penguins, *Pygoscelis adeliae* and *P. papua* respectively. They note that their data are 'not readily compared with earlier reported results in diving birds . . . because previous studies have been performed on restrained birds forcefully dived' whereas their data were obtained from 'penguins performing normal swimming and diving, *which will give results as a composite of (involuntary) diving and exercise responses*'. Radiotelemetry has since proved a useful technique for monitoring heart rate from freely diving birds and it is now clear that, as indicated by Millard, Johansen & Milsom (1973), the cardiac response during most natural dives is completely different from that obtained during involuntary submersion and may be more like that seen during exercise in air.

It has been found that externally mounted devices (of 10% the animal's cross-sectional area) affect the foraging behaviour of penguins (Wilson, Grant & Duffy 1986) so, again, care must be taken when interpreting the data from studies using such equipment.

Using totally implanted transmitters, Butler & Woakes (1979) demonstrated that, contrary to the classical hypothesis, there is no maintained bradycardia during voluntary dives in pochards and tufted ducks (*Aythya ferina* and *A. fuligula*). These birds tend to perform a number of dives in a series, and before the first dive there are increases in heart rate and respiratory frequency. Most data have been obtained from the tufted duck, and upon submersion there is an initial large fall in heart rate but it soon stabilizes at a rate which is approximately 60% *above* the resting value (Fig. 3.6). During involuntary submersion, on the other hand, these ducks show the classical diving bradycardia and heart rate reaches a value of approximately 30% the pre-dive value within 30 s of diving. This bradycardia, which is present in all other air-breathing aquatic vertebrates during involuntary submersion, is associated with a reduction in blood flow to all parts of the body except the CNS and heart, which continue to metabolize aerobically. The rest of the body, which is underperfused, metabolizes anaerobically, producing lactic acid (Scholander 1940; Andersen 1966). Monitoring heart rate by telemetry has subsequently demonstrated that there is no diving bradycardia during voluntary submersion in cormorants, *Phalacrocorax auritus* (Kanwisher, Gabrielsen & Kanwisher 1981), Humboldt penguins, *Speniscus humboldti* (Butler & Woakes 1984) and redhead ducks, *A. americana* (Furilla & Jones 1986). The absence of bradycardia during voluntary submersion raised the question as to whether or not all of the other adjustments characteristic of the classical response occur during natural dives.

Further studies on tufted ducks and Humboldt penguins have shown that there is no reduction in oxygen uptake during normal voluntary dives in these animals. In fact, in the ducks it is 3.5 times the resting value and not significantly different from that during maximum sustained swimming speed at the surface (Woakes & Butler 1983), whereas in the penguins the mean value was found to be 25% above (but not signifi-

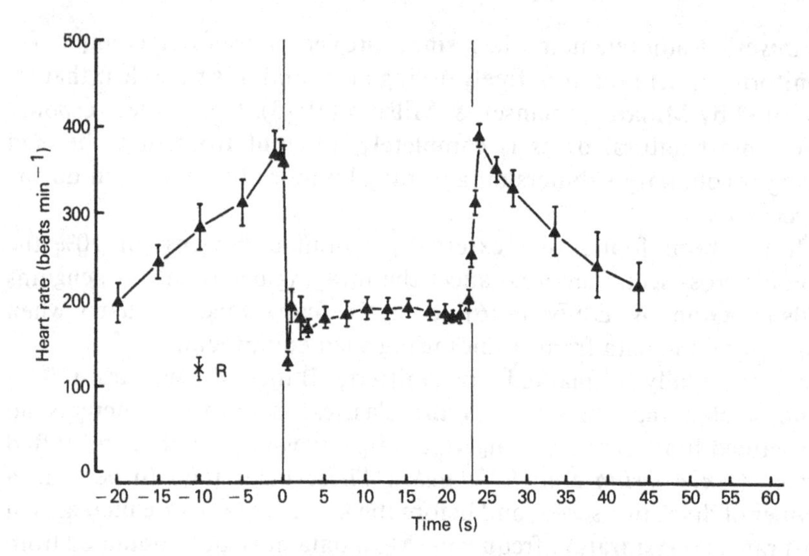

FIG 3.6 Mean changes in heart rate (±S.E.) in tufted ducks, *Aythya fuligula*, during voluntary feeding dives to a depth of 1.9 m; *R* = resting heart rate. Heart rate was determined from the ECG transmitted from an implanted PIM radiotransmitter. Times of submersion and surfacing are indicated by the vertical lines (from Butler & Stephenson 1987).

cantly different from) the resting value (Butler & Woakes 1984). The difference between the species is thought to result from two facts–the penguins are less buoyant, and they are more efficient at underwater locomotion. It appears from these values and from knowledge of usable oxygen stores and normal dive durations that these species (and presumably other birds that dive) normally remain submerged for periods that are well within their aerobic limits. They seem to metabolize completely aerobically when submerged and to replace the oxygen stores during the period at the surface between dives. There is no evidence to suggest that under normal circumstances they resort to anaerobiosis and accumulation of lactic acid (Butler & Stephenson 1987). These data from freely diving birds have clear implications for ecologists wishing to determine the energy costs of feeding in aquatic birds. Contrary to earlier suggestions, these are going to be extremely high in species such as the diving ducks.

Unfortunately it has not yet been possible to record routinely, by telemetry from freely exercising birds, physiological data other than heart rate, respiratory frequency and deep body temperature, so direct recording methods have been used to measure respiratory tidal volume, blood pressure and blood gases in flying birds (Butler, West & Jones

1977) and in swimming ducks (Woakes & Butler 1986). It is, however, extremely useful to be able to measure heart rate, via telemetry, from birds with no other form of instrumentation attached to obtain true resting values and to be able to determine the effect of different types of instrumentation.

There is a close linear relationship between heart rate and oxygen uptake in tufted ducks during steady-state swimming at different velocities (Woakes & Butler 1983), indicating that heart rate may be a useful measure of aerobic metabolism under natural conditions. However, it is clear from the study by Woakes & Butler (1986) that, although the direct recording techniques they used have little or no effect on oxygen uptake (provided the animals are given sufficient time to settle after attaching the recording equipment), they do cause an elevation of heart rate. Thus the relationship between heart rate (and, presumably, the other measured variables) and oxygen uptake is altered. Also, when feeding (diving) under water, heart rate in tufted ducks is lower than when they are swimming at the surface at the same level of oxygen usage (Woakes & Butler 1983).

The fact that the relationship between heart rate and oxygen uptake can change to quite a large extent indicates that extreme caution must be exercised when attempting to use heart rate as an indicator of aerobic metabolism in the field on the basis of one particular experimental set-up in the laboratory. This is particularly so when the relationship in the laboratory is obtained by changing environmental temperature and yet it is to be used in the field at different levels of locomotory activity, as was done by Owen (1969), Wooley & Owen (1977) and Ferns, MacAlpine-Leny & Goss-Custard (1980). It is felt that, before heart rate can be usefully employed as an indicator of aerobic metabolism in the field, more exhaustive studies on the relationship between these two variables in a number of different situations must be performed. Calibration in the field against the D_2O^{18} method (see Bryant, this volume) should also be attempted. It may still be decided that Priede & Tytler's (1977) 'quadrangle of scope' is the best that can be achieved.

MAMMALS

Running

The larger mammals are larger than most birds, and some at least can be trained more readily to carry external packages. It has also been possible to implant pressure and flow transducers into a number of larger mammals, and then to telemeter the data via a transmitter carried in a backpack by the exercising animal. It has been possible to obtain valuable information on the cardiovascular adjustments to exercise in freely running animals (Vatner *et al.* 1971, 1972). It appears that cardiac output increases sufficiently in healthy mongrel dogs to satisfy the requirements of the exercising muscles, without there being compensatory reductions in the renal and mesenteric vascular beds. As mentioned previously, the ability to record cardiac output in these animals may enable more accurate use of the Fick equation (3.1) for determining aerobic metabolism. Unfortunately the author is unaware of any attempt to do so.

Diving

The monitoring of heart rate alone via telemetry in unrestrained and freely diving mammals (including humans — Butler & Woakes 1987) has provided much important information and, as in the case of birds, has altered our views of the physiological adjustments that occur during normal diving (Harrison, Ridgway & Joyce 1972; Jones *et al.* 1973; Fedak, Pullen & Kanwisher 1988). The first of these studies indicated that, in grey seals, *Halichoerus grypus,* in a holding tank, heart rate during trained dives (i.e. when the seals dived on command signalled by an underwater light) is higher than during involuntary dives. In the second study there were variable cardiac responses during spontaneous feeding dives of harbour seals, *Phoca vitulina,* in a tank, and one seal exhibited no bradycardia at all during 20% of such dives. In the third study heart rate was measured (via acoustic transmitters) from grey seals at different swimming speeds in a swimming flume, and in harbour seals while free at sea. While ventilating their lungs heart rate was 110–120 beats min^{-1} and when the seals were swimming (or resting) under water heart rate was approximately 40 beats min^{-1}. Neither of these values varied in any systematic way with swimming speed (Fig. 3.7). However, mean heart rate (together with oxygen uptake) did increase with swimming speed as the seals spent a greater proportion of their time at the surface at higher velocities. Interestingly, heart rates much lower (<10 beats min^{-1}) were obtained when the seals were prevented from surfacing by closing off access to the respirometer in the flume. Similar patterns of heart beat were seen in the free-range harbour seals. Heart rate was approximately 120 beats min^{-1} with the animals at the surface, and was independent of dive duration, whereas during submersion it was between 45 and 30 beats min^{-1}, being lower during longer dives. Lower heart rates were only observed on very few occasions (two) when one seal engaged in aggressive behaviour with another. There were anticipatory increases in heart rate upon surfacing.

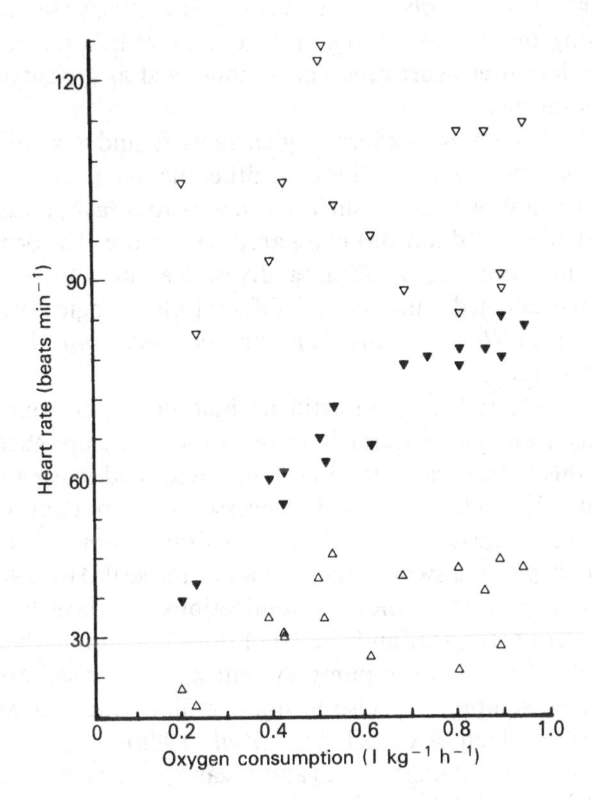

FIG 3.7 The relationship between average heart rate and oxygen consumption in a grey seal in a swimming flume: △, heart rates during dives; ▽, heart rates while at the surface breathing; ▼, average heart rates over complete dive–breathing cycles. Heart rate was obtained from an acoustic transmitter attached to the seal (redrawn from Fedak 1986).

It has been suggested, probably with some justification, that the events in the rest of the cardiovascular system, and the type of metabolism, of the seals can be deduced from the mean heart rate during complete dive/surface cycles (Fedak, Pullen & Kanwisher 1988). A similar situation also exists in diving ducks (Stephenson 1987). However, heart rate will only ever be an *indicator* of other events. Clearly, marine mammals may engage in high levels of activity when under water and 75–85% of their time may be spent below water when engaged in their normal diving behaviour. The ecology of marine mammals is different from that of other aquatic homeotherms. To a far greater extent than birds, therefore, these marine mammals lead a sub-aquatic existence with brief visits to the surface to replenish the oxygen stores (and to remove CO_2). It might be more appropriate to consider the increase in heart rate accompanying these visits to the surface as a ventilation *tachycardia* rather than the lower heart rate when submerged as a bradycardia (cf. Belkin 1964 for turtles).

The normal method of exercising in seals is under water, and the important question is whether there are differences in the cardiovascular response to swimming near the surface when there is easy access to air (or during dives of short duration) compared with those that occur during deep (and therefore longer) feeding dives. Recent use of solid-state electronics has enabled a number of different physiological variables to be collected from Weddell seals, *Leptonychotes weddellii*, diving naturally in the Antarctic.

Seals were released from an artificial hole in the ice which was far enough away from any natural hole or crack to ensure that animals returned to this hole to ventilate their lungs. Attached to each seal was a peristaltic blood sampler and a microprocessor which not only controlled the pump but also measured and stored heart rate (from the ECG), body temperature, depth and swimming velocity of the seal (Hill 1986). Fibre optic cables were used as the communications interface between the microprocessor on the seal and the laboratory computer whenever the seal surfaced. The peristaltic pump system has been used not only to withdraw blood samples, but also to inject radiolabelled metabolites in freely diving Weddell seals (Guppy *et al.* 1986). The wash-in and clearance kinetics of all metabolites and organ-specific compounds were different during spontaneous dives (whether short or long) from those seen in 'resting' animals and, according to the authors, the data from the animals when diving were consistent with the classical Irving–Scholander type of response. This implies reduced aerobic metabolism in

FIG 3.8 Changes in (a) arterial partial pressure of oxygen, P_aO_2, (b) haemoglobin concentration, Hb, and (c) concentration of oxygen in arterial blood, C_aO_2, of four Weddell seals during diving and after resurfacing. Dives were divided into those of short and long duration ($<$ and $>$ 17 min respectively). In (a) high values of P_aO_2 obtained within the first minute of descent are the result of compression of the gases in the lung, and those upon surfacing may be the result of the low body temperatures of the seals (i.e. P_aO_2 would be measured at a higher temperature than that of the blood). In (c) it is clear that, during long dives, C_aO_2 remained above the resting values of 18–19 vol.% for 15–17 min. From then on the rate of decrease was not significantly different from that measured during short dives (redrawn from Qvist *et al.* 1986).

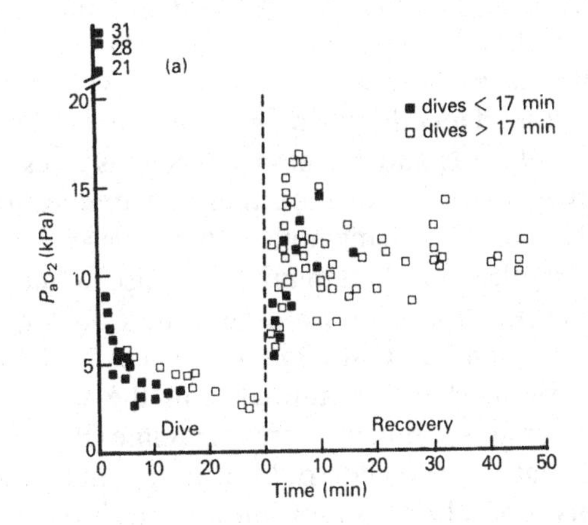
31
28
21
(a)
dives < 17 min
dives > 17 min
P_aO_2 (kPa)
Dive
Recovery
Time (min)

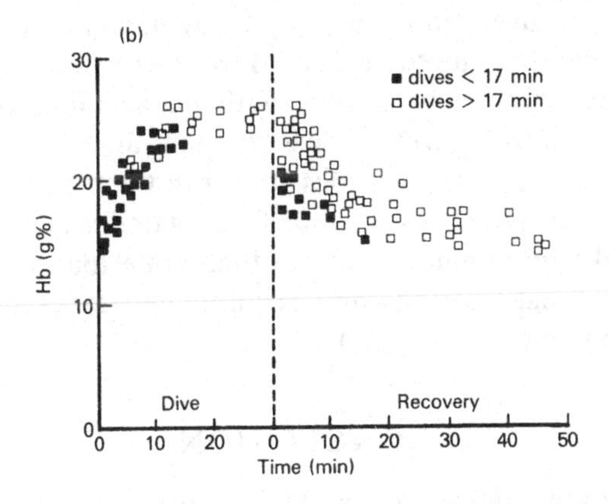
(b)
dives < 17 min
dives > 17 min
Hb (g%)
Dive
Recovery
Time (min)

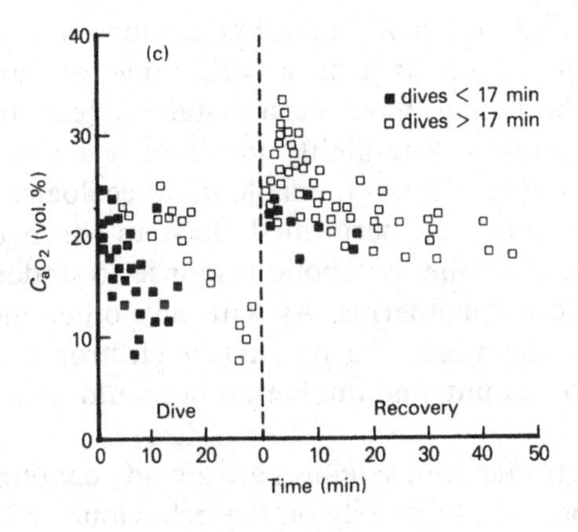
(c)
dives < 17 min
dives > 17 min
C_aO_2 (vol. %)
Dive
Recovery
Time (min)

some, if not most, tissues of the body, and the accompanying accumulation of high levels of lactic acid.

These data are, the authors say, 'incompatible with an exercise model of diving'. That rather depends on whether they were referring to exercise in normal terrestrial birds and mammals, when, as a result of matching between the delivery of oxygen by the circulatory *and* respiratory systems and the demands of the exercising mucles, blood gases are maintained at steady levels, or whether they had in mind exercise in marine animals, in which ventilation is invariably intermittent. There are large swings in the blood gases (see Fig. 3.8), so that large differences in the activity of the cardiovascular system when the animal is at the surface (or resting?) compared with when it is submerged are inevitable (Fedak 1986). It is interesting to note that blood lactate did not in fact increase substantially during or after dives of <17 min duration whereas it reached levels of up to 8 μmol ml^{-1} for dives of 24–33 min duration (Guppy *et al.* 1986). This is consistent with suggestions made by Kooyman *et al.* (1980, 1983) for Weddell seals that dives up to a certain duration (depending on the size of the animal) are completely aerobic with no accumulation of lactate, whereas during longer dives (which are few and far between) some anaerobiosis occurs and lactate does appear in the blood upon surfacing. By remaining essentially aerobic while feeding under water these animals are able to make more efficient use of their time than if they remained submerged for longer periods and accumulated excessive amounts of lactic acid (Kooyman *et al.* 1980).

CONCLUSION

There is no doubt that telemetric techniques have already given us a glimpse of the physiology of untethered and unrestricted animals, sometimes in natural and semi-natural conditions. Unfortunately it is not yet possible to obtain data on a wide range of variables so it is important that those that have been obtained (e.g. heart rate) are interpreted with caution. The ability to obtain even heart rate under natural conditions does, however, enable more ecologically meaningful laboratory experiments to be performed. Such an approach has led to a complete reappraisal of the metabolic response to underwater feeding (diving) in aquatic homeotherms. As with any other measuring technique, the effect of the measuring system (the electronics package) itself must be taken into account, and implantation should reduce these effects to a minimum.

Solid-state electronics and satellites are already enabling scientists to obtain information simultaneously on the behaviour and physiology of animals in their natural environment, and development of new transducing systems may one day enable us to monitor aerobic metabolism in free-living animals. We are, therefore, entering an exciting era. At last we are able to free ourselves and our animals from the constraints of the laboratory. Given the right conditions the next ten years will see telemetry, in the widest sense, play a dominant role in helping us understand how animals, including humans, cope with their environment, thus leading us toward a more exact ecology.

REFERENCES

Andersen, H. T. (1966). Physiological adaptations in diving vertebrates. *Physiological Reviews,* **46,** 212–43.

Armstrong, J. D. (1986). Heart rate as an indicator of activity, metabolic rate, food intake and digestion in pike, *Esox lucius. Journal of Fish Biology,* **29** (Supplement A), 207–21.

Belkin, D. A. (1964). Variations in heart rate during voluntary diving in the turtle *Pseudemys concinna. Copeia,* 321–30.

Berger, M., Hart, J. S. & Roy, O. Z. (1970). Respiration, oxygen consumption and heart rate in some birds during rest and flight. *Zeitschrift für vergleichende Physiologie,* **66,** 201–14.

Berger, M., Roy, O. Z. & Hart, J. S. (1970). The co-ordination between respiration and wing beats in birds. *Zeitschrift für vergleichende Physiologie,* **66,** 190–200.

Boyd, J. C. & Sladen, W. J. L. (1971). Telemetry studies of the internal body temperatures of adelie and emperor penguins at Cape Crozier, Ross Island, Antarctica. *Auk,* **88,** 366–80.

Butler, P. J. & Stephenson, R. (1987). Physiology of breath-hold diving: a bird's eye view. *Science Progress, Oxford,* **71,** 439–58.

Butler, P. J., West, N. H. & Jones, D. R. (1977). Respiratory and cardiovascular responses of the pigeon to sustained, level flight in a wind-tunnel. *Journal of Experimental Biology,* **71,** 7–26.

Butler, P. J. & Woakes, A. J. (1976). Changes in heart rate and respiratory frequency associated with spontaneous submersion of ducks. *Biotelemetry,* Vol. 3 (Ed. by T. B. Fryer & H. A. Miller), pp. 215–18. Academic Press, New York.

Butler, P. J. & Woakes, A. J. (1979). Changes in heart rate and respiratory frequency during natural behaviour of ducks, with particular reference to diving. *Journal of Experimental Biology,* **79,** 283–300.

Butler, P. J. & Woakes, A. J. (1980). Heart rate, respiratory frequency and wing beat frequency of free flying barnacle geese *Branta leucopsis. Journal of Experimental Biology,* **85,** 213–26.

Butler, P. J. & Woakes, A. J. (1984). Heart rate and aerobic metabolism in Humboldt penguins, *Spheniscus humboldti,* during voluntary dives. *Journal of Experimental Biology,* **108,** 419–28.

Butler, P. J. & Woakes, A. J. (1985). Exercise in normally ventilating and apnoeic birds. *Circulation, Respiration and Metabolism* (Ed. by R. Gilles), pp. 39–55. Springer, Berlin.

Butler, P. J. & Woakes, A. J. (1987). Heart rate in humans during underwater swimming with and without breath-hold. *Respiration Physiology,* **69,** 387–99.

Carey, F. G., Teal, J. M., Kanwisher, J. W., Lawson, K. D. & Beckett, J. S. (1971). Warm-bodied fish. *American Zoologist,* **11,** 135–45.

Eliassen, E. (1963). Telemetric registering of physiological data in birds in normal flight. *Biotelemetry* (Ed. by L. E. Slater), pp. 257–65. Pergamon, Oxford.

Fedak, M. A. (1986). Diving and exercise in seals: a benthic perspective. *Diving in Animals and Man* (Ed. by A. O. Brubakk, J. W. Kanwisher & G. Sundnes), pp. 11–32. Tapir, Trondheim.

Fedak, M. A., Pullen, M. R. & Kanwisher, J. (1988). Circulatory responses of seals to periodic breathing: heart rate and breathing during exercise and diving in the laboratory and open sea. *Canadian Journal of Zoology,* **66,** 53–60.

Ferns, P. N., MacAlpine-Leny, I. H. & Goss-Custard, J. D. (1980). Telemetry of heart rate as a possible method of estimating energy expenditure in the redshank *Tringa totanus* (L.). *A Handbook on Biotelemetry and Radio Tracking* (Ed. by C. J. Amlaner & D. W. MacDonald), pp. 595–601. Pergamon, Oxford.

French, J. (1986). Tracking animals by satellite. *Electronics and Power,* May, 373–6.

Furilla, R. A. & Jones, D. R. (1986). The contribution of nasal receptors to the cardiac response to diving in restrained and unrestrained redhead ducks (*Aythya americana). Journal of Experimental Biology,* **121,** 227–38.

Gessaman, J. A. & Nagy, K. A. (1988). Transmitter loads affect the flight speed and metabolism of homing pigeons. *Condor,* **90,** 662–68.

Graham, J. M. (1981). Development of ultrasonic and electromagnetic tags for heart rate monitoring in crabs. *Scottish Marine Biological Association, Marine Physics Group Report,* **17,** 1–17.

Greer Walker, M., Harden Jones, F. R. & Arnold, G. P. (1978). The movements of plaice (*Pleuronectes platessa* L.) tracked in the open sea. *Journal du Conseil Permanent International pour l'Exploration de la Mer,* **38,** 58–86.

Guppy, M., Hill, R. D., Schneider, R. C., Qvist, J., Liggins, G. C., Zapol, W. M. & Hochachka, P. W. (1986). Microcomputer-assisted metabolic studies of voluntary diving of Weddell seals. *American Journal of Physiology,* **250,** R175–R187.

Harden Jones, F. R. & Arnold, G. P. (1982). Acoustic telemetry and the marine fisheries. *Symposia of the Zoological Society, London,* **49,** 75–93.

Harrison, R. J., Ridgway, S. H. & Joyce, P. L. (1972). Telemetry of heart rate in diving seals. *Nature,* **237,** 280.

Hart, J. S. & Roy, O. Z. (1966). Respiratory and cardiac responses to flight in pigeons. *Physiological Zoology,* **39,** 291–306.

Hill, R. D. (1986). Microcomputer monitor and blood sampler for free-diving Weddell seals. *Journal of Applied Physiology,* **61,** 1570–6.

Holliday, F. G. T., Tytler, P. & Young, A. H. (1972–3). Activity levels of trout (*Salmo trutta*) in Airthrey Loch, Stirling, and Loch Leven, Kinross. *Proceedings of the Royal Society of Edinburgh,* **74,** 315–31.

Howey, P. W., Witlock, D. R., Fuller, M. R., Seegar, W. S. & Ward, F. P. (1984). A computerized biotelemetry receiving and datalogging system. *Biotelemetry,* Vol. 8 (Ed. by H. P. Kimmich & H.-J. Klewe), pp. 442–6. Pergamon, Oxford.

Jennings, J. G. & Gandy, W. F. (1980). Tracking pelagic dolphins by satellite. *A Handbook on Biotelemetry and Radio Tracking* (Ed. by C. J. Amlaner & D. W. MacDonald), pp. 753–9. Pergamon, Oxford.

Johansen, K., Franklin, D. L. & Van Citters, R. L. (1966). Aortic blood flow in free-swimming elasmobranchs. *Comparative Biochemistry amd Physiology,* **19,** 151–60.

Jones, D. R., Fisher, H. D., McTaggart, S. & West, N. H. (1973). Heart rate during breath-holding and diving in the unrestrained harbor seal (*Phoca vitulina richardi). Canadian Journal of Zoology,* **51,** 671–80.

Kanwisher, J. W., Gabrielsen, G. & Kanwisher, N. (1981). Free and forced diving in birds. *Science,* **211,** 717–19.

Kanwisher, J. W., Williams, T. C., Teal, J. M. & Lawson, K. O. (1978). Radiotelemetry of heart rates from free-ranging gulls. *Auk,* **95,** 288–93.

Kooyman, G. L., Castellini, M. A., Davis, R. W. & Maue, R. A. (1983). Aerobic limits of immature Weddell seals. *Journal of Comparative Physiology,* **B, 151,** 171–4.

Kooyman, G. L., Wahrenbrock, E. A., Castellini, M. A., Davis, R. W. & Sinnett, E. E. (1980). Aerobic and anaerobic metabolism during voluntary diving in Weddell seals: evidence of preferred pathways from blood chemistry and behaviour. *Journal of Comparative Physiology,* **B, 138,** 335–46.

Lord, R. D., Bellrose, F. C. & Cochran, W. W. (1962). Radiotelemetry of the respiration of a flying duck. *Science,* **137,** 39–40.

MacArthur, R. A. (1979). Seasonal patterns of body temperature and activity in free-ranging muskrats (*Ondatra zibethicus*). *Canadian Journal of Zoology,* **57,** 25–33.

McGinnis, S. M. & Southworth, T. P. (1971). Thermoregulation in the Northern elephant seal, *Mirounga angustirostris. Comparative Biochemistry and Physiology,* **A, 40,** 893–8.

Millard, R. W., Johansen, K. & Milsom, W. K. (1973). Radiotelemetry of cardiovascular responses to exercise in diving penguins. *Comparative Biochemistry and Physiology,* **A, 46,** 227–40.

Obrecht, H. H., Pennycuick, C. J. & Fuller, M. R. (1988). Wind tunnel experiments to assess the effect of back-mounted radio transmitters on bird body drag. *Journal of Experimental Biology,* **135,** 265–73.

Oritsland, N. A., Stallman, R. K. & Jonkel, C. J. (1977). Polar bears: heart activity during rest and exercise. *Comparative Biochemistry and Physiology,* **A, 57,** 139–41.

Owen, R. B. (1969). Heart rate, a measure of metabolism in blue-winged teal. *Comparative Biochemistry and Physiology,* **A, 31,** 431–6.

Pincock, D. G. & Church, D. W. (1984). Current trends in underwater acoustic biotelemetry systems. *Biotelemetry,* Vol. 8 (Ed. by H. P. Kimmich & H.-J. Klewe), pp. 383–6. Pergamon, Oxford.

Priede, I. G. (1977). Natural selection for energetic efficiency and the relationship between activity level and mortality. *Nature,* **267,** 610–11.

Priede, I. G. (1984). A basking shark (*Cetorhinus maximus*) tracked by satellite together with simultaneous remote sensing. *Fisheries Research,* **2,** 210–16.

Priede, I. G., French, J. & Duthie, G. G. (1984). Underwater acoustic and electromagnetic telemetry of the ecg of fish and other aquatic animals. *Biotelmetry,* Vol. 8, (Ed. by H. P. Kimmich & H.-J. Klewe). pp. 387–90. Pergamon, Oxford.

Priede, I. G. & Tytler, P. (1977). Heart rate as a measure of metabolic rate in teleost fishes: *Salmo gairdneri, Salmo trutta* and *Gadus morhua. Journal of Fish Biology,* **10,** 231–42.

Priede, I. G. & Young, A. H. (1977). The ultrasonic telemetry of cardiac rhythms of wild brown trout (*Salmo trutta* L.) as an indicator of bio-energetics and behaviour. *Journal of Fish Biology,* **10,** 299–318.

Qvist, J., Hill, R. D., Schneider, R. C., Falke, K. J., Liggins, G. C., Guppy, M., Elliot, R. L., Hochachka, P. W. & Zapol, W. M. (1986). Hemoglobin concentrations and blood gas tensions of free-diving Weddell seals. *Journal of Applied Physiology,* **61,** 1560–9.

Reinertsen, R. E. & Haftorn, S. (1986). Different metabolic strategies of northern birds for nocturnal survival. *Journal of Comparative Physiology,* **B, 156,** 655–63.

Rogers, S. C. & Weatherley, A. H. (1983). The use of opercular muscle electromyograms as an indicator of the metabolic costs of fish activity in rainbow trout, *Salmo gairdneri* Richardson, as determined by radiotelemetry. *Journal of Fish Biology,* **23,** 535–47.

Scholander, P. F. (1940). Experimental investigations on the respiratory functions in diving mammals and birds. *Hvalradets Skrifter,* **22,** 1–131.

Smith, E. N. (1975). Thermoregulation of the American alligator, *Alligator mississippiensis. Physiological Zoology,* **48,** 177–94.

Solomon, D. J. (1982). Tracking fish with radio tags. *Symposia of the Zoological Society, London,* **49,** 95–105.

Solomon, D. J. (1985). Fish in estuaries and fresh water. *Animal Telemetry in the Next Decade,* pp. 8–10. Ministry of Agriculture, Food & Fisheries, London.

Stephenson, R. (1987). *The physiology of voluntary behaviour in the tufted duck* (Aythya fuligula) *and the American mink* (Mustela vison). Ph.D. thesis, University of Birmingham.

Stevens, E. D., Bennion, G. R., Randall, D. J. & Shelton, G. (1972). Factors affecting arterial pressures and blood flow from the heart in intact, unrestrained lingcod, *Ophiodon elongatus. Comparative Biochemistry and Physiology,* **A, 43,** 681–95.

Strikwerda, T. E., Black, H. D., Levanon, N. & Howey, P. W. (1986). The bird borne transmitter. *Johns Hopkins APL Technical Digest,* **6,** 60–7.

Torre-Bueno, J. R. (1976). Temperature regulation and heat dissipation during flight in birds. *Journal of Experimental Biology,* **65,** 471–82.

Vatner, S. F., Franklin, D. Higgins, C. B., Patrick, T. & Braunwald, E. (1972). Left ventricular response to severe exertion in untethered dogs. *Journal of Clinical Investigation,* **51,** 3052–60.

Vatner, S. F., Higgins, C. B., White, S., Patrick, T. & Franklin, D. (1971). The peripheral vascular response to severe exercise in untethered dogs before and after complete heart block. *Journal of Clinical Investigation,* **50,** 1950–60.

Wang, L. C. (1978). Energetic and field aspects of mammalian torpor: the Richardson's ground squirrel. *Strategies in Cold: Natural Torpidity and Thermogenesis* (Ed. by L. C. H. Wang & J. W. Hudson), pp. 109–45. Academic Press, New York.

Weatherley, A. H., Rogers, S. C., Pincock, D. G. & Patch, J. R. (1982). Oxygen consumption of active rainbow trout, *Salmo gairdneri* Richardson, derived from electromyograms obtained by radiotelemetry. *Journal of Fish Biology,* **20,** 479–89.

Weller, C. (1984). A review of telemetric methods of patient monitoring. *Biotelemetry,* Vol. 8 (Ed. by H. P. Kimmich & H.-J. Klewe), pp. 89–92. Pergamon, Oxford.

Westerberg, H. (1984). Diving behaviour of migrating eels: studies by ultrasonic telemetry. *Biotelemetry,* Vol. 8 (Ed. by H. P. Kimmich & H.-J. Klewe), pp. 367–70. Pergamon, Oxford.

Wilson, R. P., Grant, W. S. & Duffy, D. C. (1986). Recording devices on free-range marine animals: does measurement affect foraging performance? *Ecology,* **67,** 1091–3.

Woakes, A. J. & Butler, P. J. (1983). Swimming and diving in tufted ducks, *Aythya fuligula,* with particular reference to heart rate and gas exchange. *Journal of Experimental Biology,* **107,** 311–29.

Woakes, A. J. & Butler, P. J. (1986). Respiratory, circulatory and metabolic adjustments during swimming in the tufted duck, *Aythya fuligula. Journal of Experimental Biology,* **120,** 215–31.

Wolcott, T. G. (1980). Heart rate telemetry using micropower integrated circuits. *A Handbook on Biotelemetry and Radio Tracking* (Ed. by C. J. Amlaner & D. W. MacDonald), pp. 279–86. Pergamon, Oxford.

Wooley, J. B. & Owen, R. B. (1977). Metabolic rates and heart rate–metabolism relationships in the black duck (*Anas rubripes). Comparative Biochemistry and Physiology,* **A, 57,** 363–7.

4. DETERMINATION OF RESPIRATION RATES OF FREE-LIVING ANIMALS BY THE DOUBLE-LABELLING TECHNIQUE

D. M. BRYANT
Ecology Division, School of Molecular and Biological Sciences, University of Stirling, Stirling FK 9 4LA, UK

INTRODUCTION

All animals require energy to maintain life, to grow and to reproduce. The ubiquity of energy usage encourages the view that a fuller understanding of energy flux, whether at the level of the ecosystem or the individual, can play a special role in extending our knowledge of animal function and of interrelations between animals and their environment. This notion came to the fore with the work of Lindemann (1942), and was developed more than two decades later as part of the International Biological Programme (IBP), where the role of energy in regulating ecosystem structure was given special attention. The potential of the doubly labelled water (DLW) technique (Lifson, Gordon & McClintock 1955) for accurate quantification of respiratory energy losses was recognized in several of the publications of the programme (Golley 1967; Petrusewicz & Macfadyen 1970). It offered the unique prospect of measuring respiration rates, specifically carbon dioxide production, in completely unrestrained animals. By the usual conventions of indirect calorimetry (Brody 1945), this could in turn generate estimates of energy expenditure and requirement.

The doubly labelled water technique is made possible by an exchange equilibrium between the oxygen of body water and the oxygen of expired carbon dioxide. Hence if the oxygen of body water is labelled with the stable isotope oxygen-18, loss of respiratory carbon dioxide will subsequently cause a corresponding depletion of the label. Additionally, however, water loss will cause the oxygen-18 label to be eliminated, making necessary a second label to quantify elimination of the oxygen label in the form of excretory or evaporative water. Deuterium and tritium are both suitable for this purpose, although the former is to be preferred above the radioisotope if analytical facilities are available. The difference between turnover rates of the deuterium label and the oxygen label then yields an estimate of total carbon dioxide production over the interval between an initial and final estimate of the isotopic composition of body water. In spite of its obvious applications and the recognized inadequacies of alternative approaches to estimating respiration in wild animals, the contribution of the DLW technique to the work of the IBP was negligible.

A failure to exploit the potential of the doubly labelled water technique was not confined to ecologists, who were to be pioneers in its eventual application. Even though the theory and procedures of the technique were already described by Lifson, Gordon & McClintock by 1955, with the underlying rationale established in the late 1940s (Lifson *et al.* 1949), the first study of an unrestrained animal (a homing pigeon) was not made until the 1960s (LeFebvre 1964). The first wild subjects (swallows and small rodents) were not studied until 1970 (Mullen 1970; Utter & Le Febre 1970), with work on humans waiting until the 1980s (Schoeller & Van Santen 1982; Fig. 4.1). These delays can be attributed, in part, to a number of disadvantages inherent in the technique.

1 The stable isotype oxygen-18, in the form of $H_2{}^{18}O$, is expensive, although by the standards of some other labelled compounds not unusually so, except for large animals (Mullen 1973).

2 Wild subjects must not only be captured but recaptured at an interval of about 12 h to 20 days.

3 Capture represents an intrusion into the routine of a wild animal and can cause behaviour patterns to be disrupted.

4 The current cost of founding a laboratory is around £200 000.

5 Sample analysis can be a lengthy procedure, typically taking one day per animal, and is prone to errors. Until recently, therefore, samples sent away for commercial analysis could share the twin problems of being both expensive and sometimes inaccurate.

DOUBLY LABELLED WATER TECHNIQUE

The DLW technique involves field and laboratory components. Typically, fieldwork consists of trapping the subject, introducing the labelled water (normally by injection), awaiting isotope equilibration, taking a blood sample (range 20–40 μl) and releasing the subject to resume its normal routine. This procedure takes from 1 to 2 hours for small mammals and birds. After an interval, usually of 1 to 3 days in small endotherms, but longer for larger mammals or ectotherms, the subject is

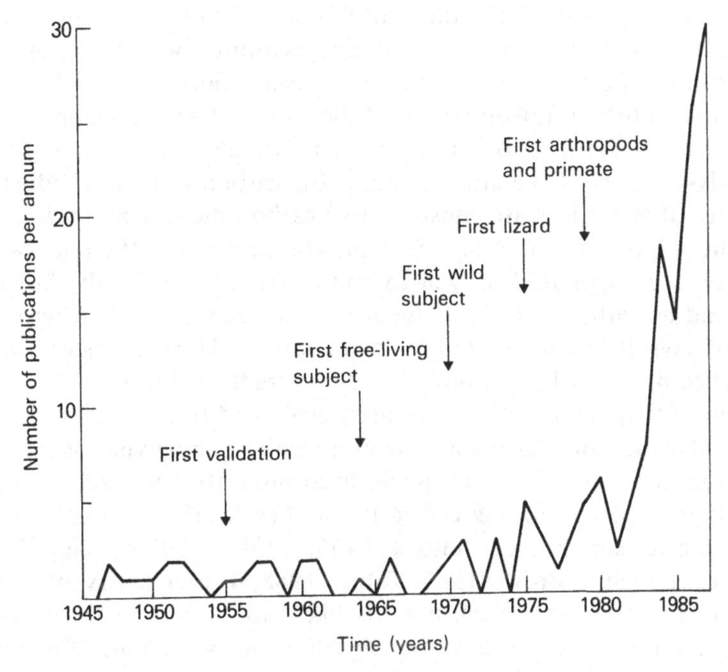

FIG 4.1 The annual number of publications concerned with the doubly labelled water technique and related theory listed in Biological Abstracts and other sources in the period 1945 to 1987.

recaptured, a second blood sample is collected, and the subject is released. General guidelines for use of the technique are similar to those for radiotags (Kenward 1987), because both involve handling the subject. If it is planned to use the DLW technique with a novel subject, then a trial involving capture, captivity for an hour or so, release, and comparisons of behaviour before and after can help to identify unsuitable subject species and individuals or suggest ways of minimizing intrusive effects inherent in the technique.

The laboratory phase involves a variety of methodologies, according to whether the water label is HDO or HTO, or if mass spectrometry is used for determination of $^{18}O/^{16}O$. To prepare samples of blood (or other body fluids) for isotope ratio mass spectrometry (IRMS), they are first vacuum-distilled and then variously treated to yield samples of hydrogen (HD) and carbon dioxide ($C^{16}O^{18}O$). The ratios of the labels in the gas samples are used to derive fractional turnover rates of the labels, from which estimates of total carbon dioxide production are calculated. Further details of the field and laboratory techniques outlined above are available from the author, or in publications cited below.

It was recognized from the outset that a number of simplifying assumptions would be necessary if the technique was to be applied successfully. These are six in number (Lifson, Gordon & McClintock 1955; McClintock & Lifson 1958a; Mullen 1973; Nagy 1980; Speakman & Racey 1988; Tatner & Bryant 1989). Briefly they assume: (i) the size of the body-water pool remains constant; (ii) carbon dioxide production and rates of water loss are constant; (iii) carbon dioxide and water lost from the subject have the same isotopic composition as the body-water pool; (iv) the introduced oxygen-18 and deuterium label only the body water and not other body constituents; (v) environmental (background) levels of oxygen-18 and deuterium do not vary; and (vi) no labelled water or labelled or unlabelled carbon dioxide enters the subject.

Each of these assumptions has been evaluated from a theoretical or empirical standpoint. Errors arising as a result of any given assumption not being met range from negligible to around 10% or, exceptionally, rather higher (McClintock & Lifson 1958b; Lee & Lifson 1960; Lifson & Lee 1960; Lifson & McClintock 1966; Mullen 1973; Nagy 1980; Schoeller, Leitch & Brown 1986; Tatner 1988; Tatner & Bryant 1989). While in isolation, therefore, none is likely to cause serious errors in estimation of respiration rates, but simultaneous violations of assumptions could be more serious. The most convincing evidence that compounding effects and other errors are small comes from the many comparisons of DLW results against conventional respirometry techniques (e.g. infra-red gas analysis) in validation trials. These suggest that a mean algebraic error of a few per cent is usual, although at the level of individual results a greater error is to be expected (Lifson, Gordon & McClintock 1955; Nagy 1980; Tatner & Bryant 1989). Some suggestions have been made for further improving the accuracy of the technique (Black, Prentice & Coward 1986; Speakman 1987; Wong *et al.* 1987). The case that some of the conventions originally adopted are unsound and the argument that a 'multi-point' method should be favoured over the 'two-point' method for estimating turnover rates (Klein *et al.* 1984; Schoeller & Taylor 1987) have been shown to be largely insubstantial (Speakman & Racey 1986). Further, the multi-point method is often unsuitable for routine application in small animals.

It may be concluded that the DLW technique yields sufficiently accurate estimates of metabolism in free-living mammals, birds and reptiles for most applications in the field of ecology (Tatner & Bryant 1989). Results for arthropods are generally more variable, although the group has not yet received sustained attention.

ENERGY EXPENDITURE IN THE FIELD: STUDIES USING DOUBLY LABELLED WATER

Extrapolation to the field of laboratory measurements of oxygen uptake or carbon dioxide production is rarely appropriate because a higher level of activity and a variable physical environment, which may itself induce increased energy costs, typically prevails under field conditions (Prinzinger 1982). Further, the metabolic costs associated with these factors are often unknown, so independent evidence to confirm the accuracy of estimates of field respiration is usually lacking. A more substantial approach is to measure all components of an energy budget in the laboratory, including the minimum metabolic cost of maintenance at thermoneutrality (called basal metabolic rate), any thermoregulatory requirement, the cost of activity and tissue production, and to allocate and sum these component costs according to activity schedules of wild animals and the nature of the field environment. Although Travis (1983) concluded that such time–energy budgets were liable to errors of 'staggering magnitude', they offer an advantage over other methods based on looser extrapolations of laboratory measurements in being capable of reflecting differences in energy use by individuals. They also offer a means of breaking down total metabolic costs, measured using DLW, into its components. Under ideal circumstances time–energy budget estimates of energy expenditure will be correlated with direct measures using DLW. Further, it is likely that a greater sophistication in measuring the costs of activity and in modelling thermal relations with the environment will encourage fuller comparability between the two methods (Buttemer *et al.* 1986).

Of three studies on aviary-held birds, the two most detailed showed close agreement between time–energy budget and DLW estimates (Table 4.1). This was to be expected where activity can be closely monitored, the behavioural repertoire is anyway restricted and the physical environment rather uniform. Less predictable was the reasonable agreement between estimates for three field studies (Table 4.1). Here the algebraic error was 6–7% as against <1% in two of the aviary studies. Yet the predictive capacity of time–energy budget models in field studies was often low, giving $r^2 \leq 10\%$ in two cases and reaching around 70% only when budgets were available for most (>75%) of the day, an ideal likely to be achieved only rarely. Further, estimation of component activity costs, such as those due to flight, will often be accessible only by using DLW. It will therefore usually not be feasible to predict energy costs using time–

TABLE 4.1. A comparison of estimates of daily energy expenditure (kJ d^{-1}) obtained using time–energy budget (TEB) and doubly labelled water (DLW) methods

	% Error[1] (algebraic)	r	n	References
Aviary studies				
Phainopepla (*Phainopepla nitens*)	−39.2[2]	0.71 (N.S)	6	Weathers & Nagy (1980)
Shrike (*Lanius nidovicianus*)	−0.1	0.79*	8	Weathers *et al.* (1984)
Budgerigar (*Melopsittacus undulatus*)	+0.3	0.85**	12	Buttemer *et al.* (1986)
Field studies				
Sparrow (*Passerculus sandwichensis*)	+6.0	0.22 (N.S.)	6	Williams & Nagy (1984)
Dipper (*Cinclus cinclus*)	+5.7	0.32*	29	Bryant, Hails & Prys-Jones (1985)
Kestrel (*Falco tinnunculus*)	+7.0	0.83**	15	Masman, Daan & Beldhuis (1988)

[1]Percentage error calculated as (TEB − DLW)/DLW × 100.
[2]See Weathers & Nagy (1980) for discussion of large error.

energy budgets and to verify their accuracy without field trials involving DLW results as a standard. In studies of energy use and demand by animals, therefore, where either accuracy of mean respiratory energy costs or an ability to distinguish such costs across a range of conditions and behaviours is required, then the DLW technique offers important advantages. This does not hold when the behaviour of the animal is markedly affected by the necessary handling of the subject. In these circumstances a naturalist's skills can be crucial in minimizing any adverse effects. Nevertheless, some subjects are likely to remain always unsuitable.

The DLW technique in ecosystem studies

A principal area of application for the DLW technique is in the study of energy flow in ecosystems to increase understanding of the scale and variability of energy flow through consumer communities. In the ideal case this could involve measuring respiration rates for all components of the system, at all stages in their seasonal cycles and yet such a task would quickly become unmanageable. Modelling will almost invariably be necessary, generally involving a greater or lesser emphasis on the system under investigation and on extrapolation from related studies.

With a view to developing an accurate predictive model for small bird communities, data were collected across a range of species, selected on the following criteria: that the sample displayed a variety of foraging habits, covered a range of body masses (10–150 g), included all stages of the annual cycle, and was selected from different environments (i.e. regarding latitude, day length and temperature). Using mean values of energy expenditure for each species at each stage in the annual cycle studied, a stepwise multiple regression analysis was used to identify the best predictors of energy expenditure in the field. No assumptions were made about ranking of variables based on experience of responses in captive animals, except inasmuch as this inevitably guided the choice of factors measured in the field or reported in the literature. Body mass, foraging guild rank, percentage flying time, latitude and temperature were used as independent variables (Table 4.2). All were significant, but with temperature only positively so, and percentage flying time only when aerial feeding birds were included. So a general model with mass, foraging guild rank and latitude as predictor variables is proposed, which explains over 80% of the variation in daily energy expenditure of the sample (Table 4.2). Mass remains the crucial factor, as in earlier findings (King 1974; Kendeigh, Dolnik and Gavrilov 1977; Walsberg 1983), but habits are invoked via feeding patterns and the environment as indicated by latitudinal position. It is likely that the twin effects of shorter activity periods and warmer temperatures at low latitudes underlie the significance of latitude in this analysis (Bryant, Hails & Tatner 1984). Nagy (1987) has also analysed a sample of DLW results for a range of animals and presents predictive regression models based on body mass alone but categorized by habitat where this has a significant effect.

An analogous approach to understanding gross patterns of variation in energy expenditure is to derive stage-specific (Table 4.3) or guild-specific estimates of metabolic rates, standardizing for body mass differences in the sample by using metabolic mass (Brody 1945). The consequences of this modelling approach for estimating community respiration can be examined using a sample data set from the literature (Table 4.4). This reveals an apparent underestimate of the order of 10–30% in the original estimate of field metabolism, which assumed all costs amounted to 2.5 × BMR (basal metabolic rate). Whether errors of this scale are important will depend on the context of the result. Of much greater concern is the dependence of broader conclusions about ecosystem function which follow from invalid assumptions about the precision of most published models of field energy requirements that are based on body mass alone. For example, Brown & Maurer (1986) examined the respective roles of smaller and larger species in terms of their energy use across a range of animal and plant communities. It is often supposed that energy use in communities is dominated by smaller

TABLE 4.2. Multiple regression analysis of factors[1] affecting daily energy expenditure (DEE)[2] of small birds[3] (10–150 g). Studies of aviary-held birds and nestlings are excluded. All estimates of DEE were obtained using DLW[4]

		β	P
(a)	Excluding foraging guild rank	β	P
	Body mass ($\log_{10}$)	+0.80	<0.001
	Latitude	+0.34	<0.001
	Temperature	+0.22	<0.05
	$r^2 \times 100$ (adjusted) = 77.5%, $F = 64.1$, d.f.$_2$ = 52, $P < 0.001$		
(b)	Including foraging guild rank	β	P
	Body mass ($\log_{10}$)	+0.93	<0.001
	Guild rank	−0.16	<0.01
	Latitude	+0.12	<0.05
	$r^2 \times 100$ (adjusted) = 80.8%, $F = 98.9$, d.f.$_2$ = 67, $P < 0.001$		

[1]Independent variables included in analysis additional to those listed above: day length, sex (♀ = 1, ♂ = 2), % 24 h in flight.
[2]Dependent variable was $\log_{10}$ (daily energy expenditure, kJ d^{-1}).
[3]Foraging guild ranks were: 1 = aerial, 2 = aquatic (diving), 3 = arboreal, 4 = pause-walk, 5 = ground, and 6 = sit and wait–sally. This was a subjective ranking placing apparently high-cost foraging techniques at the top (1) and low-cost techniques on progressively lower ranks (2–6).
[4]Species (and studies) included in analysis: dipper *Cinclus cinclus* (Bryant & Tatner 1988); house martin *Delichon urbica* (Hails & Bryant 1979; Bryant & Westerterp 1980, 1983); pied kingfisher *Ceryle rudis* (Reyer & Westerterp 1985); sand martin *R. riparia* & swallow *Hirundo rustica* (Westerterp & Bryant 1984); starling *Sturnus vulgaris* Ricklefs & Williams 1984; Westerterp & Drent 1985); savannah sparrow *Passerculus sandwichensis* (Williams & Nagy 1985; Williams 1987); blue-throated bee-eater *Merops viridis* and Pacific swallow *Hirundo tahitica* (Bryant, Hails & Tatner 1984); mockingbird *Mimus polyglottos* and purple martin *Progne subis* (Utter 1971; Utter & LeFebvre 1973); desert quail *Callipepla gambelii* (Goldstein & Nagy 1985); willow *Parus montanus*, crested *P. cristatus*, and coal tits *P. ater* (Moreno, Carlson & Alatalo 1989); phainopepla *P. nitens* (Weathers & Nagy 1980); blue *Parus caeruleus* and great tit *Parus major*, spotted *Muscicapa striata* and pied flycatchers *Ficedula hypoleuca*, wheatear *Oenanthe oenanthe*, robin *Erithacus rubecula*, blackbird *Turdus merula*, dunnock *Prunella modularis*, bullfinch *Pyrrhula pyrrhula*, (D.M. Bryant, P. Tatner *et al.*, unpublished); n = twenty-nine species/studies yielding up to seventy-one estimates of mean DEE at different stages in the annual cycle.

organisms by virtue of their greater abundance, yet Brown & Maurer showed the reverse with larger organisms dominant. Analysis of the Holmes & Sturges (1975) data (Table 4.4) showed that the apparent importance of large and small organisms depended on the model used, with the most accurate models suggesting a greater energy use by smaller birds. Thus general conclusions about energy use at the community level which imply exactness in their derivation require predictive models which have either been verified against field measurements or were derived therefrom as proposed above.

TABLE 4.3. Mass corrected energy expenditure (kJ $g^{-0.73}h^{-1}$) for different stages in the annual cycle of small birds[1]

Rank	Stage	Energy expenditure	n
(a) All species, studies and stages included			
1	Rearing young	441	33
2	Juveniles	393	2
3	Moult	385	2
4	Incubating ♀	385	14
5	Incubating ♂	383	1
6	Territorial ♂	380	5
7	Aviary	377	6
8	Winter	355	10
9	Laying ♀	281	2
10	Pullus	270	6
11	Non-breeding	240	2
(b) Excludes hirundines (swallows) and stages where $n \leq 2$[2]			
1	Rearing young	422	16
2	Territorial ♂	403	3
3	Winter	356	7
4	Incubating	345	7
5	Pullus	270	6

[1]Energy expenditures are compared by dividing kJ h^{-1} by metabolic mass, $M^{0.73}$ (exponent after Bryant & Tatner 1988). Species included as for Table 4.2.
[2]Because hirundines generally have a relatively high energy expenditure, and are well represented in some samples for each stage, they can bias the relative energy cost of the stage. Exclusion of hirundines and small samples ($n \leq 2$) therefore provides a more representative comparative data set for small birds.

TABLE 4.4. Small–bird community energetics: a comparison of different methods for estimating expenditure (kJ 10 ha^{-1} d^{-1}). Data are from Hubbard Brook study, 25 May – 9 June; Holmes & Sturges (1975)

Method	Energy expenditure	%	Relative energy demand of small and large birds[1]: Small	Large
Basal metabolic rate × 2.5	12 040	—	5630	6410
Guild means[2]	12 970	+8%	5400	7570
Regression model[3]	16 000	+33%	8420	7570

[1]A comparison of energy demands by small and large birds following the approach of Brown & Maurer (1986), who estimated energy costs by assuming they scaled as 0·67 on body mass. They concluded small species (smallest 50%) used less energy than larger species, whereas the regression model indicates an opposite conclusion.
[2]Based on foraging guild means (D. M. Bryant, unpublished); see Table 4.2 for definition of guilds.
[3]Calculated from multiple regression model (see Table 4.2; D. M. Bryant, unpublished).

The DLW technique in the study of populations and individuals

This section focuses on the causes of variation in energy expenditure at the level of individuals, and the consequences for themselves and the populations they comprise. It is here that the DLW technique makes a unique contribution because it is the only method available which provides reliable estimates of energy expenditure free from often inexplicit assumptions about their causes. Because time–energy budgets, for example, are constructed on the basis that mass differences contribute to expenditure estimates, and that a thermoregulatory component should be included, it follows that mass and temperature will be identified as key components of variation in energy expenditure (see Ettinger & King 1980). The DLW technique, in measuring total carbon dioxide production directly, makes no such assumption. Ten species have been studied in sufficient detail for concurrent data on a range of environmental, behavioural and other factors likely to influence energy expenditure to have been collected (Table 4.5). I have reduced the list of independent variables to a manageable size which in some cases accommodates similar, but not identical, factors under a single head (e.g. brood size, feeding rate and provisioning rate are all considered under nest feeds; see Bryant & Tatner 1989).

For most of the eleven independent variables examined, over half of the studies yielded significant correlations with energy expenditure (Table 4.5). In some cases an anticipated trend was not revealed, perhaps because it was overshadowed by the effects of other factors or did not vary to any extent at the time the DLW measurements were made. Correlations of opposite sign to those anticipated are particularly instructive because they may indicate novel interactions between animals and their surroundings. In only two instances was there unequivocal support for the pattern predicted on the basis of laboratory-based studies: both involved the effect of activity. Nest feeding rates and flight or ground activity were positively correlated with energy expenditure in half or more cases, with no evidence of negative trends (Table 4.5). Here, a lack of correlation with brood provisioning rates may sometimes be explained by compensatory effects of habitat or individual quality. No consistent patterns emerged amongst the remaining variables, except that body size was unexpectedly more often negatively than positively related to daily energy expenditure (Bryant & Tatner 1989). Clearly, any assumption that being large necessarily entails an increase in daily living costs is in error. Such a finding has particular relevance when the costs and benefits of being of different body sizes, or the importance of energy reserves in different-sized individuals, is being considered. It is

TABLE 4.5. Frequency of significant correlations ($P < 0.05$) for energy expenditure[1] and a range of factors in a sample of ten bird species[2]

Factors	% Yielding significant correlations	Frequency +	Frequency −	*n*
Body mass	56	3	2	9
Mass change	50	3	1	8
Body size	57	1	3	7
Food abundance	86	1	5	7
Nest feeds	50	5	0	10
Ambient temperature	50	1	3	8
Sunshine	40	0	2	5
Wind speed	25	1	1	8
% flying	83	5	0	6
Ground activity	100	2	0	2
Age	33	0	1	3

[1]Refers to daily energy expenditure (kJ d^{-1}) except for breeding house martins, where ADMR (cm^3 CO_2 g^{-1} h^{-1}) was examined (Bryant & Westerterp 1980, 1983).
[2]The following studies were included in the analysis: dipper (Bryant & Tatner 1988; Bryant, Hails & Prys-Jones 1985) house martin (Bryant & Westerterp 1980, 1983); swallow, sand martin (Westerterp & Bryant 1984); Pacific swallow, blue-throated bee eater (Bryant, Hails & Tatner 1984 and unpublished); wheatear (Tatner 1989); savannah sparrow (Williams 1987); pied kingfisher (Reyer & Westerterp 1985); starling (Westerterp & Drent 1985). These total ten species. See Table 4.2 for Latin names.

also feasible that what appear to be simply species differences in this analysis might conceal patterns which a more rigorous analysis could expose. For example, a positive relationship between house martin energy expenditure and food abundance (Bryant & Westerterp 1983) may relate to the particulate nature of the food supply, whereas other species taking larger, more scattered items could benefit by responding in a different way to changes in food supply with a consequent negative relationship between food abundance and energy expenditure.

Few weather factors were found to have a significant effect on energy expenditure, in spite of study sites ranging from a north temperate winter to the equatorial. Most striking was the relative unimportance of temperature, which on the basis of conventional treatments would be assumed to have universal significance for small homeotherms. A simple explanation might be that ambient temperature is an inadequate description of the thermal environment, and that solar heating, wind cooling and other factors must also be taken into account. To examine this point I characterized the thermal environment of the dipper, *Cinclus cinclus*, and used these observations to explore the relationship between ambient temperature and operative temperature (D. M. Bryant, unpub-

TABLE 4.6. Correlations for energy expenditure in dippers measured using doubly labelled water (DLW) and estimated using time–energy budgets[1]

	Measured energy expenditure	
	$kJ\ d^{-1}$	$kJ\ g^{-1}\ h^{-1}$
Predicted energy expenditure[1]		
T_a	0.906	0.909
T_e (bird)	0.902	0.901
T_e (river)	0.903	0.923

[1]Mean energy expenditure was predicted for each sex at all stages of the annual cycle in dippers using the model given by Bryant & Tatner (1988). T_a indicates ambient temperature was used to estimate the thermoregulatory component, whereas T_e shows operative temperature (Robinson, Campbell & King 1976; Buttemer *et al.* 1986) was used, calculated either for the microhabitat seen to be used by dippers (bird) or for a sample of stations along the river (D. M. Bryant, unpublished) (= river).
$P < 0{\cdot}01$ for all correlations.

lished). This allowed the thermal environment to be modelled with even greater precision and yet no improvement in the correspondence between time–energy budget estimates and DLW measurements of energy expenditure was evident (Table 4.6). To an extent this was expected since the dippers lived on a somewhat sheltered river system, and the impact of direct sunshine was neither frequent nor hot in the hilly ground of the study area. A more searching comparison would involve an animal routinely exposed to sun and wind. Even so, the value of thermal modelling in accounting for variation in energy expenditure in the field is always likely to be greatest when the signficance of other factors such as activity are at a minimum. While this can easily be the case for captive subjects (Weathers *et al.* 1984; Buttemer *et al.* 1986), on the basis of the present results this seems likely to be rather unusual amongst small wild birds. This should not be interpreted to suggest that thermoregulatory

FIG 4.2 The relationship between ambient temperature and energy expenditure (a) in a range of species studied using doubly labelled water. Data on temperature have either been extracted from the original papers, are the subject of personal communications or were taken from average climate tables for each study area (mean temperature of study months). Species of different mass are made comparable by dividing energy expenditure per hour ($kJ\ h^{-1}$) by metabolic mass ($M^{0.73}$). Data points refer to species and stages in the annual cycle: for this analysis different stages in the same species are treated as independent (n = 29). The quadratic equation is $y = 379.5 + 6.8x - 0.3x^2$; $F = 5.0$; $P < 0.01$; d.f.$_2$ = 68. (b) The dipper *Cinclus cinclus:* the sexes are made comparable by dividing energy expenditure per hour by metabolic mass ($M^{0{\cdot}794}$) (Bryant & Tatner 1988). Other details as for (a). The quadratic equation is $y = 5.5 + 0.15x - 0.01x^2$; $F = 5.0$; $P < 0.01$; d.f. = 74.

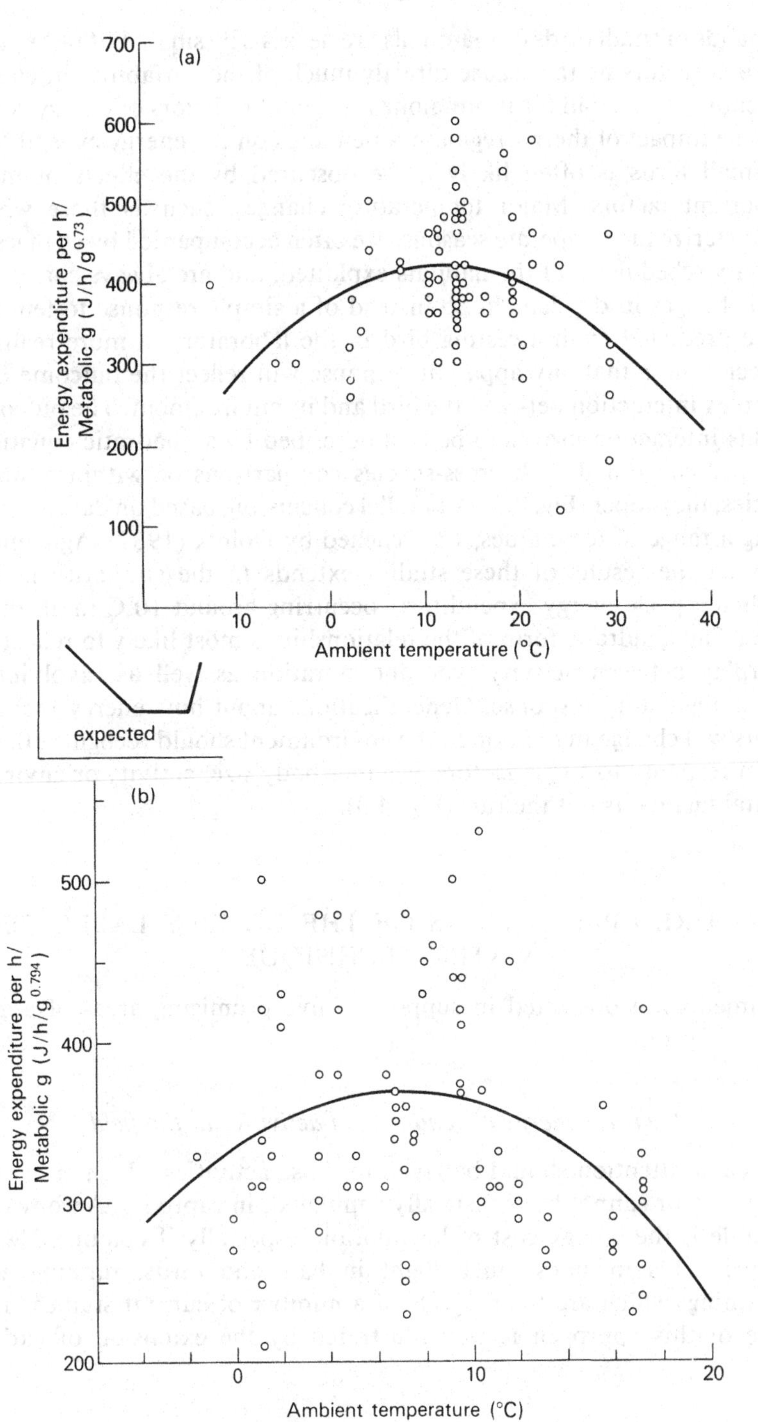
(a)
Energy expenditure per h/
Metabolic g (J/h/g0.73)
700
600
500
400
300
200
100
10
0
10
20
30
40
Ambient temperature (°C)
expected
(b)
Energy expenditure per h/
Metabolic g (J/h/g0.794)
500
400
300
200
0
10
20
Ambient temperature (°C)

demands of small birds or mammals are necessarily small, but rather that thermal factors neither cause directly much of the variability in energy expenditure of small birds nor alone are good predictors of energy costs.

The impact of thermoregulatory demands on the energy expenditure of small birds is often likely to be obscured by the effects of more important factors. Major temperature changes, such as those which characterize the temperate seasons, are often accompanied by changes in activity schedules and the habitats exploited, and are always correlated with changes in day length. So instead of a simple response to temperature predicted from a resting bird in the laboratory, a more realistic expectation is that any apparent response will reflect the outcome of a complex interaction between the bird and its environment. The outcome of this interaction proves to be best described by a quadratic equation, whether concerned with cross-species comparisons or within a single species, the dipper (Fig. 4.2). A parallel conclusion, based on data derived using a range of techniques, was reached by Dolnik (1982). Agreement between the results of these studies extends to the similarity in the predicted peak energy expenditure, occurring around 10°C in all three cases. The quadratic form of the relationship is most likely to reflect an interplay between activity type and duration as well as involving a thermoregulatory response. Generalizations about how energy requirements will change in relation to the environment should recognize that a direct response to single factors, whether body size, activity or environmental factors, is not the rule (Fig. 4.3).

FUTURE APPLICATIONS OF THE DOUBLY LABELLED WATER TECHNIQUE

Arguments are presented in support of five promising areas of application in ecology.

Measure energy expenditure of activities in the field

Particular attention should be given to those activities which either do not occur, or cannot be realistically simulated, in captivity. An obvious example is the energy cost of locomotion, especially if combined with feeding. This includes much flight in bats and birds, running and swimming, which are the subjects of a number of current studies. The value of this approach is well illustrated by the extension of earlier

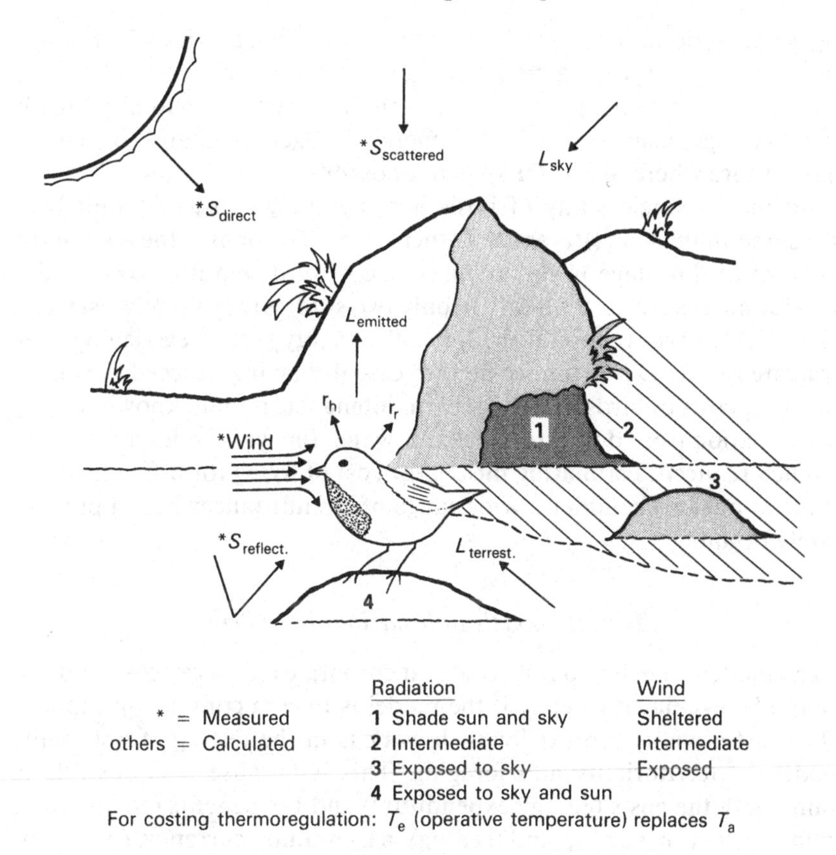

FIG 4.3 The relations between a small bird and its thermal environment. The measurements made directly or calculated to describe the thermal environment are shown. Four levels of exposure to solar radiation and three levels of exposure to wind were identified. Using the technique of Robinson, Campbell & King (1976) and Keister, Anthony & Holbo (1985) operative temperatures were derived from these data to predict the metabolic consequences of changing temperatures.

allometric flight-cost models by Masman & Klaassen (1987) to include morphometric parameters. Using flight-cost data obtained using DLW, they raised the mean correlation coefficient for the flight cost estimates from $r^2 = 0.75$ to $r^2 = 0.85$ by including mass, wing span and area as independent variables. It would be useful to extend flight-cost studies from trivial flights to extended migratory flights, although the recapture stage might present insurmountable problems. Investigation of a range of potentially high-cost field activities would yield many data of interest. Climbing, jumping, digging, building, food-handling, fighting, displaying, singing, dominance and load-carrying are all activities which might be used or occur with a frequency governed by their cost relative to alternatives. Knowledge of the cost of these activities could also greatly improve the exactness of time–energy budget estimates of energy expenditure where use of DLW is not possible.

In only a single study of birds has the energy cost of moult been measured in the field (Bryant & Tatner 1988). To confirm the apparently low cost of this stage in the annual cycle of birds requires comparable data for other species. Similarly in only two studies (Bryant & Westerterp 1980; D.M. Bryant, unpublished) has the energy cost of egg-laying been evaluated using DLW, and in neither case did laying proceed normally. So laying costs of birds, inclusive of maintenance, remain known only by extrapolation from domestic stock, for which the work of food-gathering is much reduced. Estimating the energy cost of living for different stages of the annual cycle under a wide range of circumstances has in practice barely begun.

Testing models in behavioural ecology

Many models in behavioural ecology incorporate an energy cost, and this should be evaluated directly if the model is to gain convincing support. The most obvious context for such work is in the testing of optimality models of territoriality and foraging. This is because it is possible to frame both the costs (energy expenditure) and the benefits (energy gains from territory ownership and feeding) in a common currency. One of the most fully studied territorial systems is that of the East African golden-winged sunbird *Nectarinia reichenowi* (Gill & Wolf 1975), in which territoriality was apparent when the territory was economically defendable (i.e. costs of territory defence ≤ benefits of territory holding). Yet this conclusion was reached on the assumption that territorial defence could be accurately evaluated by extrapolation from metabolism during broadly analogous activities in the laboratory. If this estimate is in error by only 10%, then territoriality was observed even where territories may not have been energetically economic to defend. Convincing evidence of the utility of this and related models must clearly demonstrate the balance of benefit to cost and also confirm the accuracy of any crucial cost parameters.

Application to 'new' organisms

As yet the bulk of studies on energy expenditure using DLW have examined small mammals and birds. The recent focus on humans has added a further large mammal useful for comparative purposes, but there remains a general dearth of studies on large mammals and large terrestrial birds (Nagy 1987). A principal constraint here is the high cost of oxygen-18-labelled water. In the case of ruminants the fate of the deuterium label in methane production needs more study (Fancy *et al.* 1986). With the more extended periods of measurement necessary for larger animals the possibility of violations of underlying assumptions (see above) also becomes greater (Nagy 1980; Black, Prentice & Coward 1986; Schoeller, Leitch & Brown 1986; Speakman & Racey 1986; Speakman 1987; Wong *et al.* 1987). Obvious applications include free-living costs of marine mammals, of different elements of large mammal assemblages, of savannah- and forest-dwelling primates, as well as comparative study of a range of life-styles in different habitats amongst other mammals and birds.

Preliminary work with arthropods has been less successful than with any other group; discrepancies between DLW measurements and conventional techniques have been unaccountably large (Cooper 1983). Sustained work with bees, dragonflies and locusts should minimize problems caused by the small size of the subjects, and ought to be readily integrated with related current work. Fish are generally considered to be unsuitable as subjects because their relatively low rate of metabolism, coupled with a likely rapid turnover of body water, suggests that isotope turnover will be overwhelmingly a consequence of losses via excretory and exchanged water. When loss of the oxygen-18 label in carbon dioxide is so small, errors can be large (Nagy 1980). Yet the water economy of fish is still relatively obscure so, with an appropriate choice of subject, useful results might be obtained. Fractionation effects at the air–skin surface make amphibians difficult candidates for study. Until this is evaluated, the ectothermic life-style is best investigated by extending studies on reptiles (Nagy, Huey & Bennett 1984) to include snakes and turtles.

Multiple-labelling techniques

While the DLW technique already involves a multiple label, there exist a number of ways in which the power of the general approach may be extended by incorporating further labels. Buscarlet & Grenot (1985) suggest using the radioisotope ^{22}Na to study sodium balance, and by inference food intake, in association with the DLW technique. Using a supplementary radiolabel in this way clearly opens up a range of possibilities. It is in the area of stable isotope applications, however, that the methodologies and equipment established for DLW work can most obviously be applied. The stable isotope labels ^{13}C, ^{15}N, ^{17}O, ^{34}S have a range of biological applications. Carbon-13 and ^{15}N might be used for monitoring lipid and amino acid metabolism in conjunction with energy use in the field, although the methods would require adaptation from those currently practised in the laboratory (Halliday & Rennie 1982; Wolfe 1984). A multi-label system giving simultaneous monitoring of energy expenditure, water turnover, protein turnover, lipid mobilization and food intake in a totally unrestrained individual is a realistic goal, largely involving a limited development of a number of established practices.

Life-history studies

Life-history theory has for long outpaced the capacity of ecologists and others to generate rigorous tests (Stearns 1976). It is only comparatively recently, for example, that evidence of a cost of reproduction has been demonstrated for field populations using a manipulative rather than a correlative approach (Nur 1984). Even so a cost of sufficient magnitude to offset the benefits of additional reproduction awaits demonstration (cf. Partridge in this volume). There is little prospect of investigating in any depth, or to any great extent, the subtleties and variety of life-history trade-offs for different species, sexes, age-classes, habitats and resource levels using most present techniques (Reznick 1985). Either estimates of cost are open to question, e.g. whether mass loss represents a cost (Bryant 1988), or quantification of cost is excessively time-consuming, e.g. measuring mortality rates for different treatments. Further, identifying which of a complex of factors is the cost-inducing agent is usually impossible in all but the most carefully controlled studies. Using DLW, however, might offer a route by which the risk linked to different behaviours could be more precisely evaluated.

To do this it would be necessary to assess the risk attached to a range of body condition states, the relationship between behaviour and effort and between effort and condition. Two essential elements are that body condition is measured with enough accuracy for variation in, especially, the lean component to be related to survival prospects, and for effort to be evaluated using a currency such as energy expenditure which facilitates direct interspecific as well as intraspecific comparisons. On the assumption that condition loss can be shown to impair survival (Marcstrom & Kenward 1981) or other fitness parameters, then a demonstration that condition loss is induced by a higher rate of energy expenditure, perhaps indirectly as a result of interactions with other factors, could be used to assess the relative risks of different behavioural and reproductive strategies across and within species.

Reasoning of this type has been used to account for clutch size variation in birds, specifically why birds often lay clutches smaller than they are able to raise to fledging. Three hypotheses which are essentially energetic in nature are potentially testable. The first suggests that effort is constrained by the capacity of the gut to assimilate energy, and the allocation of effort for brood provisioning is then limited by the ability to make energy available to metabolizing tissues (Drent & Daan 1980; Kirkwood 1983; Masman, Daan & Dijkstra 1988). It follows that effort greater than this threshold will precipitate a sharp increase in fitness costs signalled by a fall in condition, while effort below this threshold should carry no cost. By manipulating brood sizes and hence provisioning rates, the costs of raising broods of different sizes can be examined and such a threshold sought. Essential to this approach is an ability to measure accurately lean and fat condition, using ultrasound or equivalent techniques (Baldassarre, Whyte & Bolen 1980; Walsberg 1988). This is necessary because body mass changes alone are usually difficult to interpret as indicating either a cost or a benefit (Bryant 1988) and hence can be expected to yield inconsistent relationships with fitness costs.

The second hypothesis assumes that all levels of effort carry some cost, and that different levels of reproductive effort may therefore represent different trade-offs between current and future success. A crucial question here is the form of this trade-off, specifically whether fitness costs accelerate, decelerate, are linear or have some other more complex form in relation to effort (Bryant 1988). The form of the trade-off is likely to have a powerful effect on the optimal reproductive effort (Sibly & Calow 1986). The third hypothesis is that effort is unconstrained by parental risk but instead is constrained by time. On this view the parental budget is maintained continually in balance and the work of provisioning proceeds until food becomes unavailable, darkness falls or some other extrinsic constraint imposes itself. On this view energy expenditure is a consequence of time–activity budgets but does not relate in any consistent way to condition loss or fitness cost. The DLW technique could play a role in evaluating these hypotheses, even though investigations are presently hampered by an inadequate understanding of the scale and significance of changes in condition (especially lean condition) in birds.

Drent & Daan (1980) suggested that the working capacity of parents would be limited by 'physiological constraints defining a sustained work level in metabolic terms. In some situations parent birds may ignore this physiological warning level, but the penalty will be a loss of condition which will in turn entail increased mortality.' Drent & Daan (1980) observed on the basis of a small preliminary sample that rates of energy

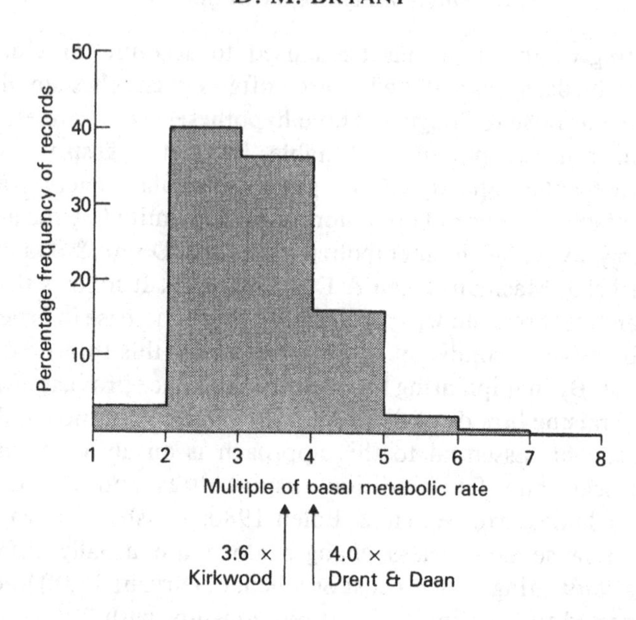

FIG 4.4 Frequency distribution of BMR multiples derived from studies using doubly labelled water on twenty-nine species of small bird. This gives a total of 501 measurements for species covering the mass range 10–150 g. The two arrows refer respectively to an estimate of maximum assimilation rate for a 10–150 g passerine bird by Kirkwood (1983) and an estimate of maximum sustainable working rate by Drent & Daan (1980).

expenditure in free-living birds reached a ceiling of 4 × BMR, an observation consistent with the first hypothesis given above assuming no survival penalty was incurred by those birds working close to this ceiling. More recent results show this to have been an underestimate and that 5 × BMR is a more representative maximum (Fig. 4.4). Across all studies, for example, 4 × BMR was exceeded in around 20% of observations and in 13 out of 26 (50%) of the species/studies. Whether those studies where BMR exceeded 4 × BMR entailed a fitness cost was not determined. Only the studies of Reyer (1984) and Reyer & Westerterp (1985) have demonstrated an increased mortality of individuals working at the highest rates, and no study has demonstrated directly that this is mediated via condition loss. Already it is clear, however, that work rates in excess of 4 × BMR are relatively common, do not necessarily involve loss of mass, and might therefore be sustainable (Bryant 1988). It follows from these observations that Kirkwood's (1983) estimate of maximal assimilation rate at around 3.6 × BMR is set too low, at least for small birds (10–150 g).

Further, the observed flexibility in gut length of birds in relation to requirements (Sibly 1981) indicates a capacity to mitigate constraint by this means, unless at the same time the cost of additional tissue maintenance and food-processing makes this unprofitable. But, if costs and benefits do trade off in this way, assimilation constraints of this kind are best viewed as a variant to be considered in trade-off hypotheses.

CONCLUSION

Two broad areas of application for the DLW technique in ecological research can be identified. On the one hand, effort can be usefully devoted to broadening the data base, in the expectation that the wider perspective will allow more exact identification of energetic constraints in different animal populations and communities. On the other hand, the technique can be used to explore the short-term energy costs associated with changing environments and alternative ecological tactics amongst individuals, whether concerned with feeding, movement, breeding or survival. Both these fields of enquiry will yield data on energy flux with an accuracy which can be obtained in no other way. To what extent the DLW technique will allow wider progress in ecological research, however, is likely to depend on the development or arguments about the role of energy as a constraint and facilitator in ecological systems, as well as on the skilful use of a range of current techniques, such as radiotelemetry and remote activity recorders, in close association with the doubly labelled water technique.

REFERENCES

Baldassarre, G. A., Whyte, R. J. & Bolen, E. G. (1980). Use of ultrasonic sound to estimate body fat deposits in the mallard. *Prairie Naturalist,* **12**, 79–86.

Black, A. E., Prentice, A. M. & Coward, W. A. (1986). Use of food quotients to predict respiratory quotients for the doubly-labelled water method of measuring energy expenditure. *Human Nutrition and Clinical Nutrition,* **40**, 381–92.

Brody, S. (1945). *Bioenergetics and Growth.* Reinhold, New York.

Brown, J. H. & Maurer, B. A. (1986). Body size, ecological dominance and Cope's rule. *Nature,* **324,** 248–50.

Bryant, D. M. (1988). Energy expenditure and body mass changes as measures of reproductive costs in birds. *Functional Ecology,* **2**, 23–34.

Bryant, D. M., Hails, C. J. & Prys-Jones, R. (1985). Energy expenditure of free-living dippers (*Cinclus cinclus*) in winter. *Condor,* **87**, 177–86.

Bryant, D. M., Hails, C. J. & Tatner, P. (1984). Reproductive energetics of two tropical bird species. *Auk,* **101**, 25–37.

Bryant, D. M. & Tatner, P. (1988). Energetics of the annual cycle of dippers *Cinclus cinclus. American Naturalist,* **130,** 17–38.

Bryant, D. M. & Tatner, P. (1989). The costs of brood provisioning: effects of brood-size and food supply. *Proceedings, 19th International Ornithological Congress, Ottawa (1988)*.

Bryant, D. M. & Westerterp, K. R. (1980). The energy budget of the house martin. *Ardea*, **68**, 91–102.

Bryant, D. M. & Westerterp, K. R. (1983). Short-term variability in energy turnover by breeding house martins *Delichon urbica:* a study using doubly-labelled water ($D_2{}^{18}O$). *Journal of Animal Ecology*, **52**, 525–43.

Buscarlet, L. A. & Grenot, C. (1985). Utilisation des isotopes stables et radioactifs dans les études bioénergétique des populations animales en milieu terrestre. *Acta Oecologica/Oecologica Generalis*, **6**, 105–34.

Buttemer, W. A., Hayworth, A. M., Weathers, W. W. & Nagy, K. A. (1986). Time-budget estimates of avian energy expenditure: physiological and meteorological considerations. *Physiological Zoology*, **59**, 131–49.

Cooper, P. D. (1983). Validation of the doubly labelled water (3H H 18D) method for measuring water flux and energy metabolism in tenebrionid beetles. *Physiological Zoology*, **56**, 41–6.

Dolnik, V. R. (1982). Methods of time and energy budgets study. *Proceedings of the Zoological Institute, USSR*, **113**, 1–36.

Drent, R. H. & Daan, S. (1980). The prudent parent. Energetic adjustments in avian breeding. *Ardea*, **68**, 225–52.

Ettinger, A. D. & King, J. R. (1980). Time and energy budgets of the willow flycatcher (*Empidonax trailii*) during the breeding season. *Auk*, **97**, 533–46.

Fancy, S. O., Blanchard, J. M., Holleman, D. F., Kokjer, K. J. & White, R. G. (1986). Validation of doubly labelled water method using a ruminant. *American Journal of Physiology*, **251**, R143–R149.

Gill, F. B. & Wolf, L. L. (1975). Economics of feeding territoriality in the golden-winged sunbird. *Ecology*, **56**, 333–45.

Goldstein, D. L. & Nagy, K. A. (1985). Resource utilization by desert quail: time, energy, food and water. *Ecology*, **66**, 378–87.

Golley, F. B. (1967). Methods of measuring secondary productivity in terrestrial vertebrate populations. *Secondary Productivity of Terrestrial Ecosystems: Principles and Methods* (Ed. by K. Petrusewicz), pp. 99–124. Warszawa-Kraków, Warsaw.

Halliday, D. & Rennie, M. J. (1982). The use of stable isotopes for diagnosis and clinical research. *Clinical Science*, **63**, 485–96.

Hails, C. J. & Bryant, D. M. (1979). Reproductive energetics of a free living bird. *Journal of Animal Ecology*, **48**, 471–82.

Holmes, R. T. & Sturges, F. W. (1975). Bird community dynamics and energetics in a northern hardwoods ecosystem. *Journal of Animal Ecology*, **44**, 175–200.

Keister, G. P., Anthony, R. G. & Holbo, H. R. (1985). A model of energy consumption in bald eagles: an evaluation of night and communal roosting. *Wilson Bulletin*, **97**, 148–60.

Kendeigh, S. C., Dolnik, V. R. & Gavrilov, V. M. (1977). Avian energetics. *Granivorous Birds in Ecosystems* (Ed. by J. Pinowski & S. C. Kendeigh), pp. 127–204. Cambridge University Press.

Kenward, R. E. (1987). *Wildlife Radio Tagging*. Academic Press, London.

King, J. R. (1974). Seasonal allocation of time and energy resources in birds. *Avian Energetics* (Ed. by R. A. Paynter), pp. 4–70. Publications of the Nuttall Ornithological Club, 15. Cambridge, Massachusetts.

Kirkwood, J. K. (1983). A limit to metabolizable energy intake in mammals and birds. *Comparative Biochemistry and Physiology*, **75A**, 1–3.

Klein, P. D., James, W. P. T., Wong, W. W., Irving, C. S., Murgatroyd, P. R., Cabrera, M., Dallosso, H. M., Klein, E. R. & Nichols, B. L. (1984). Calorimetric validation of the doubly-labeled water method for determination of energy expenditure in man. *Human Nutrition and Clinical Nutrition,* **38**, 95–106.

Lee, J. S. & Lifson, N. (1960). Measurement of total energy and material balance in rats by means of doubly-labeled water. *American Journal of Physiology,* **199**, 238–42.

LeFebvre, E. A. (1964). The use of $D_2{}^{18}O$ for measuring energy metabolism in *Columba livia* at rest and in flight. *Auk,* **81**, 403–16.

Lifson, N., Gordon, G. B. & McClintock, R. (1955). Measurement of total carbon dioxide production by means of $D_2{}^{18}O$. *Journal of Applied Physiology,* **7**, 704–10.

Lifson, N., Gordon, G. B., Visscher, M. B. & Nier, A. O. (1949). The fate of utilized molecular oxygen and the source of the oxygen of respiratory carbon dioxide, studied with the aid of heavy oxygen. *Journal of Biological Chemistry,* **180**, 803–11.

Lifson, N. & Lee, J. S. (1960). Estimation of material balance of totally fasted rats by doubly labeled water. *American Journal of Physiology,* **200**, 85–8.

Lifson, N. & McClintock, R. (1966). Theory of use of the turnover rates of body water for measuring energy and material balance. *Journal of Theoretical Biology,* **12**, 46–74.

Lindemann, R. L. (1942). The trophic-dynamic aspect of ecology. *Ecology,* **23**, 399–418.

McClintock, R. & Lifson, N. (1958a). Determination of total carbon dioxide outputs of rats by the $D_2{}^{18}O$ method. *American Journal of Physiology,* **192**, 76–78.

McClintock, R. & Lifson, N. (1958b). CO_2 output of mice measured by $D_2{}^{18}O$ under conditions of isotope re-entry into the body. *American Journal of Physiology,* **195**, 721–5.

Marcstrom, V. & Kenward, R. E. (1981). Sexual and seasonal variation in condition and survival of Swedish goshawks *Accipiter gentilis. Ibis,* **123**, 311–27.

Masman, D., Daan, S. & Beldhuis, H. J. A. (1988). Ecological energetics of the kestrel: daily energy expenditure throughout the year based on the time-energy budget, food-intake, and doubly labelled water methods. *Ardea,* **76**, 64–81.

Masman, D., Daan, S. & Dijkstra, C. (1988). Time allocation in the kestrel (*Falco tinnunculus*), and the principle of energy minimization. *Journal of Animal Ecology,* **57**, 411–32.

Masman, D. & Klaassen, M. (1987). Energy expenditure during free flight in trained and free-living Eurasian kestrels (*Falco tinnunculus*). *Auk,* **104**, 603–16.

Moreno, J., Carlson, A. & Alatalo, R. V. (1989). Winter energetics of coniferous forest tits (Paridae) in the North: implications of body size. *Functional Ecology,* **2**, 163–70.

Mullen, R. K. (1970). Respiratory metabolism and body water turnover rates of *Perognathus formosus* in its natural environment. Comparative Biochemistry and Physiology, **32**, 259–65.

Mullen, R. K. (1973). The $D_2{}^{18}O$ method of measuring the energy metabolism of free living animals. *Ecological Energetics of Homeotherms: a View Compatible with Ecological Modelling* (Ed. by S. A. Gessaman), pp. 32–43. Monograph Series, **20**. Utah State University.

Nagy, K. A. (1980). CO_2 production in animals: analysis of potential errors in the doubly labeled water method. *American Journal of Physiology,* **238**, R466–R473.

Nagy, K. A. (1987). Field metabolic rate and food requirement scaling in mammals and birds. *Ecological Monographs,* **57**, 111–28.

Nagy, K. A., Huey, R. B. & Bennett, A. F. (1984). Field energetics and foraging mode of Kalahari lacertid lizards *Eremias* sp. *Ecology,* **65**, 588–96.

Nur, N. (1984). The consequences of brood size for breeding blue tits. 1. Adult survival, weight change and the cost of reproduction. *Journal of Animal Ecology,* **53**, 479–96.

Petrusewicz, K. & Macfadyen, A. (1970). *Productivity of Terrestrial Animals: Principles and*

Methods. IBP Handbook, 13. Blackwell Scientific Publications, Oxford.

Prinzinger, R. (1982). The energy costs of temperature regulation in birds: the influence of quick sinusoidal temperature fluctuations on the gaseous metabolism of the Japanese quail *(C.c. japonica). Comparative Biochemistry and Physiology,* **71A**, 469–72.

Reyer, H.-U. (1984). Investment and relatedness: a cost/benefit analysis of breeding and helping in the pied kingfisher *Ceryle rudis. Animal Behaviour,* **32**, 1163–78.

Reyer, H-U & Westerterp, K. R. (1985). Parental energy expenditure: a proximate cause of helper recruitment in the pied kingfisher *Ceryle rudis. Behavioural Ecology and Sociobiology,* **17**, 363–9.

Reznick, D. (1985). Cost of reproduction: an evaluation of the empirical evidence. *Oikos,* **44**, 257–67.

Ricklefs, R. E. & Williams, J. B. (1984). Daily energy expenditure and water turnover rate of adult European starlings *Sturnus vulgaris* during the nesting cycle. *Auk,* **101**, 707–16.

Robinson, D. E., Campbell, G. S. & King, J. R. (1976). An evaluation of heat exchange in small birds. *Journal of Comparative Physiology,* **105**, 153–66.

Schoeller, D. A., Leitch, C. A. & Brown, C. (1986). Doubly-labeled water method: in-vivo oxygen and hydrogen isotope fractionation. *American Journal of Physiology,* **251**, R1137–R1143.

Schoeller, D. A. & Taylor, P. B. (1987). Precision of the doubly labeled water method using the two-point calculation. *Human Nutrition Clinical Nutrition,* **41**, 215–24.

Schoeller, D. A. & van Santen, E. (1982). Measurement of energy expenditure in humans by doubly labelled water method. *Journal of Applied Physiology,* **53**, 955–9.

Sibly, R. M. (1981). Strategies of digestion and defecation. *Physiological Ecology: an Evolutionary Approach to Resource Use.* (Ed. by C. R. Townsend, & P. Calow) pp. 109–39. Blackwell Scientific Publications, Oxford.

Sibly, R. M. & Calow, P. (1986). *Physiological Ecology of Animals: an Evolutionary Approach.* Blackwell Scientific Publications, Oxford.

Speakman, J. R. (1987). Calculation of CO_2 production in doubly-labelled water studies. *Journal of Theoretical Biology,* **126**, 101–4.

Speakman, J. R. & Racey, P. A. (1986). Measurement of CO_2 production by the doubly labeled water technique. *Journal of Applied Physiology,* **61**, 1200–2.

Speakman, J. R & Racey, P. A. (1988). The doubly labelled water technique for measurement of energy expenditure in free-living animals. *Science Progress,* **72**, 227–37.

Stearns, S. C. (1976). Life-history tactics: a review of the ideas. *Quarterly Review of Biology,* **51**, 3–47.

Tatner, P. (1988). A model of the natural abundance of oxygen-18 and deuterium in the body water of wild animals. *Journal of Theoretical Biology,* **133**, 267–80.

Tatner, P. (in press). Energetic demands during brood rearing in the wheatear *Oenanthe oenanthe* using the doubly labelled water method. *Ibis,* **131**.

Tatner, P. & Bryant, D. M. (in press). The doubly labelled water technique for measuring energy expenditure. *Techniques in Comparative Respiratory Physiology* (Ed. by C. R. Bridges & P. J. Butler). Cambridge University Press.

Travis, J. (1983). A method for the statistical analysis of time–energy budgets. *Ecology,* **63**, 19–25.

Utter, J. M. (1971). *Daily energy expenditure of free-living purple martins* (Progne subis) *and mockingbirds* (Mimus polyglottos) *with a comparison of two northern populations of mockingbirds.* Ph.D. thesis, Rutgers University.

Utter, J. M. & LeFebvre, E. A. (1973). Daily energy expenditure of purple martins (*Progne subis*) during the breeding season: estimates using $D_2{}^{18}O$ and time budget methods. *Ecology,* **54**, 597–604.

Utter, J. M. & Le Febre, E. A. (1970). Energy expenditure for free flight by the purple martin. *Comparative Biochemistry and Physiology*, **35**, 713–9.

Walsberg, G. E. (1983). Avian ecological energetics. *Avian Biology*, Vol. 7 (Ed. by D. S. Farmer, J. R. King & K. C. Parkes), pp. 161–220. Academic Press, New York.

Walsberg, G. E. (1988). Evaluation of a nondestructive method for determining fat stores in small birds and mammals. *Physiological Zoology*, **61**, 153–9.

Weathers, W. W., Buttemer, W. A., Hayworth, A. M. & Nagy, K. A. (1984). An evaluation of time-budget estimates of daily expenditure in birds. *Auk*, **101**, 459–72.

Weathers, W. W. & Nagy, K. A. (1980). Simultaneous doubly labeled water ($^3H\ H^{18}O$) and time-budget estimates of daily energy expenditure in *Phainopela nitens*. *Auk*, **97**, 861–7.

Westerterp, K. R. & Bryant, D. M. (1984). Energetics of free-existence in swallows and martins (Hirundinidae) during breeding: a comparative study using doubly labelled water. *Oecologia*, **62**, 376–81.

Westerterp, K. R. & Drent, R. H. (1985). Flight energetics of the starling *Sturnus vulgaris* during the parental period. *Proceedings, 18th International Ornithological Congress, Moscow*, pp. 392–9.

Williams, J. B. (1987). Field metabolism and food consumption of savannah sparrows during the breeding season. *Auk*, **104**, 277–89.

Williams, J. B. & Nagy, K. A. (1984). Daily energy expenditure of savannah sparrows: comparison of time-energy budget and doubly-labeled water estimates. *Auk*, **101**, 221–9.

Williams, J. B. & Nagy, K. A. (1985). Daily energy expenditure by female savannah sparrows feeding nestlings. *Auk*, **102**, 187–90.

Wolfe, R. R. (1984). *Tracers in Metabolic Research: Radiosotope and Stable Isotope Mass Spectrometry Methods.* Laboratory and Research Methods in Biology and Medicine, 9. Liss, New York.

Wong, W. W., Butte, N. F., Garza, C. & Klein, P. D. (1987). Energy expenditure of term infants determined by the doubly-labeled water method, indirect calorimetry and test-weighing. *Pediatric Research*, **21**, 282A.

III. CONTROL OF POPULATION SIZE

By and large ecology is not an especially controversial subject, with scholarly arguments competing for time in common rooms and symposia. One topic stands out as the stimulant for more studies and as the basis for exciting polemic: the question of control of population size. For many years this was largely the province of animal ecologists who led the way in trying to understand the factors and processes determining population change and bringing about regulation. Opposing views spearheaded by Nicholson and Varley in the density-dependent camp, by Andrewartha and Birch in the 'climatic school' and by Milne in the pragmatic central ground sometimes degenerated into semantics which threatened violent increases in intellectual entropy. Plant ecologists were fortunate in two respects. They joined the fray at a time when the discussion was beginning to be resolved, and they had a clear-sighted lead from Harper.

Nevertheless we are still in need of incisive thinking about terminology and of a clear theoretical framework against which models and field studies can be tested. This is provided by the first paper in this section, by Murdoch and Walde. As emphasis has shifted from temporal to spatial considerations, this paper deals with the problem of handling subpopulations, with emphasis on control of population size in insects.

The paper by Potts and Aebischer, in contrast, takes a particular case study, that of the grey partridge, and shows how a mobile, larger animal with complex behaviour may have equilibrium densities varying over almost two orders of magnitude in different parts of the species' range. Their take-home message is that simultaneous study of different populations of the same species is a fruitful approach, and that the value of long-term field studies should not be under-rated.

The third paper in this section, by van der Meijden, also chooses to emphasize the significance of spatial heterogeneity, with explicit treatment of local extinction and recolonization, this time in plant populations. The case is argued for field experiments designed in a strong framework of theory and observation.

Nowhere is the need for iteration of good field observation, sophisticated modelling and carefully designed experimentation more necessary than in this area.

5. ANALYSIS OF INSECT POPULATION DYNAMICS

W. W. MURDOCH[1] AND S. J. WALDE[2]
[1]*Department of Biological Sciences, University of California, Santa Barbara, CA 93106, USA, and* [2]*Department of Biology, Dalhousie University, Halifax, Nova Scotia B3H 4JI, Canada*

INTRODUCTION

The assignment to discuss insect population dynamics led us to review many of the now abundant long-term studies of insect populations. Some of these studies arose from the practical consideration of pest control. Others were directed by an interest in one of the central arguments of ecology in the 1950s and 1960s, that of the role of density dependence. Behind most of the studies was the notion that the careful collection of population estimates over long periods, often supplemented by life tables and estimates of the mortality contributed by different sources, would illuminate the causes of population fluctuation and regulation.

The larger framework implied by the symposium title *Toward a More Exact Ecology* naturally leads one to consider the power of this descriptive approach. While useful conclusions and insights have indeed been gained, we are struck also by its limitations. In this paper we try to illustrate the greater power that can be achieved by combining long-term studies with field experiment and modelling of suspected mechanisms.

Given the flavour of the symposium title, we have not reviewed the literature with the goal of inducing general results. Rather, we have focused on three major problems that seem central to analysing population dynamics in the field, and have tried to suggest possible approaches or solutions.

The first problem is testing for density dependence in field data. Some theoretical background (section 5.1) is needed for discussing actual tests (section 5.2), whose difficulties lead to a broader view of the problem (section 5.3). The second problem is distinguishing between 'classical' regulation of a single local population and regulation of collections of weakly coupled subpopulations (here termed 'ensembles') in a heterogeneous environment (sections 5.4 and 5.5). Finally, section 5.6 deals with the third problem, namely the need to use few-species models to analyse the dynamics of populations that exist in many-species communities.

On several occasions we allude to the potential usefulness of models tailored to particular field populations, and refer briefly to one such model in the discussion.

5.1 REGULATION AND DENSITY DEPENDENCE — THEORY

Any review of studies of insect population dynamics immediately runs into an argument about density dependence and regulation that is more than 30 years old but still consumes ecologists' energy. It is highlighted by the following quotations from recent papers: 'populations are ultimately regulated by density dependent processes' (Latto & Hassell 1987) and 'density dependence is the most plausible mechanism by which population size can be restricted within relatively narrow limits' (Latto & Hassell 1987). 'We partly or completely replace the principle of density dependence by the antagonistic statistical principle of spreading the risk . . . heterogeneity . . . and asynchrony . . . [lead to] increased stability (= keeping density within limits) irrespective of whether any of the processes are basically density dependent' (den Boer 1987).

Two definitions are indispensable in reaching a position on this argument. First, both sides seem to accept that *regulation of a population means long-term persistence and fluctuations within limits, presumably with the lower limit* >0. Regulation thus includes not only stable populations (i.e. those tending to return after perturbation to a point equilibrium), but also all formally unstable populations that nevertheless respond to a bounded attractor whether it be cyclic or chaotic.

Second, density dependence seems to have been given various meanings. We have found it impossible to state a definition that combines the following attributes: (i) corresponds to the field ecologist's verbal description, which often implies that vital rates are explicitly a function of population density, (ii) includes all potentially stabilizing mechanisms, which is the general view of its role in the field, and (iii) is consistent with its use by theoreticians in recent demonstrations relating density dependence and regulation.

We have accepted the definition that appears to be most commonly used by theoreticians since it both is unambiguous and corresponds with usage in recent theoretical work on the relation between density dependence and regulation (e.g. Reeve, in press): *density dependence implies there is some dependence of per capita growth rate on present and/or past population densities.* This definition includes, for example, the simplest and neutrally stable Lotka–Volterra predator–prey model:

$$dH/dt = aH - bHP$$
$$dP/dt = cHP - dP. \qquad (5.1)$$

Here the *HP* terms provide the essential density dependence because through them the per capita death rate of the prey depends ultimately on prey density (the per head death rate is b*P* and *P* is affected by *H* via c*HP*). Thus density dependence exists even though the vital rates, as expressed in the constants a, b, c and d, do not depend explicitly on *H* or *P*.

Given these definitions of regulation and density dependence, theoretical work points to the clear conclusion that regulation cannot occur in the absence of density dependence: such a population in a stochastic environment will drift at random to zero or infinity. Density dependence, as defined, prevents this drift by creating an equilibrium.

The conclusion that regulation requires density dependence applies not only to single populations, unconnected to others (referred to hereafter as 'isolates'), but also to collections of subpopulations linked by movement of individuals between them (henceforth 'ensembles'). That is, linking together randomly drifting subpopulations (each of which lacks an equilibrium) merely produces a randomly drifting, non-regulated ensemble (Chesson 1981; Klinkhamer, de Jong & Metz 1983; Metz, de Jong & Klinkhamer 1983; and for a review see Taylor 1988).

Two features of the relation between regulation and density dependence need to be clarified.

1 Although all density dependence as defined above is equilibrium-creating, only a subset is regulating. Consider, for example, the simplest Nicholson–Bailey model

$$H_{t+1} = FH_t\exp(-aP_t)$$
$$P_{t+1} = H_t[1-\exp(-aP_t)]. \qquad (5.2)$$

The density dependence is again indirect (as in the Lotka–Volterra model) and, as is well known, is delayed by a generation (Varley, Gradwell & Hassell 1973). When perturbed, the populations undergo ever-increasing fluctuations, which in the real world would lead inevitably to extinction; i.e. the populations are not regulated although the system is density-dependent. Density dependence, as defined, is thus a necessary but not a sufficient condition for regulation.

This example highlights the semantic problem that seems to explain some of the disagreement on population regulation. Field ecologists tend to use 'density-dependent' and 'stable' as synonyms, whereas the rigorous mathematical definition of density dependence implies only the existence of an equilibrium, which need not be stable or regulated.

2 Field ecologists are in fact interested in a subset of density-dependent processes: those that lead to regulation. There seems little point in detecting the type of density dependence in the Nicholson–Bailey model if it does not explain why the population is regulated. However (as noted above), there seems to be no simple way to specify this subset, other than the circular definition that it includes all mechanisms that are potentially stabilizing. This has implications for field tests for density dependence (section 5.2).

The large body of theoretical work done in recent decades points to at least the following types of regulating density dependence that belong in this subset:
(i) direct effects on birth- and death-rates — for example in predator–prey models, type-3 functional responses, a logistic term in the prey, interference among predators (e.g. Murdoch & Oaten 1975; Hassell 1978), (ii) recruitment from a source that is to some extent 'external', e.g. a refuge, or an invulnerable reproductive stage (e.g. Murdoch *et al.* 1987), and (iii) movement between subpopulations each of which, as an isolate, may be unregulated but contains equilibrium-creating density dependence (e.g. Reeve, in press).

5.2 DETECTING DENSITY DEPENDENCE IN FIELD DATA

The above introductory remarks on density dependence lead to the first problem in the analysis of field populations: density dependence can be shown to be essential to regulation, yet it has been extremely difficult to demonstrate the occurrence of density dependence in persistent field populations by analysing long runs of sampling data. Two techniques have been commonly used. One is analysis of birth- rates or stage-specific mortality from life-table data; Dempster (1983) provides many cases that fail this test. The other technique is based on the idea that the proportionate rate of change in density should be greater the further the population is away from equilibrium. The latter tests are all based on the following model of population dynamics:

$$\log N_{t+1} = \text{b} \log N_t + r + e_t, \tag{5.3}$$

where r is the density-independent rate of increase, and e_t is a random component of the rate of change between sampling times whose expected value is 0. In the deterministic version e_t is a constant = 0. The system is density-independent if b = 1 and is density-dependent if b < 1. A population with b = 1 and e_t random undergoes random walk (random walk with drift if $r > 0$). (Bulmer's (1975) test assumes $r = 0$ in the density-independent case.)

Because it requires only population estimates over time, the second approach is applicable to a wider data set and we shall concentrate on it. Gaston & Lawton (1987) illustrate the failure of this approach. They analysed eleven field populations, taken from studies in which the authors claimed to have shown that density-dependent factors were operating, yet Gaston and Lawton could detect density dependence in none or only four, using one or other of two recent statistical tests (Table 5.1). Table 5.1 also shows that a third test in this genre, due to Pollard, Lakhani & Rothery (1987), also fails to find density dependence in many of the cases.

The standard response to such failures is to point to statistical deficiencies in the tests, or to a possible lack of statistical power, i.e. it is argued that random environmental variation produces noise that swamps the density-dependent signal (e.g. Hassell 1986). There may, however, be additional and more biological reasons for these failures. For example, (i) density dependence may be only intermittent, or 'vague' (Strong 1986), (ii) the equilibrium (whatever its nature) may trend through time in response to changes in the environment, (iii) the

TABLE 5.1. Populations showing density dependence (+) in eleven examples brough together by Gaston & Lawton (1987); these authors used Bulmer's (1975, $R < R_L$ test and Slade's (1977) test, and obtained no positive result with the latter. We have added analyses using the test of Pollard, Lakhani & Rothery (1987). A number of life stages were analysed in several populations. They are recorded here as showing density dependence if any stage showed density dependence

	Test	
Species	Bulmer	Pollard, Lakhani & Rothery
Budmoth:		
Delicious apple	—	—
McIntosh apple	—	—
Codling moth	—	+
Cinnabar moth	+	—
Spruce budworm	—	—
Pine looper	—	—
Winter moth	+	—
Whitefly	+	—
Great tit	+	+
Blue tit	+	+
Tawny owl	—	—

stabilizing mechanism(s) may lead to dynamic behaviour (e.g. cycles, quasi-cycles, chaos) that, because of its complexity, is not likely to be detected by tests based on model (5.3), and (iv) regulation may result from 'ensemble dynamics' in which case it may be detectable only at one particular spatial scale. In fact, these are all examples of a situation in which the underlying dynamic model in the test (eqn 5.3) does not match the dynamics of the field population.

Possibilities (iii) and (iv) deserve amplification. Complex equilibrium behaviour (iii) can be illustrated by a simple example corresponding to type (ii) stabilizing mechanisms listed in section 5.1. Consider the simplest (and unrealistic) representation of a predator–prey interaction in which the Lotka–Volterra model is modified by adding an invulnerable adult stage (Smith & Mead 1974):

$$\begin{aligned} \mathrm{d}Y/\mathrm{d}t &= aA - bYP - mY \\ \mathrm{d}A/\mathrm{d}t &= mY - gA \\ \mathrm{d}P/\mathrm{d}t &= cHP - dP\,, \end{aligned} \qquad (5.4)$$

where Y and A are, respectively, the densities of the vulnerable young and invulnerable adult stage, P is the predator density, a and g are, respectively, the adult reproductive and death-rates, m is the maturation rate of the young, b is the attack rate of the predator and d is its death-rate.

This system is always (oscillatorily) stable. The resulting population fluctuations for one set of parameter values, following a single perturbation, are shown in Fig. 5.1(a). These twenty 'years' of data are comparable to the sort of information that might be available for analyses of density dependence in a field population, except that in real populations many other mechanisms would be operating and it would typically not be easy to determine that the population had gone through two oscillations. (A generation here is taken to be the sum of the expected lifetimes of the juvenile and adult stages, and is set equal to 1 year.)

A variety of tests has been developed for the hypothesis $b < 1$, based on model (5.3) (e.g. Bulmer 1975; Pollard, Lakhanid & Rothery 1987). These all admittedly have statistical weaknesses. Nevertheless, no test of whatever statistical purity will find density dependence in the data from Fig. 5.1 (a) as replotted in Fig. 5.1 (b); clearly the data would not allow rejection of the hypothesis $b = 1$, even though the population is stable. (The reason, as shown in Fig. 5.1 (a) is that the period of the damped oscillation is much longer than one generation and is unrelated in any obvious way to the life-history of the prey species). There is of course a

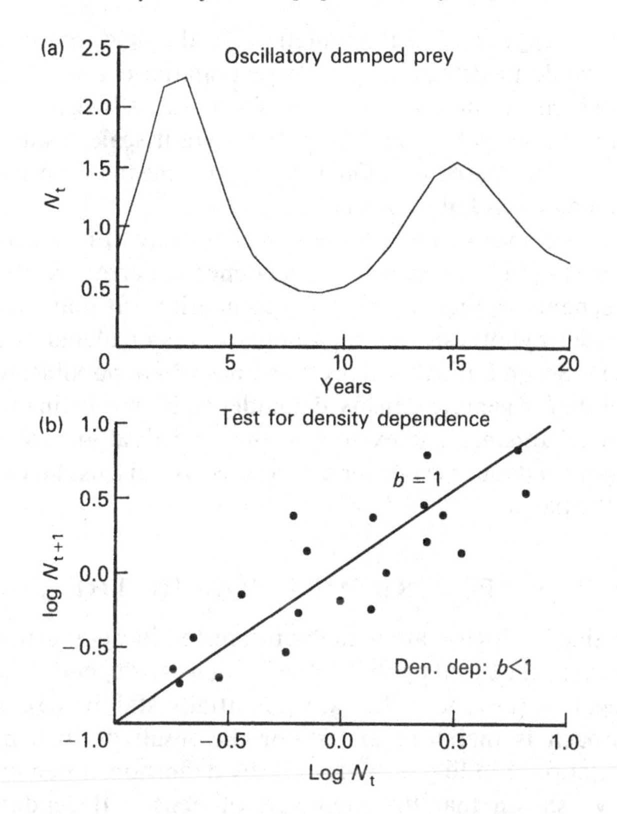

FIG 5.1 (a) Deterministic changes in the density of the juvenile prey population of eqn. (5.4), in which the adults are invulnerable. (b) A test for density dependence in the data in (a), which requires $b < 1$.

(long) lag that can be substituted in the ordinate of Fig. 5.1 (b) that will demonstrate density dependence. But that is clear only because we have the model (eqn 5.4) and the deterministic data it generates—the exigencies of nature will frequently render it impossible to know the model, determine whether there are regular oscillations, or measure their period. That is, frequently a mere time series will not be adequate to tell us that model (5.3) is the wrong model, or what the right one might be.

Turning to point (iv), Reeve (in press) has analysed regulated parasitoid–host ensembles in which each isolate obeys the negative binomial modification of the Nicholson–Bailey model, with a value of the parameter k (= 0.5) that renders it stable. He has shown, nevertheless, that density dependence can be detected in this particular case only if dynamics averaged across subpopulations are analysed; density dependence does not appear in a subpopulation. In the field, however, it will rarely be possible to determine the correct spatial extent of the population over which we should be looking. Thus, unless by good fortune or unusual circumstances we know the right spatial scale, using tests for density dependence based on model (5.3) again means we are using the wrong model and will fail to detect it.

We draw two morals from these considerations. First, because the model underlying tests for density dependence is narrow relative to the range of mechanisms that can regulate populations, it may be better to test the broader hypothesis that the population is regulated. We turn to this idea next. Second, it will usually be better, where possible, to replace the above tests for generic density dependence by modelling particular stabilizing mechanism(s) and examining the model's ability to explain or predict long-term data or experimental results. We discuss an example at the end of the paper.

5.3 A TEST FOR REGULATION IN THE FIELD

We suggest that ecologists are actually interested in demonstrating that populations are influenced by that subset of density-dependent processes that can regulate (i.e. those that are potentially stabilizing). An alternative approach is therefore to test for the result of such processes, namely regulation. Finding regulation is, by definition, a demonstration that we have shown that the right sort of density dependence is in operation (section 5.1).

As noted in the introduction, regulated populations by definition fluctuate between limits; i.e. the fluctuations do not grow with time, at least after some initial period. The following is therefore a test for regulation. For a sequence of population densities $N_i = N_1, N_2, \ldots N_k$, the cumulative variance up to some time t is

$$s^2(t) = [1/(t-1)] \sum (N_i - n)^2 , \qquad (5.5)$$

where n is the mean over t sampling dates. If the fluctuations in density are bounded, so is the cumulative variance. The cumulative variance plotted against time must therefore level off or become asymptotic. The great advantage of this test is that it applies to any bounded population regardless of how complicated the underlying dynamic model is.

Populations failing this test (i) are undergoing random walk, or (ii) have fluctuations about an equilibrium which are increasing through time (for example, as in the Nicholson–Bailey model), or (iii) are regulated but around a trending equilibrium. It might be possible to treat this last case by removing the trend statistically.

The particular form of $s^2(t)$ versus time will depend upon the underlying dynamic model. For example, in model (5.3) it can be shown that, if $b = 1$ (the population is undergoing random walk), the expected cumulative variance (i.e. obtained from many randomly walking populations) will increase linearly with time. If $b < 1$, the cumulative variance may increase gradually to a limit, or may show an initial 'overshoot' owing to the fact that the first observed density is different from the long-term mean (R. M. Nisbet, personal communication).

We calculated the cumulative variance for most of the populations examined by Gaston & Lawton (1987). All but two of these populations show clear evidence of cumulative variances that level off, and several overshoot initially as expected. Three examples are shown in Fig. 5.2. Two populations fail this test for regulation. Southern's tawny owl population trended over most of the sampling period, and it also fails all the tests in Table 5.1. The winter moth population shows level variance for a long period, but then shows a sharp increase at the end of the data set corresponding with an increase in the amplitude of fluctuations (Fig. 5.3).

We have applied this procedure to other insect populations and find, as might be expected, widespread evidence for regulation. By contrast, a population shown by Hassell, Southwood & Reader (1987) to be well

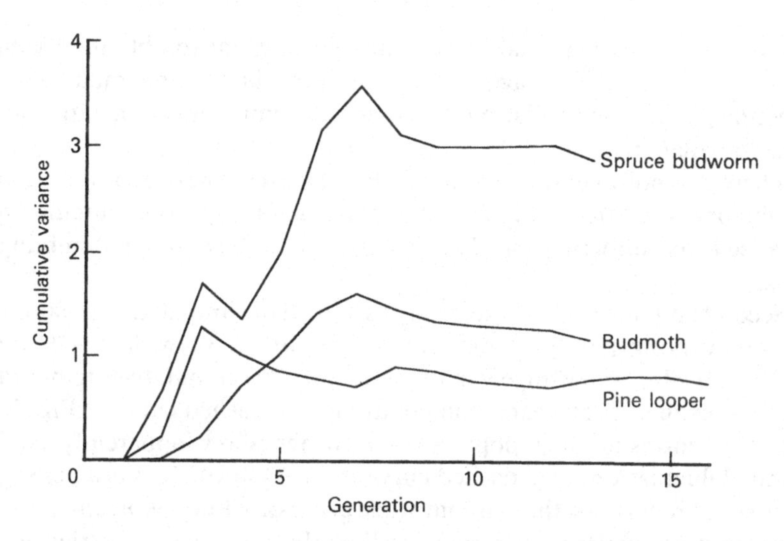

FIG 5.2 Cumulative variances of three insect populations. See text for details.

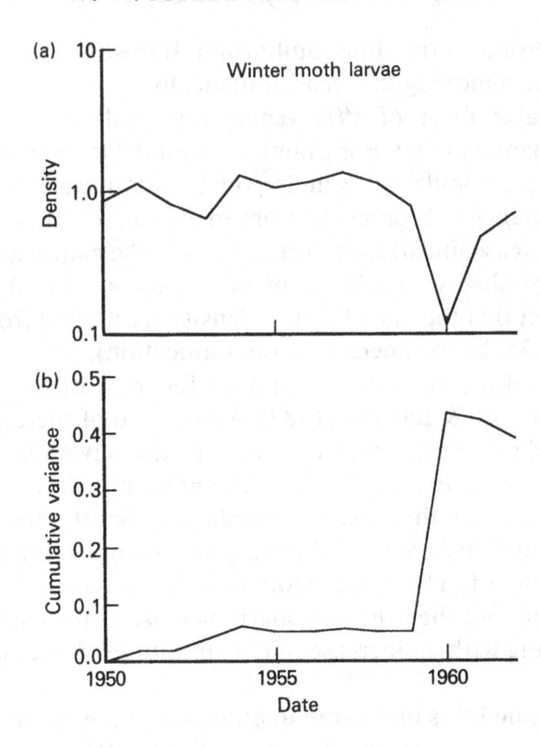

FIG 5.3 (a) Population estimates of the winter moth over time (from Varley, Gradwell & Hassell 1973). (b) Cumulative variance of these data.

described by an exponential growth model shows inexorably increasing variance. These results suggest that this may be a good method for detecting population regulation, and that many populations in nature are indeed regulated.

There are, however, two problems. First, we have not yet developed a satisfactory statistical test for the hypothesis that the cumulative variance levels off with time. This is not likely to be a major stumbling-block.

Second and more serious, as many as a third of simulated populations that are varying purely at random (as in eqn (5.3), with $b = 1$ and $e_t \sim N(0,1)$) also show nicely flattened plots of cumulative temporal variance, even over 50 years of fluctuations (see dashed curve in Fig. 5.4 (b)). The corresponding population histories show apparently well-bounded fluctuations (e.g. dashed curve in Fig. 5.4(a)). This is a perhaps surprising property of the random-walk process: while the *mean* cumulative temporal variance does increase linearly with time, the variance of

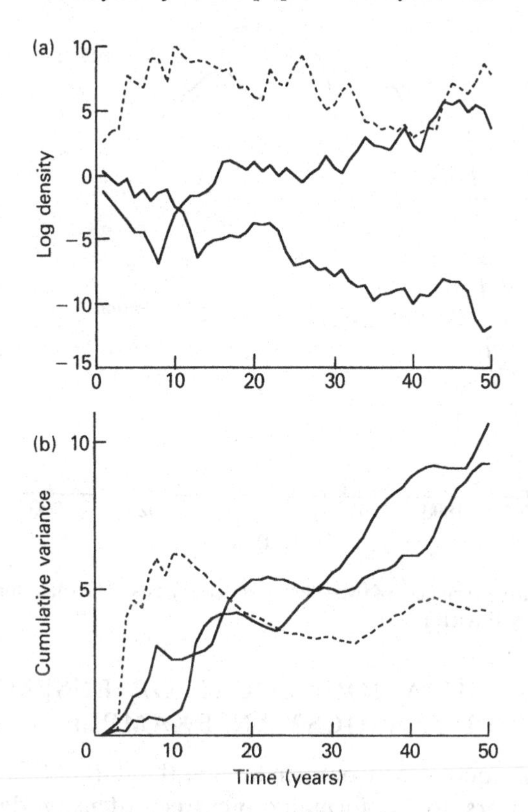

FIG 5.4 (a) Three random-walk populations. (b) Cumulative variances for the populations in (a).

single replicates often behaves badly and very unlike the mean (A. Stewart-Oaten, W. W. Murdoch & S. J. Walde, unpublished). Eventually, of course, all single realizations of the process will accumulate variance with time, but it seems that 'eventually' can be distressingly far away.

The cumulative variance technique appears to have promise, but clearly its results need to be treated with caution when they are applied to any particular population. The random-walk results reinforce our earlier conclusion (section 5.2) that it is better to use tests of specific mechanisms and models than generic tests.

Fig. 5.5 illustrates a point worth stressing: a regulated population may be extremely variable. The data in this figure are estimates of German forest pests as plotted by Varley (1949). *Dendrolimus* changed 20 000-fold in about 20 years, yet judged by its cumulative variance it appears to be a good example of regulation.

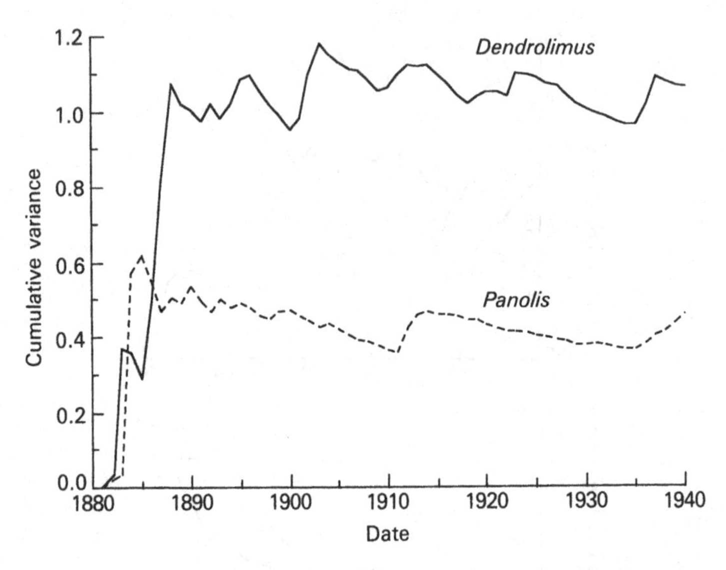

FIG 5.5 Cumulative variances of two species of insect pests in German forests (data from Varley, Gradwell & Hassell 1973).

5.4 REGULATION: LOCAL OR ENSEMBLE DYNAMICS? AN EXAMPLE

We return to the quotation from den Boer at the start of this paper. While den Boer appears to be formally incorrect (density dependence *is* necessary for regulation), the central ideal seems to be valid. Indeed this once heterodox notion has become generally accepted. It poses an interesting problem for the analysis of field populations.

The essential claim is that ensembles can be regulated even if they are composed of subpopulations that, as isolates, are unregulated. The claim has now been demonstrated frequently in models of ensembles, e.g. those of Hastings (1977), Nachman (1987a,b) Zeigler (1977) and Reeve (in press).

The essential features of regulated ensemble dynamics appear to be: (i) the isolates need to contain equilibrium-creating density dependence, (ii) the isolates need not be regulated, however, and can be diverging from equilibrium toward extinction: the immigration terms caused by movement between subpopulations, which are added when the subpopulations are linked to each other, act as potentially stabilizing factors on each local population, and (iii) the various parts of the environment need to be sufficiently different to cause den Boer's out-of-synchrony fluctuations (see Taylor 1988 for a useful summary).

The simplest demonstration of the mechanism is obtained by linking subpopulations obeying the Lotka–Volterra model in eqn (5.1) (Murdoch & Oaten 1975). Local stability is obtained via the appearance of negative terms along the diagonal of the stability matrix, of the form: $-m_j(H_j)^*/(H_i)^*$, representing immigration from patch *j* to patch *i*.

The problem we address is this. How does one determine *in the field* when regulation is the property of an ensemble rather than the result of stabilizing mechanisms occurring within a local population? Obviously, demonstrating regulation or density dependence using any of the techniques discussed in sections 5.2 and 5.3 does not answer this question.

It will generally be difficult to demonstrate regulation via ensemble dynamics when the only data available are long-term population estimates, mainly because it will be hard to show that subpopulations would be unregulated if they were isolated from the rest of the ensemble. We have found one example (not an insect) that appears to fit the bill, although the very features that make it exemplary underscore the difficulties likely to be encountered in most studies of insect populations.

Giant kelp (*Macrocystis pyrifera*) occurs near the shore on rocky bottoms in the southern California Bight in water about 2–20 m deep. The holdfast of the adult plant is attached to rocks and reef outcrops, and the fronds form a canopy on the sea surface. Each adult produces millions of spores throughout most of the year. The vast majority of these spores move no more than a few metres from the adult plant (Anderson & North 1966, 1967; Dayton *et al.* 1984), though some may move long distances (A.W. Ebeling, personal communication). In periods of adequate light on the bottom, which occur mainly in spring, the spores give rise to a microscopic sexual generation which leads to a tiny sporophyte, which grows into the adult plant over a period of about 6 months.

The ocean floor in the southern Bight is a sandy plain interrupted by reefs and rocky outcrops. The kelp beds studied along the 50-km stretch of shore between latitudes 33° 18′ and 33° 35′ are separated from each other by not more than a few km. Outside the area the nearest beds are 25 km to the south and about 40 km to the north, so this 50-km stretch seems to hold a fairly isolated collection of beds.

There are numerous records of extinction of kelp beds. The causes are not often known, but include severe winter storms, El Niño (warm water) conditions, sea urchin plagues, and sand movement (Rosenthal, Clarke & Dayton 1974). When a bed goes extinct, it is reconstituted from spores dropped by drifting adult plants that have been uprooted from extant beds, usually by storms. The kelp beds in this area were recorded at intervals between 1955 and 1974 by surveys from boats and aerial photography (North 1977).

This set of isolated kelp beds seems to be a good candidate for regulation via ensemble dynamics. Although there was no time when the collection as a whole went extinct, about two-thirds of the fifteen beds went extinct in this 20-year period (Table 5.2). Furthermore, there is no evidence that the five persistent beds are fundamentally different or form a permanent refuge: each of them was reduced to extremely low numbers at one time or another in the 20-year period. The dynamics of different beds are not entirely independent — for example, the El Niño event of 1957–8 caused general reductions which were followed by broad recovery. But there were independent extinctions and recoveries.

This example has even broader implications. There is good evidence that, between extinctions, the beds are dominated by density-dependent forces, and show good regulation at the local level. Studies by Dayton *et al.* (1984) and Dean, Thies & Lagos (1989) show that recruitment to the adult population is intensely density-dependent, the dependence upon adult density operating with a lag of up to a year. The result appears to be either 3- or 4-year cycles (e.g. Fig. 6 of Dayton *et al.* 1984) or (in models) stable dynamics (Nisbet & Bence, in press). Regulation of kelp populations thus appears to be a mixture of local and ensemble phenomena, and we might expect this to be true of many populations.

TABLE 5.2. Presence (+) and absence (−) of kelp beds at fifteen sites in a north–south sequence along an isolated 50 km stretch in the southern California Bight (after North 1977)

	Year										
Site	55	58	63	67	68	69	70	71	72	73	74
1	+	−	−	−	−	−	−	−	+	+	+
2	+	+	+	+	+	−	+	+	+	+	+
3	+	+	+	+	+	−	+	+	+	+	+
4	+	+	+	+	+	−	+	+	+	+	+
5	+	+	+	+	+	+	+	+	+	+	+
6	+	+	+	−	−	−	−	−	−	−	+
7	+	+	−	+	+	−	+	+	+	+	+
8	+	+	+	+	+	+	+	+	+	+	+
9	+	+	+	+	+	+	+	+	+	+	+
10	+	+	+	−	−	−	−	−	−	−	−
11	+	+	+	+	+	+	+	+	+	+	+
12	+	+	+	+	+	+	+	+	+	+	+
13	+	+	−	−	−	−	+	+	+	+	+
14	+	+	−	−	−	−	−	−	+	+	+
15	+	+	+	−	−	−	−	−	+	+	+

This example illustrates why it is likely to be difficult to demonstrate regulation via ensemble dynamics in populations of insects. Insect subpopulations are rarely so distinctly separated. Nor can we be sure, even when some local extinctions have been documented, that all of the interacting subpopulations are unregulated. For example, overall persistence of a set of butterfly populations in California seems to depend on a large and persistent refuge population that permits re-establishment of smaller populations when these go extinct (Harrison, Murphy & Ehrlich 1988). Insects typically are also better dispersers than kelp, so subpopulations seem much less likely to show such dramatic evidence of non-regulation as extinction; even when extinction occurs, it will be more difficult to document. In the next section we present some ways of testing for ensemble dynamics in insect populations.

5.5 REGULATION, ENSEMBLE DYNAMICS AND TEMPORAL VARIABILITY

While most ecological theory concerns stability and/or regulation, field ecologists are also interested in explaining the great range of temporal variability observed in natural populations. For example, some bird populations have fluctuated less than threefold in 50 years (Lawton 1988), while we referred above to insects changing by 20 000-fold in 20 years. One would like to know what causes one population to fluctuate a lot and another to be tightly bounded. (Indeed, if we are interested in the biological control of a pest insect, we are perhaps more interested in temporal variability than we are in whether or not the population is stable or regulated (Murdoch, Chesson & Chesson 1985; Murdoch, in press).

We expect ensemble dynamics to reduce temporal variability, i.e. the variability of a collection of subpopulations should be less than that of the average subpopulation. A test for ensemble dynamics would therefore be to show that temporal variability declines as a larger and larger collection of subpopulations is included in the total population sampled (e.g. Murdoch, Chesson & Chesson 1985; Fig. 5.1).

Reeve (in press) has recently provided theoretical support for this approach. He analysed the dynamics of model ensembles whose Nicholson–Bailey-type subpopulations, when isolated, were unregulated and showed ever-increasing fluctuations ($k = 2$). The coupled subpopulations were in a spatially heterogeneous and randomly varying environment. He showed that temporal variability measured in a single (but not isolated) subpopulation is greater than that observed when densities are averaged across subpopulations. He also showed that such ensembles are more likely to be regulated (in this case stable) and to fluctuate within narrower bounds: (i) the more heterogeneous the environment is, and (ii) the more weakly the subpopulations are connected.

We need now a brief digression on the measurement of temporal variability (details from A. Stewart-Oaten and colleagues, unpublished). The usual measure of temporal variability is the standard deviation of the logarithm of successive population *estimates* (or, more crudely, the estimated N_{max}/N_{min}). This measure is inappropriate, however, because it confounds spatial with temporal variability. Each estimate of density at time t (i.e. the mean of a set of spatially distributed samples) is different from the actual population density at that time. Consequently, the variance of the estimated densities over time contains both the temporal variance of the true population mean plus a component contributed by the sampling error on each date. The latter component is bound to decrease with the number of samples used to estimate the mean on each date (and hence will always give a false impression of ensemble dynamics in operation); also, it will increase with spatial variance. A. Stewart-Oaten and colleagues (unpublished) provide a solution to this problem, which is to estimate the temporal variance of the true population density by substituting spatial variance from total temporal variance.

By way of illustration, W. W. Murdoch, A. Stewart-Oaten & S. J. Walde (unpublished) analyse the temporal variability of populations of the red scale insect controlled by the parasitoid *Aphytis* in citrus groves in southern California. This is a prime example of biological control. In addition, the system seems to fit the classical notion of a stable interaction (Murdoch, Chesson & Chesson 1985; Reeve & Murdoch 1986); the populations undergo quite narrow fluctuations, particularly for a pest species, and we have found no evidence for local extinction. So far, however, we have been unable to detect the operation of density-dependent mechanisms, and this might therefore be a candidate for regulation by ensemble dynamics.

We have failed to find evidence that ensemble dynamics are important in this case. We used the individual tree as a subpopulation. Each one contains several hundred thousand red scale, and the scale move very little: the youngest instar is the only mobile stage and typically crawls at most only about 1 m. We have analysed the temporal variability of the populations on individual trees, then of populations on trees grouped by pairs, then grouped four at a time, and so on. The prediction is that temporal variability will decrease as we average over more subpopulations (trees). In fact, temporal variability hardly declined with ensemble size up to the maximum of eight trees (Table 5.3). The reason that ensemble

TABLE 5.3. Effect of increasing ensemble size (number of trees) on temporal variability of a red scale population in a lemon orchard (original results — see text)

Number of trees	Estimate of temporal variability (SD)
1	0.205 (0.070)
2	0.203 (0.028)
4	0.203 (0.017)
6	0.196 (0.013)
8	0.195

dynamics are unimportant in this grove is that the fluctuations on different trees are largely in synchrony.

With R. F. Luck we are also in the middle of an experimental test of ensemble dynamics for the red scale insects. Trees in one set in a grapefruit grove have been individually almost completely isolated by fine-mesh cages, while populations in control trees are left fully linked to the rest of the grove. If ensemble dynamics are important we would expect the temporal variability of the caged trees to increase above that shown by the controls. This experiment has been going for only a few months, but so far almost no difference between the temporal variability of caged and uncaged trees has appeared, giving support to the above analysis of sampling data.

The apparent stability and low temporal variability of the red scale/*Aphytis* system, combined with the absence of evidence for ensemble dynamics, thus suggests that classical, locally operating density-dependent mechanisms are important. This conclusion is consistent with earlier speculation that the red scale/*Aphytis* system might exemplify the classical paradigm of biological control (Murdoch, Chesson & Chesson 1985).

5.6 MANY-SPECIES COMMUNITIES, FEW-SPECIES MODELS

We claimed earlier that it is better to substitute, for generic measures of density dependence and regulation, tests of models incorporating the regulatory mechanism(s) thought to be operating in a particular situation. This implies that models of the population dynamics can be developed that are tractable. This leads us to our third problem: communities typically contain many interacting species, yet analytical models of population dynamics of necessity contain very few species — often only two. We need to use such few-species models because it is difficult to analyse the dynamics of populations in models with many species, and in particular the connection between local and global stability becomes more tenuous (Goh 1979).

The problem could be circumvented if we knew that communities consist of weakly coupled subsystems of few species (Lawton 1976; May 1977). Then we could analyse the dynamics of these simple subsystems separately. In this section we present evidence for weak coupling in some freshwater communities.

We first need operational definitions for weakly and strongly coupled. Suppose we are interested in the dynamics of a 'target' species — species 1 — which interacts with species 2. From the point of view of species 1, species 2 is weakly coupled if we can describe the dynamics of species 1 by an equation for species 1 only, i.e. we do not need to write a coupled equation for species 2.

Two conditions that allow this simplification are as follows: (i) if species 2 has a small effect on species 1, it can be ignored, (ii) if there is an incomplete feedback loop from species 1, through species 2, and back to species 1, we can treat species 2 as part of the independently fluctuating environment of species 1. These conditions can be defined as follows

1 Let

$$dN_1/dt = N_1 f_1(N_1) + a_1 N_1 g_1(N_2),$$

where the function f_l describe the intraspecific interaction and the function g_l (modified by a constant a_1) describes the effect of species 2 on the rate of change of species 1. Then if a_1 is small, we need not write an equation for species 2 because its effect is negligible.

2 The more interesting case is where a_1 is not small. We then need to write an equation for species 2 if its effect on species 1 at time t in turn depends on the density of species 1. This occurs if species 1 has a large effect on species 2, i.e. if

$$dN_2/dt = N_2 f_2(N_2) + a_2 N_2 g_2(N_1)$$

and a_2 is large. A diagram of such strong coupling is in Fig. 5.6 (top). A standard example is a closed predator–prey system, in which each species determines the dynamics of the other.

Now consider the case where a_2 is small. In this case, although species 2 strongly affects species 1 (a_1 is large), the feedback loop is incomplete (Fig. 5.6, bottom). The effect of species 2 on species 1 in this case can be strong and variable, but it is like any other part of the independent and variable environment (such as temperature), and its effect on species 1

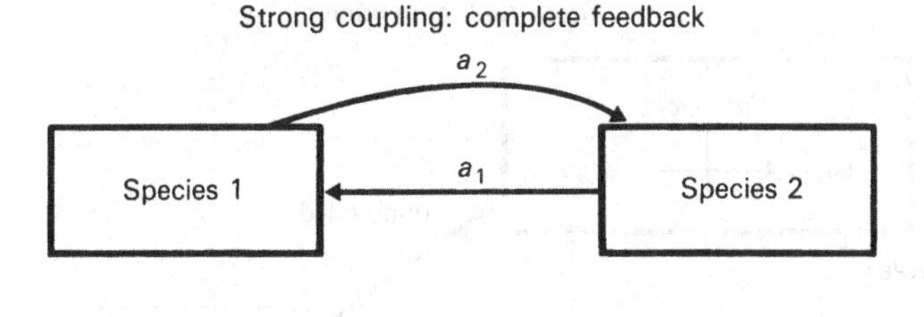

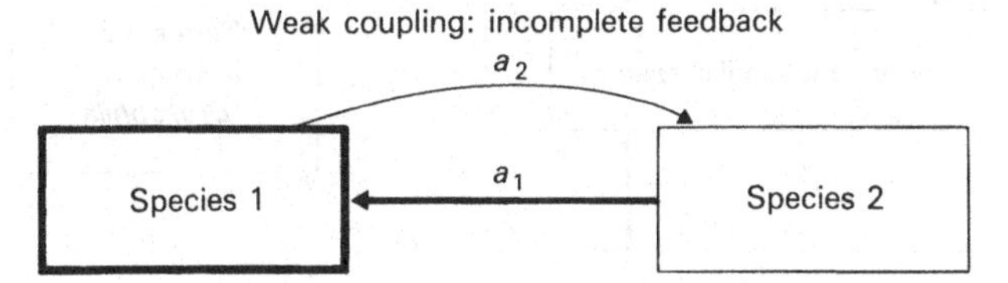

FIG 5.6 A diagrammatic representation of strong and weak coupling between two species. See text for details.

can be incorporated through its influence on parameter values. This is the case of weak coupling.

Murdoch (1979) reported on a community of freshwater organisms in stock tanks and ponds in southern California that appeared too complex to yield to a description by few-species models. Since then two sets of studies have been done on this system. Both of them strongly support the idea that, in fact, the community can be subdivided into weakly coupled subsystems.

The community in the experimental stock tanks consists of *Notonecta,* a backswimming bug, two or three species of zooplankton dominated by *Daphnia,* and terrestrial surface prey whose rate of supply can be experimentally manipulated. This set of organisms is typical of the interaction in ponds in the area, except that there the terrestrial prey arrive at random (Murdoch, Scott & Ebsworth 1984).

First, we have analysed the dynamics of *Notonecta,* the dominant predator (Orr, Murdoch & Bence, in press). As shown in Fig. 5.7 the potential complexity is quite large. The smaller *Notonecta* instars prefer zooplankton — almost entirely *Daphnia.* The larger stages eat *Daphnia,* but prefer surface prey. The larger stages also cannibalize, especially the first two immature instars — the probability of cannibalism increasing with the number of cannibals and decreasing with the amount of alternative prey. The zooplankton in turn could interact with each other and with several species of algae on which they feed.

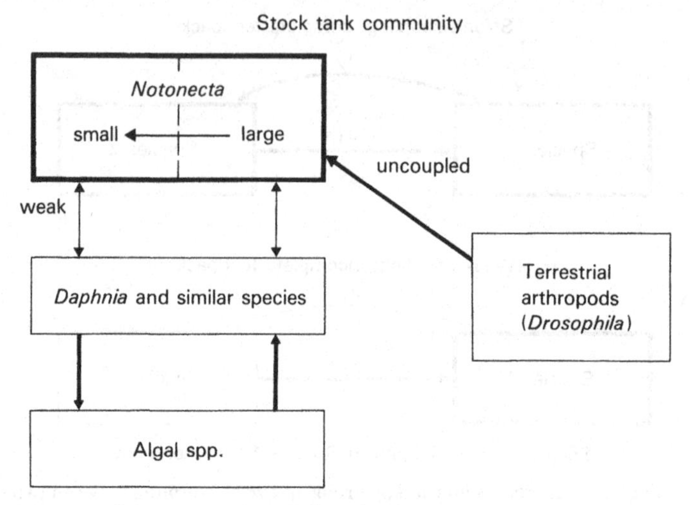

FIG 5.7 Diagram of the community relations in an experimental stock tank.

The dynamics of *Notonecta* were analysed in an experiment with a two-way design, in which the level of surface prey (*Drosophila*) was kept constant at either a high or a low level. This was crossed with a convergence experiment, in which initial adult *Notonecta* density was raised above or below what we considered to be the appropriate equilibrium density. The experiment was started in early summer with adult *Notonecta*. They produce eggs and these develop through five immature stages to form the overwintering population of adults.

In spite of the potential for complex interactions, and because of cannibalism, the system turns out to be dynamically very simple: *Notonecta* dynamics were determined almost entirely by the rate of supply of surface prey.

First, the final overwintering *Notonecta* populations converged to an equilibrium density set by the rate of supply of surface prey. In fact, convergence to equilibrium occurred by the second or third instar (Orr, Murdoch & Bence (in press). Convergence occurred because (i) The fecundity of *Notonecta* was strongly food- and density-dependent, and (ii) cannibalism was also strongly density- and food-dependent. In each case the relevant food was surface prey. Essentially all mortality of small *Notonecta* was caused by cannibalism, so that, although *Daphnia* was the major food supply of the small instars, and though its density fluctuated through time and varied among tanks, the dynamics of the smaller instars were related only to the food supply of the larger instars of *Notonecta*. Thus *Notonecta* dynamics can be described without regard to the dynamics of the zooplankton.

The combination of (i) different food supplies for large and small instars, and (ii) cannibalism, greatly simplifies the dynamics of *Notonecta*. In fact, since the rate of supply of the controlling terrestrial food supply is wholly unaffected by the density of *Notonecta*, the food supply becomes a part of the environment, and the dynamics can be modelled as a single-species system. Features (i) and (ii) are common in freshwater predators, and the third is typical of some, such as many fish species.

The second major component of the community is zooplankton, particularly *Daphnia*. McCauley & Murdoch (1987) have recently investigated the dynamics of this genus in a broad range of habitats.

The coupling between *Daphnia* and *Notonecta* in the experimental stock tanks was very weak — essentially *Notonecta* had no effect on the dynamics of *Daphnia* (W. W. Murdoch, unpublished data). However, this weak coupling might be an artefact of the conditions in the experimental tanks. The studies of McCauley & Murdoch (1987) on *Daphnia*/algae dynamics in natural systems, however, suggest that *Daphnia* and its predators may be typically weakly coupled (Fig. 5.8).

McCauley & Murdoch (1987) have analysed the dynamics of *Daphnia* and algal populations following the spring decline and preceding the winter collapse in about thirty habitats spread broadly across the north temperate region, including lakes, ponds and reservoirs. In spite of the

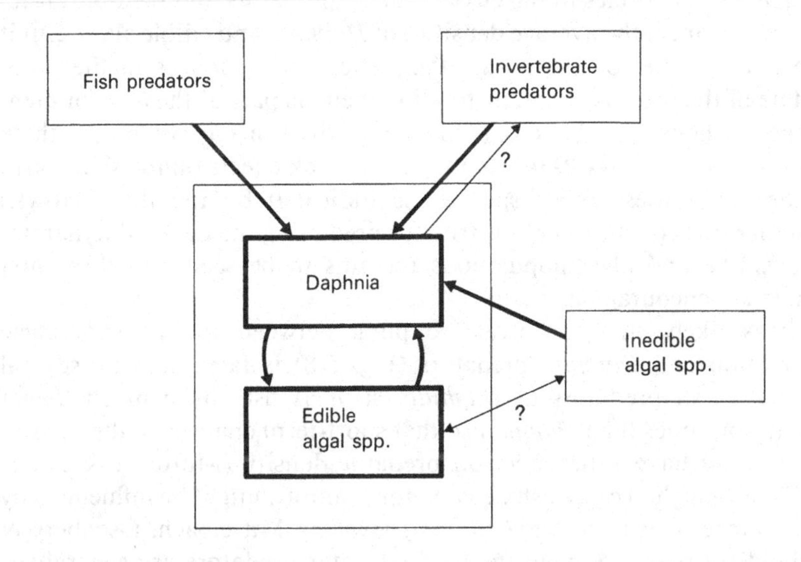

FIG 5.8 Diagram of the simplified dynamics of the interactions between *Daphnia* and algae, and between this system and the rest of the community, in lakes. See text for further details.

fact that these habitats contain invertebrate predators of *Daphnia,* and in some instances fish predators, it appears that the dynamics of *Daphnia* are determined by its interaction with its food supply — the edible algae. Furthermore, even though there may be dozens of edible algal species in these natural environments, it seems that the algae can be treated as a uniform 'prey species'.

In essence, *Daphnia*/algae dynamics can be adequately described as a two-species predator–prey interaction. *Daphnia* suppresses the density of edible algae far below the level set by nutrients. In turn, *Daphnia* is severely food-limited and its density is set by the turnover rate of the algae. The small amplitude cycles or quasi-cycles that are commonly found in *Daphnia* populations appear to be close analogues of those found in the laboratory when populations of *Daphnia* are given a constant supply of algae. The amplitude (less than fourfold) and period (usually 20–40 days) of the cycles and the detailed demography of the populations (bursts of reproduction followed by the march of a cohort through the age classes and suppression of subsequent reproduction) in the field are very similar to those found in the laboratory.

McCauley, Murdoch & Watson (1988) have also shown that the qualitative behaviour of the average densities of edible algae and *Daphnia*, as one moves from oligotrophic to nutrient-rich lakes, are also in accord with predictions of the simplest two-species predator–prey models. Other species in the environment (predators and inedible algae) do seem to affect the average densities of *Daphnia* and edible algae. But it appears possible to incorporate their effects via changes in the parameters of the models — i.e. by treating them as part of the environment of the predator–prey system. (This is indicated in Fig. 5.8 by the thick unidirectional arrows. *Daphnia* may have weak effects on the short-term dynamics of these other species, as indicated by the thin arrows.) Whether this conclusion holds true for predicting the detailed dynamics of *Daphnia* and algal populations remains to be seen, but these first results are encouraging.

It is likely that the weak coupling between the *Daphnia*/algae interaction and *Daphnia* predators (Fig. 5.8) is facilitated by several factors. First, predators of *Daphnia*, such as fish, have much longer generation times than *Daphnia* so that short-term changes in the density of *Daphnia* have little effect on predator density (Murdoch & Bence 1987), although *average* fish density, for example, might be influenced by the average density of *Daphnia*, as shown by Mittelbach, Osenberg & Leibold (in press). Second, many freshwater predators are generalists, which will tend to uncouple their dynamics from those of any one prey species. Third, the diets of such predators are typically size-dependent, so any one class of prey tends to be important only for short periods of the predator's life-cycle.

These results provide experimental evidence that at least one apparently complex community can be decomposed into subunits, the dynamics of whose constituent populations we can describe with few-species models. The species are similar to those that occur in many freshwater communities, and provide hope that this property might be quite general in such environments. However, freshwater communities tend to have fewer insect species, and therefore fewer interactions, than terrestrial communities. It would therefore be interesting to see similar analyses and experiments done in terrestrial communities to see if these also are less complex dynamically than seems at first sight to be the case.

DISCUSSION

At various points in this paper we have suggested that it will be useful to develop and test models that incorporate the actual mechanisms believed to be operating in particular situations. There is, of course, a long history of the use of population models in ecology. But these are mostly what May (1974) has termed 'strategic' models. While they may provide broad insight, they are of limited use for distinguishing between competing hypotheses seeking to explain a particular set of population dynamics, because it is usually possible to account for the qualitative features of populations (e.g. stability or cycles) with any one of a large number of such models.

Recent progress in modelling techniques suggests we should be able to move beyond this stage. For the past several years we (W. S. C. Gurney, E. McCauley, W. W. Murdoch and R. M. Nisbet) have been pursuing the goal of explaining and predicting the dynamics observed in the laboratory and field in the *Daphnia*/algae system described above. Our approach has been, first, to develop a detailed model of individual *Daphnia* energetics, since the interaction with food is crucial. This model, using the wealth of data available on physiology of individuals, is able to predict the growth and fecundity schedule of *Daphnia* in a broad range of food regimes. These results in turn have served as the basis for a stage-structured population model of *Daphnia* feeding on algae. The functions and parameters in this model are all derived from studies of individuals, independently of any population data.

We have focused on the oscillations and quasi-cycles (hereafter just 'cycles') found in the laboratory and field, as discussed above. Two paradigms suggest themselves as potential explanations for the cycles: non-linear 'predator–prey' dynamics of the sort found in Lotka–Volterra models, and time lag/response time relationships (May 1976). The most likely mechanism fitting the first paradigm is 'paradox of enrichment' cycles arising from the interaction between the destabilizing type-2 functional response of *Daphnia* and the stabilizing logistic growth of the algae (Rosenzweig 1971). Our hypothesis has been that it is the second paradigm that explains the observed cycles, because of the demographic details described above (and, in particular, the fact that the period of the cycle equals the generation time of *Daphnia*).

Our stage-structured model does well in predicting the behaviour of *Daphnia*, given a constant rate of supply of algae in the laboratory (Nisbet *et al.*, in press). It gets correct the period and amplitude of (damped) cycles seen in the laboratory, as well as the demographic pattern associated with dominance–suppression (section 5.6). But it is on field populations that it has thrown the most useful light.

When we substituted a dynamic algal population for the constantly supplied food in the laboratory, we obtained paradox-of-enrichment-type oscillations that arose from the interaction between the logistic algae and the type-2 functional response of the *Daphnia*. However, these cycles have none of the features seen in real populations (the period is much too long and the amplitude far too large) (Fig. 5.9 (a)) We also found, however, that a small change in one parameter of the functional response (an increase in the half-saturation constant) suppresses the paradox-of-enrichment cycles, and allows the standard field cycles to emerge (Fig. 5.9 (b)). These have the correct period and amplitude, and the results seem clearly to reinforce our view that the cycles in the field are time lag/response cycles, not paradox-of-enrichment cycles. (We do not suggest that the parameter change in the functional response is necessarily the correct one. But it is certain that the parameter values in the field are not exactly those that obtain in the laboratory. Our problem now is to find the key parameter changes that do occur in the field.)

The discouraging aspect of these results is that they suggest that one may need to get the model's parameter values close to correct to distinguish between competing hypotheses. Probably no insect is as well studied as the crustacean *Daphnia* in both field and laboratory, and at both individual and population levels. However, especially in the area of biological control of insect pests, there are species for which enormous amounts of high-quality information are available.

In general, however, we find these results, in combination with those

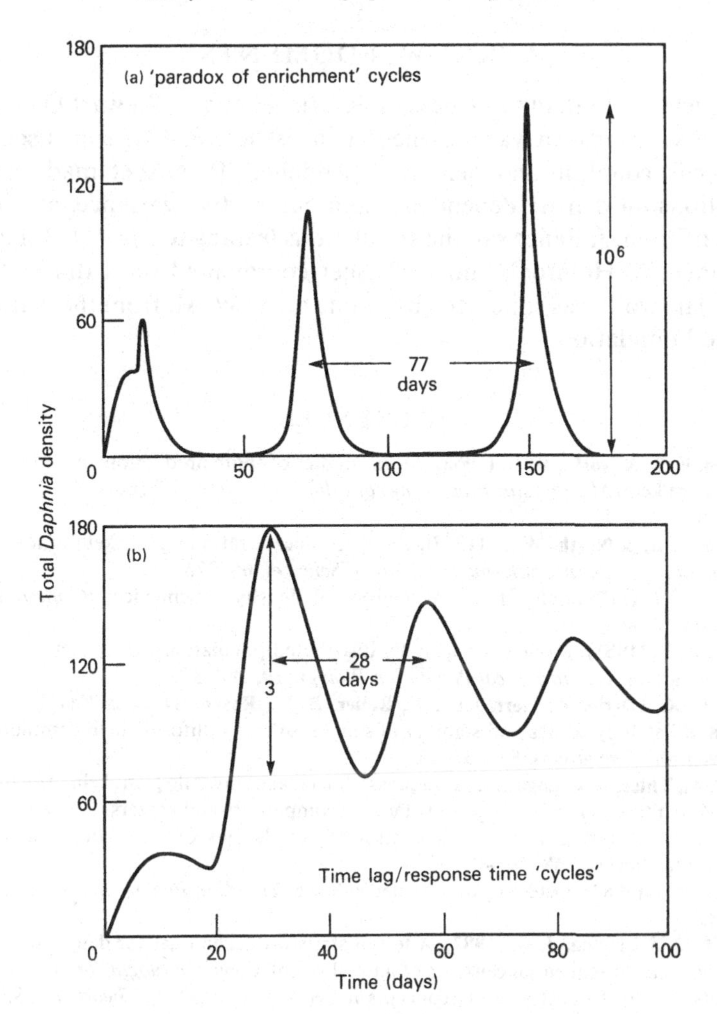

FIG 5.9 Output from a stage-structured model of a *Daphnia* population feeding upon logistically growing algae (from Nisbet *et al.*, in press). (a) 'Paradox of enrichment' cycles; parameter values obtained from lab studies of individual *Daphnia.* (b) 'Time lag/response time' damped oscillations obtained by increasing the half-saturation constant in the functional response of *Daphnia*.

of section 5.6, encouraging. They tell us that we may indeed be able to decompose complex natural communities and get their dynamics correct, using few-species models. They also argue that the combination of long-term population studies, field experiments and modelling of particular mechanisms can be a powerful one.

ACKNOWLEDGMENTS

It is a pleasure to thank P. Chesson, R. Nisbet and A. Stewart-Oaten for useful discussions on various theoretical issues. P. Chesson discussed both weak coupling and density dependence. R. Nisbet made useful suggestions on density dependence and cumulative variance, as did A. Stewart-Oaten. J. Bence ran the simulations leading to Fig. 5.1. J. Bence, J. Connell, S. Holbrook and R. Nisbet commented on a draft of the paper. The work was supported by grants to W. W. M. from the National Science Foundation.

REFERENCES

Anderson, E. K. & North, W. J. (1966). *In situ* studies of spore production and dispersal in the giant kelp, *Macrocystis. Proceedings of Fifth International Seaweed Symposium,* **5**, 73–86.

Anderson, E. K. & North, W. J. (1967). Zoospore release rates in giant kelp *Macrocystis. Bulletin of Southern California Academy of Sciences,* **66**, 223–32.

Bulmer, M. G. (1975). The statistical analysis of density dependence. *Biometrics,* **31**, 901–11.

Chesson, P. L. (1981). Models for spatially distributed populations: the effect of within-patch variability. *Theoretical Population Biology,* **19**, 288–323.

Dayton, P. K., Currie, V., Gerrodette, T., Keller, B. D., Rosenthal, R. & Ven Tresca, D. (1984). Patch dynamics and stability of some southern California kelp communities. *Ecological Monographs,* **54**, 253–89.

Dean, T. A., Thies, K. & Lagos, S. L. (in press). Survival of juvenile giant kelp, *Macrocystis pyrifera:* the effects of demographic factors, competitors and grazers. *Ecology.*

Dempster, J. P. (1983). The natural control of populations of butterflies and moths. *Biological Reviews,* **58**, 461–81.

Den Boer, P. J. (1987). Detecting density dependence. *Trends in Ecology and Evolution,* **2**, 77–8.

Gaston, K. J. & Lawton, J. H. (1987). A test of statistical techniques for detecting density dependence in sequential censuses of animal populations. *Oecologia,* **74**, 404–10.

Goh, B. S. (1979). Robust stability concepts for ecosystem models. *Theoretical Systems Ecology* (Ed. by E. Halfon), pp. 468–87. Academic Press, San Francisco.

Harrison, S., Murphy, D. D. & Ehrlich, P. R. (1988). Distribution of the bay checkerspot butterfly, *Euphydryas editha bayensis:* evidence for a metapopulation model. *American Naturalist,* **132**, 360–82.

Hassell, M. P. (1978). *The Dynamics of Arthropod Predator Prey Systems.* Monographs in Biology, 13. Princeton University Press.

Hassell, M. P. (1986). Detecting density dependence. *Trends in Ecology and Evolution,* **1**, 90–3.

Hassell, M. P., Southwood, T. R. E. & Reader, P. M. (1987). The dynamics of the viburnum whitefly (*Aleurotrachelus jelinekii*): a case study of population regulation. *Journal of Animal Ecology,* **56**, 283–300.

Hastings, A. (1977). Spatial heterogeneity and the stability of predator–prey systems. *Theoretical Population Biology,* **12**, 37–48.

Klinkhamer, P. G. L., de Jong, T. J. & Metz, J. A. J. (1983). An explanation for low dispersal rates: a simulation experiment. *Netherlands Journal of Zoology,* **33**, 532–41.

Latto, J. & Hassell, M. P. (1987). Do pupal predators regulate the winter moth? *Oecologia*, **74**, 153–5.

Lawton, J. H. (1976). Mathematical models in ecology. *Nature*, **264**, 138–9.

Lawton, J. H. (1988). More time means more variation. *Nature*, **334**, 563.

McCauley, E. & Murdoch, W. W. (1987). Cyclic and stable populations: plankton as paradigm. *American Naturalist*, **129**, 97–121.

McCauley, E., Murdoch, W. W. & Watson, S. (1988). Simple models and variation in plankton densities among lakes. *American Naturalist*, **132**, 383–403.

May, R. M. (1974). *Stability and Complexity in Model Ecosystems*, 2nd edn. Princeton University Press.

May, R. M. (1976). Models for single populations. *Theoretical Ecology: Principles and Applications* (Ed. by R. May), pp. 4–25. Blackwell Scientific Publications, Oxford.

May, R. M. (1977). Thresholds and breakpoints in ecosystems with a multiplicity of stable states. *Nature*, **269**, 471–7.

Metz, J. A. J., de Jong, T. J. & Klinkhamer, P. G. L. (1983). What are the advantages of dispersing: a paper by Kuno explained and extended. *Oecologia*, **57**, 166–9.

Mittelbach, G. G., Osenberg, C. W. & Leibold, M. A. (in press). Trophic relations and ontogenetic niche shifts in aquatic ecosystems. *Size-structured Populations: Ecology and Evolution* (Ed. by B. Ebenman & L. Persson). Springer, Berlin.

Murdoch, W. W. (1979). Predation and dynamics of prey populations. *Fortschritte der Zoologie*, **25**, 295–310.

Murdoch, W. W. (in press). Biological control in theory and practice. *Critical Issues in Biological Control, Aggregation and Population Dynamics* (Ed. by M. Mackauer & L. Ehler), Intercept, Wimborne.

Murdoch, W. W. & Bence, J. R. (1987). General predators and unstable prey populations. *Predation in Aquatic Communities: Direct and Indirect Effects* (Ed. by C. Kerfoot & A. Sih), pp. 17–29. University Press of New England, Hanover, New Hampshire.

Murdoch, W. W., Chesson, J. & Chesson, P. L. (1985). Biological control in theory and practice. *American Naturalist*, **125**, 344–66.

Murdoch, W. W., Nisbet, R. M., Blythe, S. P., Gurney, W. S. C. & Reeve, J. D. (1987). An invulnerable age class and stability in delay-differential parasitoid–host models. *American Naturalist*, **129**, 263–82.

Murdoch, W. W. & Oaten, A. (1975). Predation and population stability. *Advances in Ecological Research*, **9**, 1–131.

Murdoch, W. W., Scott, M. A. & Ebsworth, P. (1984). Effects of the general predator *Notonecta* upon a freshwater community. *Journal of Animal Ecology*, **53**, 791–808.

Nachman, G. (1987a). Systems analysis of acarine predator–prey interactions. I. A stochastic simulation model of spatial processes. *Journal of Animal Ecology*, **56**, 247–65.

Nachman, G. (1987b). Systems analysis of acarine predator–prey interactions. II. The role of spatial processes in system stability. *Journal of Animal Ecology*, **56**, 267–81.

Nisbet, R. M. & J. R. Bence. (in press). Alternative dynamic regimes for canopy-forming kelp: a variant on density vague population regulation. *American Naturalist*.

Nisbet, R. M., Gurney, W. S. C., Murdoch, W. W. & McCauley, E. (in press). Structured population models: a tool for linking effects at individual and population level. *Biological Journal of the Linnean Society*.

North, W. J. (1977). *Annual Report of the Kelp Habitat Improvement Project*, 1 July 1974–30 June 1975, with addendum for July 1975 to July 1977.

Orr, B. K., Murdoch, W. W. & Bence, J. R. (in press). Populations regulation, convergence and cannibalism in *Notonecta* (Hemiptera). *Ecology*.

Pollard, E., Lakhani, K. H. & Rothery, P. (1987). The detection of density-dependence from a series of annual censuses. *Ecology*, **68**, 2046–55.

Reeve, J. D. (in press). Environmental variability, migration and persistence in host–parasitoid systems. *American Naturalist*.

Reeve, J. D. & Murdoch, W. W. (1986). Biological control by the parasitoid *Aphytis melinus*, and population stability of the California red scale. *Journal of Animal Ecology*, **55**, 1069–82.

Rosenthal, R. J., Clarke, W. D. & Dayton, P. K. (1974). Ecology and natural history of a stand of giant kelp, *Macrocystis pyrifera*, off Del Mar, California. *Fishery Bulletin*, **72**, 670–84.

Rosenzweig, M. L. (1971). Paradox of enrichment: destabilization of exploitation ecosystems in ecological time. *Science*, **171**, 385–7.

Slade, N. A. (1977). Statistical detection of density dependence from a series of sequential censuses. *Ecology*, **58**, 1094–102.

Smith, R. H. & Mead, R. (1974). Age structure and stability in models of predator–prey systems. *Theoretical Population Biology*, **6**, 308–22.

Strong, D. R. (1986). Density-vague population change. *Trends in Ecology and Evolution*, **1**, 39–42.

Taylor, A. D. (1988). Large-scale spatial structure and population dynamics in arthropod predator–prey systems. *Annales Zoologici Fennici*, **25**, 63–74.

Varley, G. C. (1949). Population changes in German forest pests. *Journal of Animal Ecology*, **18**, 117–22.

Varley, G. C., Gradwell, G. R. & Hassell, M. P. (1973). *Insect Population Ecology: an Analytical Approach*. Blackwell Scientific Publications, Oxford.

Zeigler, B. P. (1977). Persistence and patchiness of predator–prey systems induced by discrete event population exchange mechanisms. *Journal of Theoretical Biology*, **67**, 677–86.

6. CONTROL OF POPULATION SIZE IN BIRDS: THE GREY PARTRIDGE AS A CASE STUDY

G. R. POTTS AND N. J. AEBISCHER
The Game Conservancy, Fordingbridge, Hampshire SP6 1EF, UK

INTRODUCTION

The ideas most ornithologists have on the control of population size are in large part formed by studies of insects. This is not surprising considering that case studies on the control of population size in birds have accounted for less than a tenth of those on animals generally. For instance, out of thirty studies included by Stubbs (1977) and of forty-five reviewed by Connell & Sousa (1983), the majority concerned insects and only six referred to birds. A recent review restricted to insects alone cited sixty-three case studies (Stiling 1988).

The numerical dominance of insects — 22 500 species in Britain, compared to 1700 for vascular plants, 210 for birds and 58 for mammals — is obvious, but, in many respects, the study of birds is ahead of studies on other organisms. After all, birds have clear advantages: they are for the most part diurnal, are attractive to look at, and have family lives that we can relate to. Moreover, they are easy to count or even to census, especially when nesting, and they can be individually marked or radio-tracked with relative ease. Estimates of bird population density which reach back over more than 20 years have been available for some time; for example Tanner (1966) mentions thirteen species. We take the view that ornithologists could contribute far more towards studies of the control of population size than, in general, they do. Rather than survey the whole order, even if that were possible, we give the example of the grey partridge *Perdix perdix*. It is the one we know best, and we believe principles arise which are of general application.

In 1913, when the British Ecological Society was founded, partridges were more than ten times as numerous in Britain as they are now. There must have been over seven million partridges in the country at the end of the breeding season, because they provided a sustainable yield of about 2.5 million shot per year (Potts 1986). None of this was achieved with the benefit of ecological science as we know it. There had not even been an official 'inquiry' on the partridge, as there had been on the red grouse (Lovat 1911). Nor was it the outcome of *laissez-faire* or ignorance. Partridges were managed for their yield over about half the area of the countryside of Britain, and there was a great store of management expertise, accrued from a tradition of practical experimentation by estate owners.

A Rip van Winkle amongst these owners, having slept for the past 75 years, would, on waking, surely marvel at our Land-Rovers, D-vacs and computers. We feel, though, than he would easily understand our current recommendations for partridge management. In passing we note that he would understand them better now than if he had woken in the 1960s when predation was considered of little significance. Maynard-Smith (1978) wrote, 'ecology is still a branch of science in which it is usually better to rely on the judgement of an experienced practitioner rather than on the predictions of a theorist'. We like to think that we have at last caught up!

The purpose of this paper is twofold. First, we summarize some of our research into the control of partridge population size, with particular attention to annual variation, to population trends and to the question of what determines the density of a population at equilibrium; the partridge offers a good subject here because in a number of detailed studies its equilibrium densities have varied from less than 1 to more than 50 pairs per km^2. Second, we consider how the study of bird populations has advanced since the review of Lack (1966), and we propose a way forward for the future.

METHODS

Basic natural history of the partridge

The partridge is a bird typical of open landscapes in what were originally the temperate grassland ecosystems of the steppes. It starts breeding in the year after hatching. The pairs usually form in February and then search for permanent grassy nesting cover, preferably on a dry bank. The partridge lays the largest clutch of any species of bird, averaging fifteen to sixteen eggs, but predation of the hen and eggs is high. The chicks hatch in mid- to late June and, during the first 2 weeks of life, feed mainly on insects in cereals or in other tall grasses. By late summer, adults and young form groups of about twelve birds known as 'coveys'; they feed on waste grain and on seeds of stubble-weeds such as black bindweed

Polygonum volvulus. In winter and spring most partridges feed by grazing on growing cereals or alternatively on pastures, especially clover and grass leys. The species is remarkably sedentary and the home range may involve only two or three fields.

Sussex study

The Sussex study area covers 62 km^2 of the South Downs — indicated by the broken line in Fig. 6.1. Some data were collected annually from 1955 but the main study began in March 1968 and is continuing. Partridge numbers, breeding success and losses were recorded annually, using the same methods, on five farms totalling 29 km^2 (also indicated in Fig. 6.1). The farms were privately owned and, because we were monitoring, we sought no influence over any aspect of farm or game management. Full details of the study area have been given elsewhere (Potts & Vickerman 1974; Potts 1980). A key feature was the choice of the outer boundaries of the study area: the areas beyond the boundaries were relatively unsuitable for grey partridges, which reduced the influence of ingress and egress.

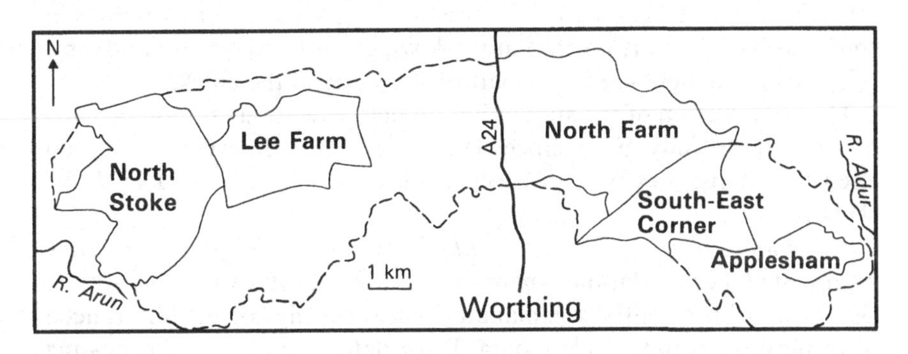

FIG 6.1 The study area on the South Downs, Sussex, showing the five farms where partridge populations have been monitored since March 1968.

Having the advantage of previous intensive studies on the ecology and breeding biology of the partridge, this study concentrated on three different but complementary approaches. First the long-term *monitoring* was used to generate data which defined the overall problem — too few partridges — in terms of the life-history dynamics. *Modelling* was then used to simulate the population processes, and to design experimental tests of solutions in terms of partridge population *management*.

Monitoring, essentially, was a census of partridges on stubble-fields in

August or early September, in which birds were aged and sexed. The methods of converting sex ratio and young : old ratio to measures of survival have been given by Potts (1986). The precision of the monitoring in terms of measuring population size, mortality rates, etc., has been estimated elsewhere, most recently by Aebischer & Potts (in press). This work concluded that sampling errors were not hindering the analysis and interpretation of the monitoring data, except that in recent years the number of broods available for estimating chick survival rate has been too low on some of the individual farms. This is an inevitable consequence of the general decline of partridges throughout the study period.

Mortality rates were expressed as *k*-factors, i.e. as log (initial number) — log (number surviving). The *k*-factors were k_1, loss of eggs when hen survived; k_2, loss of eggs caused by loss of hen; k_3, chick losses to age 6 weeks; k_4, shooting; k_5, losses between shooting and nesting. There was no k for failure to hatch when fully incubated; this is rare and constant. Eggs in the above analysis are therefore, strictly, potential chicks. In this paper k_1 and k_2 were combined where necessary to make the data comparable with those obtained in other studies. Probit survival was selected as being suitable for investigating the causes of variations in chick survival. This is normal in work which seeks to quantify adverse effects of pesticides — a major part of the study of the chick.

The construction of the simulation models, essentially minor variants of one successfully used since 1975, has been described in detail elsewhere (Potts 1980), with a complete print-out published by Potts (1986).

The model was considered *validated* when the outputs accurately represented the population dynamics on the Sussex study area. However, such a model is essentially explanatory, intrinsically correct but in need of verification by independent data. These data were generated by testing key processes in carefully controlled experiments: density-dependent nest predation on Salisbury Plain, Wiltshire (Tapper, Brockless & Potts 1988) and the influence of insect food supply on chick survival at Manydown, northern Hampshire (Rands 1985).

A search of the literature and various archives revealed thirty-nine studies of the partridge which were sufficiently detailed to reconstruct basic population numbers and to estimate breeding stock density, $k_1 + k_2$, k_3 and $k_4 + k_5$. These data were published by Potts (1986, Table 10.4), and since then a further three sets have become available (Delacour 1987; Pegel 1987; John Carroll, University of North Dakota, Bismarck, unpublished); together they will henceforth be referred to as the 'world data'.

RESULTS

Density-independent forms of mortality: chick losses

In the first modern analysis of partridge population dynamics, chick mortality was established as the key factor responsible for change (Blank, Southwood & Cross 1967). Unfortunately the original data, from the Game Conservancy's experimental estate, were biased by upward trends in stock density (due to predator control instigated at the start) and in chick mortality (due to the introduction of pesticides). Thus the key factor also emerged as a significant density-dependent factor. Indeed, further analysis of the data by Manly (1977) suggested that, because of this density-dependence, the net effect of chick mortality was to offset variation in chick production. Subsequent work, however, has shown that chick mortality is not affected by the density of chicks hatching, the prime reason being that the mortality is attributable, mostly, to a shortage of insect food (Potts 1986). Normally partridge chicks are unable to depress supplies of their insect food, and cannot compete for it with other chicks; the annual variation in the insect supply was mostly attributable to the weather in the previous six weeks (Potts 1986). This lack of density-dependence in chick survival must be emphasized since the conclusions of Blank, Southwood & Cross (1967) have been quoted uncritically in recent years.

The relationship between the annual chick survival rate (probits) in the Sussex study area and the mean annual density of preferred insects in cereals in the same area is shown in Fig. 6.2. The densities of five preferred insect taxa were weighted by their individual regression coefficients on probit chick survival; these taxa were, in descending order of importance, small diurnal ground beetles, caterpillars (sawflies and Lepidoptera), leaf beetles (Chrysomelidae), plant bugs and leaf hoppers, and aphids. These were then combined to produce the index of insect abundance (density) shown in Fig. 6.2. The densities of these preferred insects explained 52% of the variation in chick survival rates (probits).

This correlative work was much strengthened by the results of monitoring the diet and survival of eighteen radio-tracked broods (Potts & Aebischer, in press). The outcome suggested, on the one hand, that the complete loss of insects would reduce chick survival to a level below that necessary to replace adult losses even when these were at their lowest observed. This would inevitably cause equilibrium levels (see later) to trend (*sensu* Murdoch & Walde, this volume) towards extinction. On the other hand, doubling insect numbers would restore chick survival rates and, through them, the equilibrium levels to those of the pre-pesticide era.

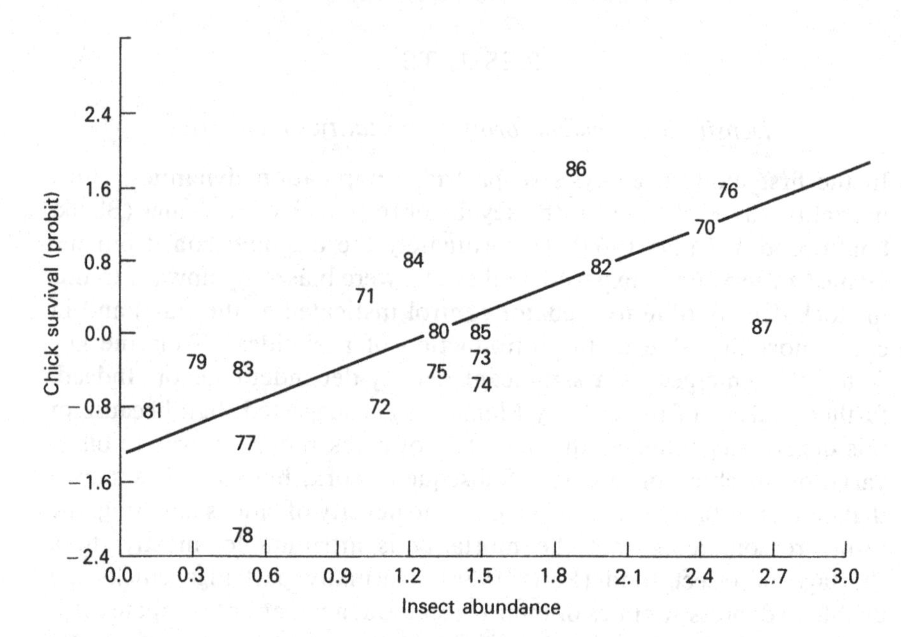

FIG 6.2 Survival of chicks of grey partridge in the Sussex study area as a function of insect abundance: probit of chick survival to 6 weeks in relation to the sum of the preferred taxa of insect (logarithm of the number of individuals per D-vac sample from about 0.5 m^2) at the time of hatching in the years 1970 to 1987, each taxon being weighted for its effect on chick survival (see text).

In order to verify the effects of insect food supply on chick losses, an experiment was begun at Manydown, Hampshire, in 1982. Cereal headlands were selectively sprayed in order to improve the chick food supply. The densities of preferred insects were raised by a factor of 2.5 (Rands 1985) and the chick survival by a factor of 1.8 (Potts 1986) with the consequent effects on population stock size indicated in Fig. 6.3 (a); the effects were in accordance with the predictions from the models.

To summarize, the key factor determining population change was chick survival, this in turn being determined by the weather and pesticide-dependent supply of insect food in cereal crops. Chick survival was the only stochastic input into the simulation model which enabled it to reproduce the annual variation in partridge density (Potts 1980). Excluding the occasional loss of clutches in heavy rain, chick survival rates are the only external sources of annual variation identified to date. This is not to say that the chick survival rate was varying at random (it was insect-dependent), only that it was not determined by population density.

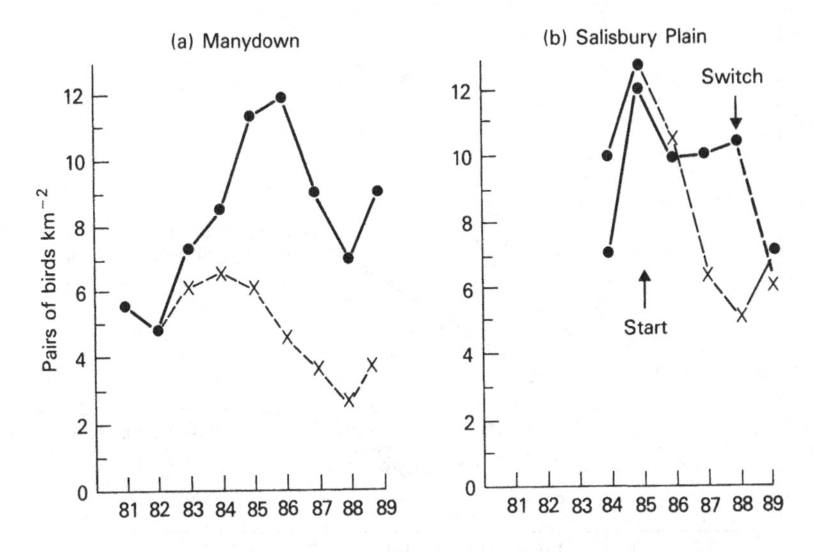

FIG 6.3 The results of two experiments in the management of populations of grey partridge. (a) Stock densities at the Manydown Estate (11 km^2) in northern Hampshire, where 'conservation headlands' (selectively sprayed field margins) were instituted in half the cereal fields from 1982: the continuous line shows the density in these fields, and the pecked line shows the levels that would have been found in them without conservation headlands, based on data for chick survival in the control fields. (b) Stock densities on two similar areas, each of 5.5 km^2, on Salisbury Plain, Wiltshire, one where nest predators were controlled (continuous line) and one where they were not (pecked line).

Density-dependent forms of mortality

Nest losses

The mortality attributable to k_1 was for the most part due to egg predation, most typically by carrion crows *Corvus corone*. This k factor represented the proportion of hens that survived incubation but which did not hatch their eggs (from the first or a replacement clutch). Accounting for most of the variation in nest losses, it was clearly density-dependent where predators were not controlled (Fig. 6.4; see also Potts 1986).

The mortality attributable to k_2 is predation of the incubating hen, most typically by foxes. It has increased steadily through the past 20 years in line with the increase of the fox population (Fig. 6.5). This trend masked the presence of density dependence, as detected by simple regression of k_2 against density. For four of the five farms the effects of

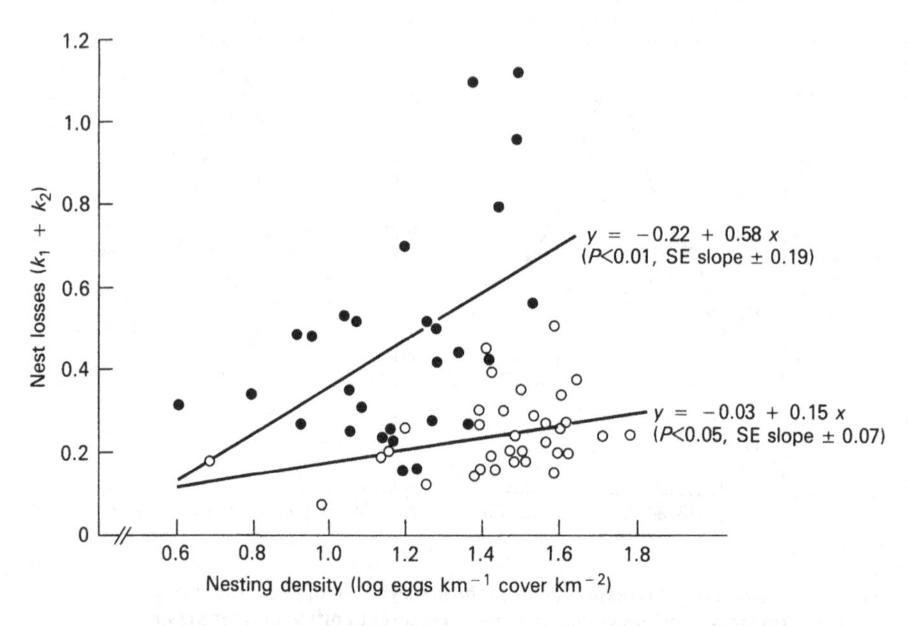

FIG 6.4 The relationship between nest losses of grey partridges ($k_1 + k_2$) and nest density in areas in Sussex with gamekeepers (o) and without (•): each point represents one area-year (from Potts 1986).

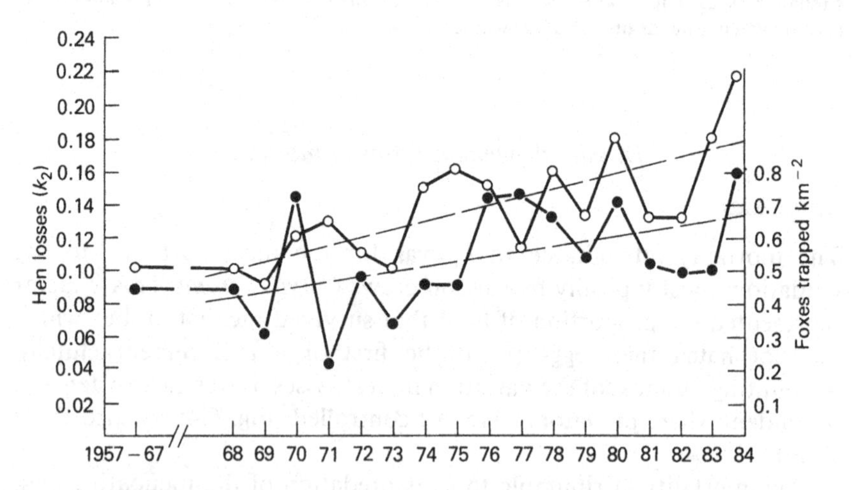

FIG 6.5 Changes with time in the losses of grey partridge hens (k_2) in part of the Sussex study area (o) (mostly resulting from predation by foxes) and in the numbers of foxes trapped km^{-2} in the same area between 1 March and 1 August (•). The upward trends in both measures (pecked lines) are significant ($P < 0.05$), as is the correlation between them ($P < 0.02$). Control of foxes in this area ceased after 1984.

density on k_2 itself were statistically significant only when the time (= fox) trend was taken into account:

$$k_2 = -0.240 - 0.022x_i - 0.289x_{ii} \ (r^2 = 0.17, \text{d.f.} = 46, P < 0.01),$$

where x_i was year (1968 = 1; 1987 = 20) and x_{ii} was log density of nests per linear km of nesting cover per km^2. On the fifth farm, k_2 was related to density even without the time-trend ($P<0.02$). In a separate experimental study area on Salisbury Plain the effects of nest predation on the partridge population have now been verified (Fig. 6.3(b)).

To summarize, k_1 and k_2 are normally strongly density-dependent, but the amount of mortality is a function of both partridge nesting density and predator density. With no foxes there is no fox predation at whatever density the partridges reach! We return to this trending of density dependence later — it has far too often been ignored, for example in k-factor analysis. In our case the problem became clear only after the study had progressed for 18 years and, in the case of the pied flycatcher (*Ficedula hypoleuca*), a similar problem was encountered in a 17-year study (Stenning, Harvey & Campbell 1988).

Losses through shooting, winter cold and predation

The relationship between shooting rate, k_4, and density has been explored elsewhere in some detail (Potts 1980). Higher rates of shooting occur when the population is high and a worthwhile bag is expected; shooting is difficult to organize and not worthwhile at low densities. The relationship is sigmoid because after about half the birds have been shot the survivors become more wary and more difficult to shoot; in practice shooting is also curtailed because of the need to conserve next year's stocks.

Winter losses, k_5, are of two kinds. First, there is density-dependent emigration resulting from the search for adequate nesting cover (Church, Harris & Stiehl 1980) and, secondly, there is predation in late winter in areas with prolonged snow cover (Potts 1986). The data from Russia (Fig. 6.6), where the winters are most severe, show that these losses, though caused by obviously density-independent factors (snow and ice) nevertheless resulted in density-dependent mortality, probably because the actual mortality was attributable to predators.

The causes of year-to-year variation in density

The Sussex study yielded eighty-eight estimates of spring stocks averaging 9.6 pairs km^{-2} but varying from two to 27 pairs km^{-2}. Estimates of

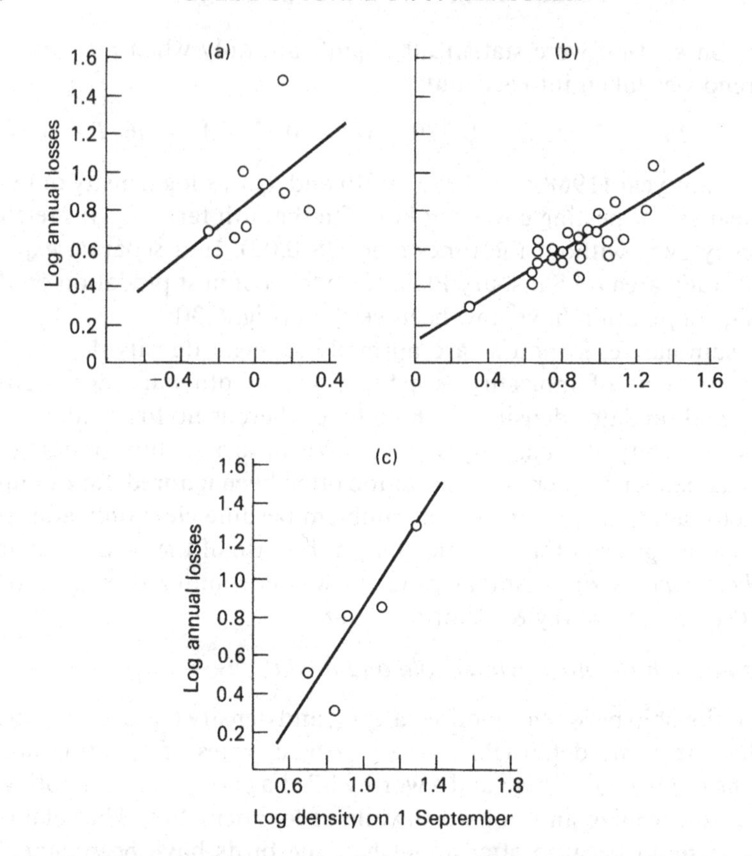

FIG 6.6 The relationship between density of grey partridges on 1 September (i.e. before shooting) and adult losses ($k_2 + k_{4+5}$) in subsequent year (mostly in winter) at three sites in the USSR; (a) Caucasian foothills, (b) Imperial Hunting Reserve at Gatchina, and (c) Trans-Baikal region (from Potts 1986).

log change in stock, i.e. log (N_{t+1}/N_t), averaged −0.032: this was equivalent to a decline of 7% per annum.

These changes were almost equally attributable to nest predation (k_{1+2}), to chick losses (k_3) and to shooting and other non-breeding losses (k_{4+5}). Overall 67% of the variation was accounted for in the equation:

$$y = -0.470 + 0.358\, k_{1+2} + 0.280\, k_3 + 0.760\, k_{4+5}$$
$$(r^2 = 0.67, \text{d.f.} = 84, P < 0.001), \qquad (6.1)$$

where y is the log proportional change in stock from one year to the next. The average contribution to change made by each k factor, obtained by multiplying each mean k by its matching regression coefficient, was about the same for all k factors:

$$k_{1+2} \quad 0.358 \times 0.392 = 0.140$$
$$k_3 \quad 0.280 \times 0.631 = 0.177$$
$$k_{4+5} \quad 0.760 \times 0.250 = 0.190$$

Much of the 33% not explained is attributable to sampling errors. The variance of the residuals in the eighty-eight cases decreased as stock numbers (= n) rose, according to the relationship:

$$r^2 = 0.86n/\,(10.72 + n)$$

Discounting these sampling errors, the relationship (eqn 6.1) explained 86% of the annual changes in partridge stock.

In the world data the picture was much the same, with a clear relationship between overall mortality and changes in partridge stocks (Fig. 6.7).

The changes described above occurred about density levels at which the populations were historically in equilibrium. Attention is now given to the levels about which the changes take place.

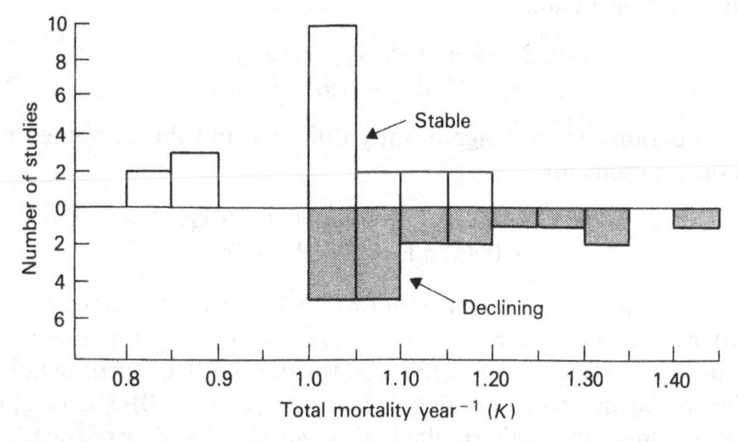

FIG 6.7 The frequency distribution of studies in the world data set for grey partridge in relation to the total mortality per year, shown separately for stable and declining populations.

The control of equilibrium density

Equilibrium densities are defined here as those series without a time trend. The world data contain twenty such series. If the 20-year Sussex data for the five farms are split into means for the two 10-year periods this yields a further ten series; it so happens that the main change in Sussex took place half-way through the study.

In both data sets — world and Sussex — the best determinant of equilibrium density was k_{1+2}; in neither set was k_3 significant. Chick mortality, k_3, did not affect equilibrium values in this study for two reasons. First, in the world data declining populations were excluded and amongst the stable populations there was no significant variation in k_3. Likewise in the Sussex data there was no further decline in k_3 after the study started. In the world data k_5 was very important in three cases (those in Fig. 6.6). The data in the figure come from areas where the winter is particularly harsh. They illustrate that, although the life-cycle of the partridge in such areas was dominated by an obvious density-independent factor (severe weather), most of the adult mortality was density-dependent. Because there was always an inverse relationship between k_{1+2} and k_{4+5} (shown in Fig. 6.8) we needed to deal with the two together. With y denoting the equilibrium level, we found for Sussex:

$$y = 1.602 + 1.584k_{1+2} + 0.150k_{4+5}$$
$$(r^2 = 0.70, \text{d.f.} = 7, P < 0.01)$$

and for the world data:

$$y = 2.589 + 3.095k_{1+2} + 2.004k_{4+5}$$
$$(r^2 = 0.43, \text{d.f.} = 19, P < 0.01).$$

These equations are not significantly different and the combined data yield the relationship

$$y = 2.489 + 2.799k_{1+2} + 1.924k_{4+5}$$
$$(r^2 = 0.44, \text{d.f.} = 29, P < 0.001).$$

The relationship between equilibrium density and the combined density-dependent factors ($k_{1+2} + k_{4+5}$) is shown in Fig. 6.9; the equilibrium levels were determined by the density-dependent factors.

This is shown also by simulation modelling (Fig. 6.10). High and low predation pressures both resulted, through density dependence, in a population at equilibrium, but the levels of equilibrium were very different. This was because of a trade-off between density-dependent reproductive losses and density-dependent non-breeding losses. Figure 6.8 clearly shows how the two were linked within the Sussex and the world data. Thus the trade-off resulted in equilibrium; as the compensation was not one-to-one, the level of equilibrium varied. The results from the model, in terms of changes in the population parameters, are given in Table 6.1. At equilibrium, the population suffering high nest predation had a lower recruitment rate than the one with low predation; to compensate, the adult expectation of life was greater.

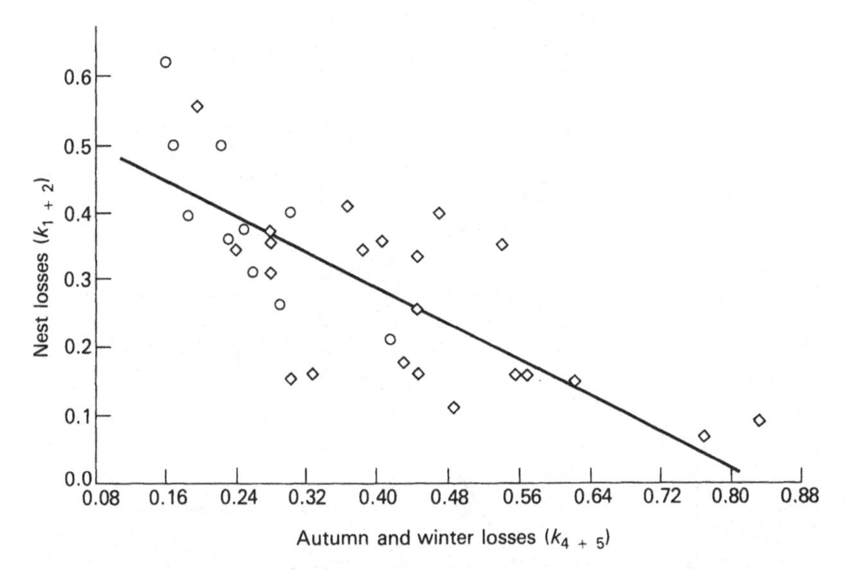

FIG 6.8 The negative relationship between nest losses (k_{1+2}) and autumn and winter losses (k_{4+5}) in a variety of populations: o, in Sussex; ◇, other studies (stable populations). Chick losses (k_3) were not related to any of the other losses.

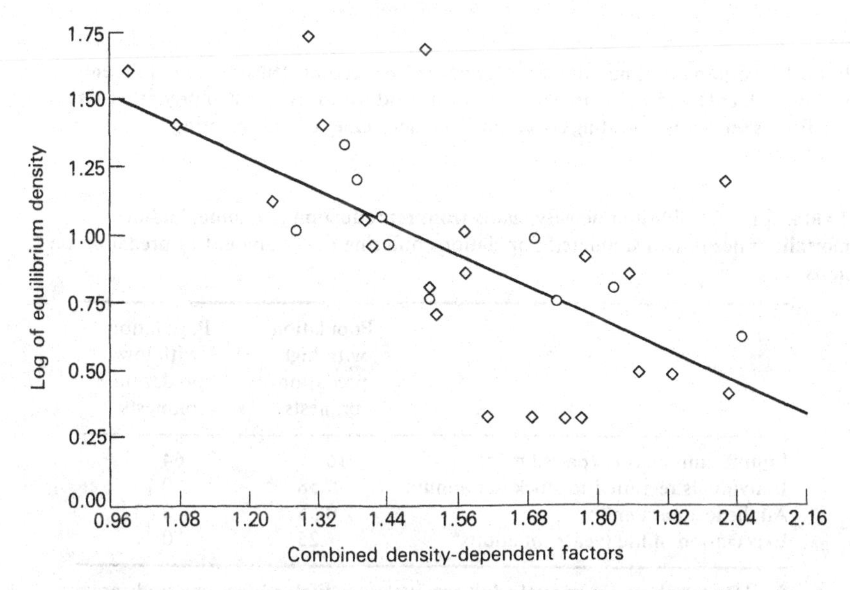

FIG 6.9 The relationship between density-dependent mortalities, combined by using the regression coefficients to weight them, and the density of breeding pairs in stable populations of grey partridge (as in Fig. 6.8).

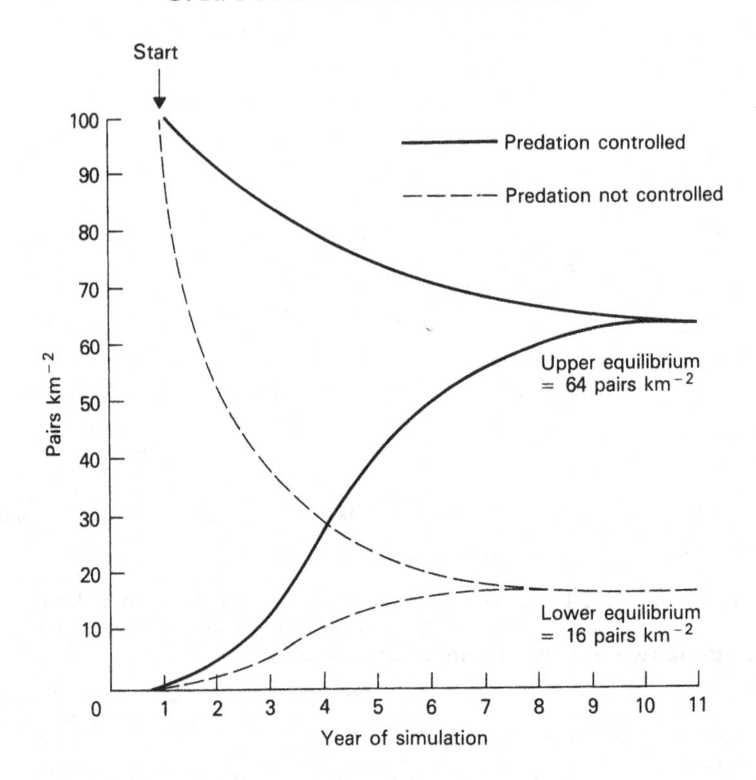

FIG 6.10 Computer simulations of grey partridge breeding densities with different starting values (1 and 100 pairs km^{-2}) and with and without control of predators. Other conditions set: 9 km of nesting cover km^{-2}, no herbicides and no shooting.

TABLE 6.1. Equilibrium density, gains from reproduction and annual adult mortality rates in two simulated populations differing in the amount of predation on nests

	Population with high predation on nests	Population with low predation on nests
Equilibrium density (pairs km^{-2})	16	64
Individuals recruited to stock per annum	0.58	0.76
Adults lost per annum	0.55	0.73
Expectation of life (years) of adults*	1.25	0.77

* $-1/\ln(s)$ where s = annual adult survival rate; 'life' = life in the study area, i.e. it excludes net emigration.

To summarize, the equilibrium density levels were determined by the outcome of the interplay of the density-dependent factors, and by the density-independent factor, chick mortality. The latter, however, did not explain a significant amount of the variation in equilibrium levels when the populations were not trending.

DISCUSSION

Determination of population size, using the partridge as an example

Density-dependent mortality and equilibrium levels

Haldane (1952) considered it 'blindingly obvious' that density-dependent mortalities are higher at high than at low densities. We, however, consider it equally obvious that high mean densities cannot be reached unless density-dependent mortality is relaxed relative to the low-density areas. Our data show that density-dependent mortalities are *lower* at higher densities, and modelling revealed that the expectation of life is not the same for all equilibrium levels. Our view is thus the opposite of that of Krebs (1986), who went so far as to say that one does not necessarily need to measure birth and death-rates in order to explain differences in mean density. This is because in most other studies densities have been, in effect, expressed relative to their equilibrium level. It has usually not been possible to investigate, within the same study, the factors controlling the densities of many different populations of the same species. At lower levels of equilibrium density, the adults live longer but have high density-dependent losses; at high levels the adults have a much shorter expectation of life but have reduced density-dependent mortalities.

Surprisingly the density-independent mortality was not higher at low densities. We have not found anywhere where partridges persist in the absence of density-dependent factors, as is the case, for example, with the trout (*Salmo trutta*) at Wilfin Beck, where abiotic density-independent factors were found sufficient to depress numbers indefinitely (Elliott 1987). We saw, earlier, that winter losses in partridge populations can be density-dependent even where the winter weather is extreme (Fig. 6.6), with a given bout of severe weather causing a higher proportion of loss to predation at high densities than it does at low ones. The phenomenon is not restricted to partridges. In the case of the shag (*Phalacrocorax aristotelis*), it was found that as densities rose on the Farne Islands a greater proportion of birds nested in poor nest sites, poor because they were exposed to the weather and to a lesser extent to predation. Thus a given amount of bad weather caused losses which rose proportionately with the number of shags nesting (Potts, Coulson & Deans 1980). In the partridge, predation and food shortage were exacerbated by the snow and ice — the birds did not die of cold *per se*. Darwin (1859) put it better than we can, 'when we travel from south to north, we invariably see some species gradually getting rarer and rarer, and finally disappearing; and the change of climate being conspicuous, we are tempted to attribute the whole effect to its direct action. But this is a false view'.

Balance of density-dependent and density-independent mortalities

The broadly inverse relationships amongst the different density-dependent factors, and between the density-dependent and density-independent factors, we believe to be typical of most populations. Their most important feature is not in their precise balance but in the net effect on the equilibrium level. Thus in partridges, nest predation or, alternatively, autumn shooting is completely offset by reduced emigration, but the net effect on the population is a clear reduction in its mean size (Fig. 6.9). Parasitism in partridges by the nematode *Trichostrongylus tenuis* does not occur to a significant extent in partridge populations which are limited by predation, but in any case the long-term reduction in the population caused by parasitism would probably not be as severe as that caused by predation (Potts 1986). However, the principle is clear enough and probably widely applicable. For example, the density-dependent mortality caused by a tachinid parasitoid in the winter moth can apparently become redundant after the effects of density-dependent predation by staphylinids (Hassell 1980). Evans & Pienkowski (1985), however, proposed for shore-birds a model with two equilibrium levels, low density and high density. The lower of the two levels had a higher density-independent mortality and the density-dependent mortalities were assumed to be 'not altered by the change in density-independent factors'; this remains to be investigated.

Existence of density-dependent mortality

It has to be said that our experience of partridge populations and of the other species we have studied lead us to be sceptical about recent attempts to minimize the importance of density-dependent factors in the control of population size, e.g. those of den Boer (1987), Elliott (1987) and Stiling (1988). Indeed we support those who continue to demonstrate the importance of density-dependent mortalities, e.g. Hassell (1980), or who claim that density-dependent mortalities are elusive for statistical reasons (Royama 1981; Hassell 1986) or for biological reasons (Solomon 1985). Solomon clearly recognized the problem when he wrote of density-dependent mortalities in salmonids 'whose parameters are in part determined by density-independent factors (e.g. stream discharge dictating carrying capacity)'. The same principle arose in our study as fox numbers rose, or when hedgerows were removed.

Relevance of deterministic chaos

Schaffer & Kot (1985) took the view 'that what passes for fundamental concepts in ecology is as a mist before the fury of the nonlinear storm'. However, their evidence for deterministic chaos caused by non-linearity, the phenomenon of 'strange attractors', may be generated by our partridge model with a little added random noise. The view that population sizes might often be unpredictable or determined simply by 'butterfly effects' (Gleick 1988) is, in our opinion, not only wrong but seriously misleading for conservationists. In contrast, we echo Hassell (1986), who considered that 'populations that persist do so, not by a juggling act balancing their rate of increase against a pot-pourri of unpredictable environmental factors, but rather by one or more processes that are broadly density-dependent'. Schaffer & Kot (1985) hope for a reappraisal of ecology and for models which are no longer 'caricatures'. We believe that our model, far from being such a caricature, gives a fair representation of essential reality. How else could it have been verified experimentally?

This said, we have one important caution. We are arguing strongly that the partridge population is balanced and adjusted to its resources, but the balance is not perfect. It is misleading to write 'If the balance of nature exists it has proved exceedingly difficult to demonstrate' (Connell & Sousa 1983). The density-dependent factors change with the resources themselves; unlike Connell & Sousa, we would not expect long-term stability.

The ultimate determinant of population size

None of this should minimize the role of density-independent factors. Both types of factor are essential to the control of population size. Moreover, the density-independent factors always have the final say, as was pointed out by Milne (1984), who wrote 'The ultimate control of decreases in numbers rests with the density-independent factors alone, for unless the density-independent factors stop causing a decrease then the remnants of the individuals that are left by the density-dependent factors must continue down towards zero.'

This is certainly the case with the partridge, where changes in the density-independent factor — the insect supply — caused the equilibrium levels to trend. Unless we can control the adverse effects of density-independent pesticides there is little hope of stemming the decline in population — whatever the original equilibrium density. It is because we see competition for resources as the basic mechanism of the control of population size that we feel confident about the ability of ecologists to change population sizes — up or down as necessary. We can provide resources or limit them, but we are unable to do much about the weather itself or about systems with deterministic chaos.

Evolution of population studies of birds

In many ways the partridge study is typical of those on bird populations generally, and many of these have been of even longer duration or covered larger areas. In one way, however, our study has been unusual: it has employed long-term monitoring, simulation modelling and experimentation in an integrated approach to an applied problem with a clear single objective — to restore partridge numbers. In our view there is considerable merit in tackling the same problem in several different and integrated ways. We believe that the strategy of combining monitoring, modelling and experimentation must be the way forward in understanding what controls population size in birds generally. Monitoring ensures that trends or unusual changes in population size may be picked up and analysed; modelling is the means whereby hypotheses and predictions may be made more exact for subsequent testing by experimentation. So how have population studies of birds evolved over the years, and how closely does their pattern conform to our proposed way forward?

Lack (1966) used the criteria for his review of population studies on birds 'that each should have continued for at least 4 years and should consist of more than just an annual census'. On this basis five species had been studied by 1954 and nineteen by 1966. Even after doubling Lack's time criterion, we estimate the current total as upward of seventy species. There has therefore been a phenomenal growth in the amount of basic population data on birds — some of it now very long term. However, we note that, although 59% of bird species are passerines, only 21% of the long-term studies mentioned above were on passerines. Part of the discrepancy is certainly due to the authors' greater familiarity with non-passerines, but some at least must reflect the larger size, higher conspicuousness and greater ease of study of the latter group. So the first part of our suggested three-pronged strategy — monitoring — is in a healthy state, albeit with a strong bias towards the larger, showier species.

The growth in the number of species studied has led directly to only a very modest growth in the number of population simulation models, the second of our three approaches. The first such model appeared only 20 years ago for the great tit *Parus major* (Pennycuick 1969) and populations of only about twelve of the seventy species mentioned above have been subject to any kind of computer modelling. Surprisingly these account for only half the total number of bird population models of which we are aware, the other half not being based on long-term monitoring data. Even more surprisingly, emphasis on the third approach — experimentation — has actually fallen: 40% of studies in 1954 (Lack 1954), 26% in 1966 (Lack 1966) and, we estimate, 12% in 1988. This is true despite some early pleas for the experimental approach (Watson & Jenkins 1968). It is difficult to estimate just how many studies have used all three approaches in an integrated way and have replicated them, although the studies on the red/willow grouse and the great tit are important examples.

It is clear from the above that the monitoring side of our integrated strategy is by far the strongest at the present time. Looking to the future, much of the excellent data already collected by monitoring should be synthesized in the form of simulation models, which would make possible quantification of the hypotheses concerning the control of population size, ready for experimental testing. It seems to us that for the time being the best policy would thus be to capitalize on the data from existing studies rather than amass further new data. The long-term studies should be given particular attention, and a representative selection of them should be continued indefinitely.

REFERENCES

Aebischer, N. J. & Potts, G. R. (in press). Sample size and area: implications based on long-term monitoring of partridges. *Proceedings of the International Workshop on Terrestrial Field Testing of Pesticides* (Ed. by L. Somerville & C. H. Walker). Taylor & Francis, London.

Blank, T. H., Southwood, T. R. E. & Cross, D. J. (1967). The ecology of the partridge 1. Outline of population processes with particular reference to chick mortality and nest density. *Journal of Animal Ecology*, **36**, 549–56.

Church, K. E., Harris, H. J. & Stiehl, R. B. (1980). Habitat utilisation by grey partridge (*Perdix perdix* L.) pre-nesting pairs in east-central Wisconsin. *Proceedings of Perdix II Grey Partridge Workshop* (Ed. by R. Peterson & L. Nelson), pp. 9–20. University of Idaho.

Connell, J. H. & Sousa, W. P. (1983). On the evidence needed to judge ecological stability or persistence. *American Naturalist,* **106**, 789–824.

Darwin, C. (1859). *The Origin of Species.* Murray, London.

Delacour, G. (1987). *Statut de la perdrix grise en Alsace: étude menée entre 1975 et 1985.* Unpublished *mémoire,* Université de Bourgogne, Dijon, France.

Den Boer, P. J. (1987). Density dependence and the stabilization of animal numbers, 2. The pine looper. *Netherlands Journal of Zoology,* **37**, 220–37.

Elliott, J. M. (1987). Population regulation in contrasting populations of trout *Salmo trutta* in two Lake District streams. *Journal of Animal Ecology,* **56**, 83–98.

Evans, P. R. & Pienkowski, M. W. (1985). Are shorebird populations regulated? *Behaviour of Marine Animals,* Vol. 5. *Shorebirds: Breeding Behaviour and Populations* (Ed. by J. Burger & B. L. Olla), pp. 110–17. Plenum Press, New York.

Gleick, J. (1988). *Chaos: Making a New Science.* Heinemann, London.

Haldane, J. B. S. (1952). Animal populations and their regulation. *New Biology,* **15**, 9–24.

Hassell, M. P. (1980). Foraging strategies, population models and biological control: a case study. *Journal of Animal Ecology,* **49**, 603–28.

Hassell, M. P. (1986). Detecting density dependence. *Trends in Ecology and Evolution,* **1**, 90–7.

Krebs, C. J. (1986). Are lagomorphs similar to other small mammals in their population ecology? *Mammal Review,* **16**, 187–94.

Lack, D. (1954). *The Natural Regulation of Animal Numbers.* Clarendon, Oxford.

Lack, D. (1966). *Population Studies of Birds.* Clarendon, Oxford.

Lovat, Lord (1911). *The Grouse in Health and in Disease.* Smith, Elder, London.

Manly, B. F. J. (1977). The determination of key factors from life table data. *Oecologia,* **31**, 111–17.

Maynard-Smith, J. (1978). *Models in Ecology.* Cambridge University Press.

Milne, A. (1984). Fluctuation and natural control of an animal population, as exemplified in the garden chafer *Phyllopertha horticola* (L). *Proceedings of the Royal Society of Edinburgh,* **82B**, 145–99.

Pegel, M. (1987). Das Rebhuhn (*Perdix perdix* L.) im Beziehungsgefüge seiner Um- und Mitweltfaktoren. *Systematische Untersuchungen über die Existenz- und Gefahrdungskriterien einheimischer Wildtiere, Teil 2,***18**, 1–200. Enke, Stuttgart.

Pennycuick, L. (1969). A computer simulation of the Oxford great tit population. *Journal of Theoretical Biology,* **22**, 381–400.

Potts, G. R. (1980). The effects of modern agriculture, nest predation and game management on the population ecology of partridges *Perdix perdix* and *Alectoris rufa. Advances in Ecological Research,* **11**, 2–79.

Potts, G. R. (1986). *The Partridge: Pesticides, Predation and Conservation.* Collins, London.

Potts, G. R. & Aebischer, N. J. (in press). Modelling the population dynamics of the grey partridge: conservation and management. *Bird Population Studies: Their Relevance to Conservation and Management* (Ed. by C. M. Perrins, J. -D. Lebreton & G. J. M. Hirons). Oxford University Press, Oxford.

Potts, G. R., Coulson, J. C. & Deans, I. R. (1980). Population dynamics and breeding success of the shag, *Phalacrocorax aristotelis,* on the Farne Islands, Northumberland. *Journal of Animal Ecology,* **49**, 465–84.

Potts, G. R. & Vickerman, G. P. (1974). Studies on the cereal ecosystem. *Advances in Ecological Research,* **8**, 107–97.

Rands, M. R. W. (1985). Pesticide use on cereals and the survival of grey partridge chicks: a field experiment. *Journal of Applied Ecology*, **22**, 49–54.

Royama, T. (1981). Evaluation of mortality factors in insect life table analysis. *Ecological Monographs*, **51**, 495–505.

Schaffer, W. M. & Kot, M. (1985). Do strange attractors govern ecological systems? *BioScience*, **35**, 342–50.

Solomon, D. J. (1985). Salmon stock and recruitment, and stock enhancement. *Journal of Fish Biology*, **27**, 45–57.

Stenning, M. J., Harvey, P. H. & Campbell, B. (1988). Searching for density-dependent regulation in a population of pied flycatchers *Ficedula hypoleuca* Pallas. *Journal of Animal Ecology*, **57**, 307–17.

Stiling, P. (1988). Density-dependent processes and key factors in insect populations. *Journal of Animal Ecology*, **57**, 581–93.

Stubbs, M. (1977). Density dependence in the life-cycles of animals and its importance in *K*- and *r*- strategies. *Journal of Animal Ecology*, **46**, 677–88.

Tanner, J. T. (1966). Effects of population density on growth rates of animal populations. *Ecology*, **47**, 733–45.

Tapper, S. C., Brockless, M. & Potts, G. R. (1988). The predation control experiment: the turning point. *The Game Conservancy Annual Review*, **19**, 105–11.

Watson, A. & Jenkins, D. (1968). Experiments on population control by territorial behaviour in red grouse. *Journal of Animal Ecology*, **47**, 595–614.

7. MECHANISMS IN PLANT POPULATION CONTROL

E. VAN DER MEIJDEN
Department of Population Biology, University of Leiden, PO Box 9516, 2300 RA Leiden, The Netherlands

INTRODUCTION

The increase, during the last two decades, in knowledge about plant populations has been impressive (Harper 1977; Solbrig 1980; White 1985; Crawley 1986). At the same time there has been a gradual but characteristic change in the nature of population studies on plants from being exploratory and descriptive in a rather general way toward the unravelling of mechanisms and testing of theoretical concepts. Even so, the proportion of experimental studies is still limited (cf. Partridge, this volume). In this chapter I shall consider the present state of knowledge in the field of population control in plants. That knowledge is biased in that there are more studies on species that are easily observable than on species that are difficult to find or count, due to their size, abundance, life-form, position in the vegetation, etc. I will restrict myself mainly to short-lived plants. Population control in long-lived plants, or in rare plants, may be fundamentally different because within-species interactions may be less frequent than in short-lived, abundant species that tend to grow in almost pure stands (Shaw & Antonovics 1986). Although the former species are more difficult to study in the field, there is much basic information available. Nevertheless, the same bias in selection of short-lived abundant plants is found in population modelling. I would like to stimulate the construction of models in which parameters concerning these differences can be studied.

WHAT IS POPULATION CONTROL?

Population control covers all ways in which population numbers are being affected, and in that sense it is no more than a collective noun for several different processes (cf. Fig. 7.1). Population control limits numbers and mass at, or below, their maximum possible values under a specific set of environmental conditions. That set of values is equivalent to the theoretical carrying capacity and to the collection of 'safe sites' in

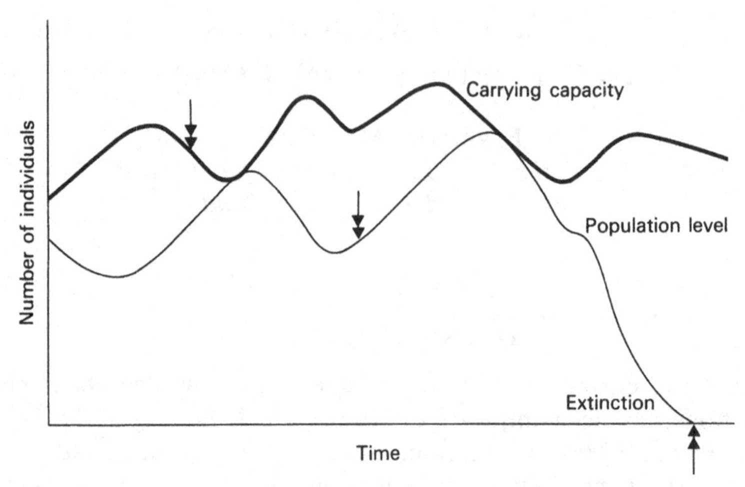

FIG 7.1 Aspects of population control: carrying capacity, population fluctuations below carrying capacity and extinction. Arrows indicate the processes leading to the different levels (see text).

population models. The level of the carrying capacity is determined by the combination of all biotic and abiotic components affecting the environment of a species; it includes the presence of other plant species, herbivores and disease, and it includes the effects of weather and soil on the suitability of the environment for germination, growth and survival.

Time delay in recovery through individual growth, reproduction and dispersal may lead to population levels or cyclic or stochastic fluctuations below the maximum level. Carrying capacity in monocultures that cover the whole surface of an area is well defined. A surplus of seeds will eventually lead to a strict interdependence between plant size at maturity and plant numbers. Soil fertility does not affect the general relation, but it determines which combinations of plant size and numbers will eventually be realized. These latter rules were already established 35 years ago by the Japanese school from Osaka (Yoda *et al.* 1963, elaborated by White 1980). If the whole surface of an area is not suited for plant growth and especially when we are not dealing with a monoculture, and normally we are not, carrying capacity becomes one of the most difficult population parameters to measure, judging from the minute number of publications in which estimates are published.

Carrying capacity will not always be reached, perhaps almost never, although we are still far from making an intelligent estimate of how often it is not reached. Estimates can be made through carefully designed experiments on seed addition, but perhaps even more so by the development of theory relevant for plant populations that tells experimenters how and what to measure. Several studies on seed addition, although not all, demonstrate an increase in population level, indicating that the population is below carrying capacity (Schenkeveld & Verkaar 1984; Shaw & Antonovics 1986).

Information on carrying capacity is relevant for both empirical and theoretical studies, to judge the importance of density dependence.

Population control also involves the ability of a population to persist for a certain period, the mechanisms that buffer against extinction, and finally extinction itself. Some of the buffer mechanisms are well known; clearly seed dispersal and delayed germination are important, but so also is delayed development, as seen in the aged saplings forming the sapling bank under mature trees. How often are these mechanisms of vital importance? How frequent are extinction events? At what scale do plants go extinct?

The complicating factors, that many plant ecologists are aware of in studying population control are the patchy spatial distribution of plants and the fact that they stand still and do not move around freely as animals do. Undoubtedly somebody will have said once that it is exactly that characteristic of plants which makes them easy to study compared with animals. However, by moving around freely, animals meet many other animals and they may all influence each other in food intake, transfer of disease, etc. Density estimates may therefore relate to intensity of competition. When competing, plants influence each other almost exclusively as neighbours. The importance of that observation becomes clear when comparing the distribution pattern of suitable sites for growth, and the distribution pattern of individuals over those sites. In some sites many individuals are found, in others only a few. Although total density may be far below carrying capacity, competition may be severe in some sites and absent in others. This point was stressed by Antonovics & Levin (1980) and in several later publications (Ågren & Fagerström 1984; Pacala & Silander 1985) but, nevertheless, has not been taken up properly in the experimental field. There is, however, a collection of descriptive studies that is based on this thought: the nearest-neighbour approach, in which the effect of neighbours on each other is examined (Pielou 1962; Yeaton & Cody 1976; Phillips & MacMahon 1981). Like many other descriptive studies, it contains a pitfall: plants may be small not only as a result of competition but also because, for example, they grow on poor soils or have suffered from a pest or pathogen. In populations in general it will be impossible to estimate the importance of density dependence if the spatial distribution is unknown.

It will, moreover, be impossible to judge which population models are relevant, and even which evolutionary models are relevant.

ARE PLANT POPULATIONS DENSITY-REGULATED?

Density effects, through intraspecific competition, have been shown to be important in a number of plant species in closed monocultures, and in a number of perennial plant species that form the dominant element of local vegetation (see reviews by Harper 1977 and Antonovics & Levin 1980). This is not so in all, however. Hubbell & Foster (1986) studied a species-rich tropical rain forest. Out of forty-eight common tree species less than 50% displayed negative density dependence when juveniles per hectare were plotted against the number of adults. With one exception the negative density effects proved too weak to regulate (to compensate for) the number of adults. Hubbell & Foster suggest that 'being in the right place at the proper time (a gap at the moment when it comes into being) might be more important to a tree's success in reaching the canopy than the taxon to which the tree belongs'. Because trees are surrounded by random combinations of other tree species, interspecific competition is diffuse and is not expected to lead to specialization but, instead, to greater generalization and similarity. The number of trees per hectare is almost constant.

Fowler (1986) studied density-dependent processes experimentally in natural populations of the perennial bunchgrass species *Bouteloua rigidiseta* , both by removing 50% of the mature individuals with more than five tillers or by increasing the number of seeds per quadrat to about 3.4 times that of the control plots (Table 7.1). The shoots of adults were removed by cutting their bases from the roots in order not to disturb the soil environment. Seeds were sown evenly over the quadrats and mixed into the vegetation. The magnitude of the responses to changes in density were smaller than the changes themselves, indicating that density-dependent effects had no regulating strength although they were present. The number of recruits increased in nearly direct proportion to seed input, instead of staying at the same level as without addition. Consequently the number of recruits was not density-regulated but limited by seed fall in the previous year. Removal of adult plants was only partly compensated for by an increase in survival and in tiller numbers (19%). Density-dependent regulation was weak, or absent in this 2-year study. Fowler suggested that the strength of density-dependent and independent factors may vary among years, in this case due to population-reducing droughts. Several other experimental studies on perennial herbs, e.g. that

TABLE 7.1. Experiments on survival of the grass *Bouteloua rigidiseta* in an arid grassland in Texas 1 year after plant removal and seed addition (from Fowler 1986)

	Number of plants (per 0.4 m^2)		
	Initial number	Surviving after 1 year	Percentage
(a) Removal experiment			
50% removal of individual plants	396	158	39.9
Control	445	276	62.0
	Number of recruits after 1 year		
(b) Addition experiment			
Seed addition (3.4 × control)	207.9		
Control	78.1		

of Goldberg (1987) on *Solidago* and that of Shaw & Antonovics (1986) on *Salvia lyrata* demonstrate strong density-dependent responses, but do not allow one to judge whether regulation actually takes place because of the experimental set-up. In *Salvia* the main negative effect was through adults on seedlings. Interestingly, a similar effect could not be found in a demographic study on the spontaneous population of the same species that was carried out simultaneously in the same area.

An unusual result was obtained by Symonides and colleagues in their study on *Erophila verna* (Symonides, Silvertown & Andreasen 1986). This annual grows at extremely high densities, almost in monocultures, in sand dunes in Poland. Density had an extremely strong effect on plant numbers, even stronger than expected when it was regulating, in the majority of 1-dm^2 plots used for censusing. Density acted in an overcompensating way leading to cycles of abundance that were not synchronous over all patches. The mechanism involved is an overcompensatory reduction in seed production per plant with density.

De Jong & Klinkhamer (1988) found a strong density-dependent survival of seedlings of *Cynoglossum officinale* in 2-dm^2 patches underneath the parent plants. Multiplying the mid-value of density with survival demonstrates that there is density-dependent regulation of seedling numbers on this scale. Multiplying the number of surviving seedlings with average weight even suggests overcompensation in biomass (Table 7.2). In this case, regulation is brought about by competition between sibs. Sib competition among seedlings, but also among mature

TABLE 7.2. Density-dependent survival of seedlings of *Cynoglossum officinale* at various natural seedling densities in plots of 2 dm^2 (after de Jong & Klinkhamer 1988)

Initial seedling density per dm^2	Number of surviving seedlings per dm^2	Biomass per plant (g)
1–5	1.50	0.234
6–10	3.68	0.692
11–15	4.55	0.473
>15	3.20	0.192

plants, is likely to be a common phenomenon. One wonders whether there may be positive effects involved for survival of the parent genotype through the occupation of space followed by competition leading probably to the best-adapted offspring.

The role of density-dependent regulation has been questioned particularly for species in sparse populations. Grubb (1986) studied *Rhinanthus minor* in chalk grassland and found that density effects at a scale of 10 × 10 cm plots, were obvious only in one out of five years. Kelly's (1982) study, in a similar area, in smaller plots (5 × 5 cm) on three different sparse species of short-lived plants (*Euphrasia officinalis, Gentianella amarella* and *Linum catharticum*) showed that density dependence in survival and performance occurred over less than 5% of the study area.

Kelly (1982) also reported on density-dependent effects in four dune annuals, using study plots of 10 × 10 cm around each plant. Despite the apparently sparse populations, density-dependent effects in either mortality or fecundity were found in all four species (cf. Grubb, Kelly & Mitchley 1982).

We have studied a number of biennial plant species of which some certainly are sparse. To see whether individuals of these species, growing in an open sand dune habitat, influenced each other, nearest-neighbour pairs were selected along random transects in spring. Short distances between nearest neighbours might lead to competition. Larger distances between plants should demonstrate less effect (Fig. 7.2). This method overcomes the surface and scale problems, but on the other hand certainly does not allow one to judge regulation. In almost all species and all populations density-dependent effects occurred (Table 7.3).

As mentioned earlier, the main problem in determining whether density-dependent regulation through competition occurs in populations of common, but also especially of sparse species, is that it is usually not

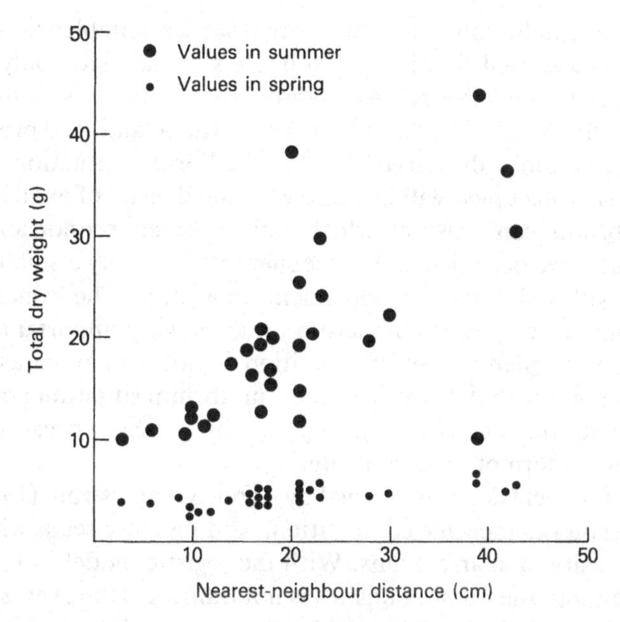

FIG 7.2 Nearest-neighbour analysis of thirty pairs of plants of *Cynoglossum officinale* in June (•) and September (●) of the same year. Total dry weight in June was estimated from a regression of weight on length of the longest leaf times leaf number. In September plants were sampled and weighed.

known when the carrying capacity of the area locally has been reached or what its level is.

Klinkhamer and de Jong (1989) constructed a deterministic model to test whether density-dependent mortality may have a significant effect in keeping sparse populations sparse. The model is based on the assumption

TABLE 7.3. Correlation coefficients between nearest-neighbour distance and total dry weight per pair in five populations of *Senecio jacobaea* and *Cynoglossum officinale*. Thirty pairs were measured per population in June and September

	Population				
	1	2	3	4	5
Cynoglossum officinale					
June	0.43**	0.17	0.69***	0.40*	0.51***
September	0.77***	0.64***	0.65***	0.80***	0.69***
Senecio jacobaea					
June	0.67***	0.27	0.55***	−0.17	0.16
September	0.80***	0.91***	0.88***	0.68***	0.56***

* $P < 0.05$; ** $P < 0.01$; *** $P < 0.005$.

that a habitat can be subdivided into sites that are suitable and sites that are not. It is assumed that in each suitable site (safe site) only a single individual can establish itself and reproduce. Parameters in the model are the density of safe sites, their total surface area, and seed production. Seeds are randomly dispersed. In the equilibrium situation the proportion of sites occupied will increase with the density of available sites. An equilibrium can exist at which only a small proportion of the available sites are occupied and consequently when only a small number of plants will suffer density-dependent mortality. The conclusion is therefore that intraspecific competition may be very important, even at low densities of plants when competition is not very obvious. A very interesting point is that, because in the equilibrium situation not all sites are occupied, the carrying capacity is not reached because of the distribution pattern of seeds and sites.

A similar conclusion was reached by Ågren & Fagerström (1984) from their model on interspecific competition. Adding more seeds will lead to a higher density of sparse plants. With the logistic model in mind, that would mean that the number of sites is not limiting. However, seeds and safe sites appear to be limiting at the same time in this situation.

So, if density-dependent effects do occur in sparse species, we know they have the strength to regulate even if they are infrequent.

In sparse species density effects may be underestimated because plant numbers have not been related to the appropriate surface area for these species. Perhaps more important, however, is the scale of the arena for competition or other density-dependent processes. Because plants influence only neighbours, competition effects should be studied on that level. It is not total density, but local density that is experienced. The problem disappears, to a large extent, when we are dealing with monocultures. Density dependence may then indeed become almost perfect if all individuals are surrounded by equal numbers of even-sized conspecifics.

The extremely diverse and often incomplete approach of the studies of density dependence in plants that is recorded in the literature emphasizes the need for more agreement between theoretical and empirical studies.

ARE POPULATIONS WITHOUT DENSITY-DEPENDENT PROCESSES DOOMED TO EXTINCTION?

The affirmative statement can be found in many textbooks. Unfortunately it is not followed usually by information on mechanisms of persistence, or extinction frequencies of species. Strong (1986) envisaged a ceiling and a floor density where density-dependent mechanisms should be most pronounced. Floors to density would be more poorly understood than ceilings simply because sparse populations are more difficult to find and study. Yet it is not very difficult to suggest density-dependent mechanisms that could possibly lead to persistence. One of these has to do with the distribution of degrees of resistance in plants against herbivores. If resistance follows a bell-shaped distribution, like for instance alkaloid concentration in ragwort (van der Meijden *et al.* 1984), and if herbivores prefer plants with low levels of resistance, then an increase in herbivory would lead to an increase in toxicity of the surviving plants. This in turn would lead to a decrease in herbivory. Similarly bell-shaped distributions of resistance of plants against drought, frost and other environmental factors would buffer.

How frequently do populations become extinct? Extinction and persistence are related to scale (MacArthur & Wilson 1967). Local subpopulations will disappear more quickly than metapopulations. It is on the level of the local subpopulation that intra- and interspecific interactions, and also the effects of the abiotic environment, are to be studied. It is also between these subpopulations that dispersal is relevant. Therefore, it is on this level too that extinction should be studied. Botanists with their knowledge about succession are familiar with changes in vegetation composition involving local extinction of plant populations. However, not all extinctions are part of succession. Grubb (1986) describes local extinction of species of short-lived plants in quadrats of 0.25m^2 in a chalk grassland. The species do not disappear from the study area, but show definite spatial dynamics which Grubb (1982) called 'shifting clouds of abundance'. Schat (1982), working on short-lived plants on beach plains, found similar spatial changes in abundance and could relate them to environmental circumstances such as flooding and drought. We are dealing here with local subpopulations that are separated from other subpopulations (or 'interaction groups'). Gene flow between these groups may be very low. In some species a small seed bank may be present that can lead to recovery, while in others recovery is possible only through immigration. Such dynamics are not limited to plant populations. The theory of 'spreading of risk' by den Boer (1968) was inspired by the observation of local extinction of carabid beetles, together with another observation, namely that those species that went extinct locally had more winged individuals than the more persistent species, indicating a higher level of migration.

Data on a group of 'biennial' species of which discrete local subpopulations have been followed since 1980 by monitoring permanent

25-m^2 samples show that, in this group of species, extinction is a very frequent event (Table 7.4). Apparently, different species are not behaving in the same way. Sites are repopulated by the species of *Cirsium, Cynoglossum* and *Echium* whereas the species of *Arctium, Carlina* and *Verbascum* tend to stay unoccupied on this time scale. We have gained the impression that the observed fluctuations represent the situation over a much larger area. A similar picture was found in ragwort (*Senecio jacobaea*), which has been monitored in 102 local populations in the same dune area of about 10 km^2 (Fig. 7.3). Populations were sampled in 4-m^2 quadrats. If not a single mature rosette or seedling plant was found in a subpopulation sample area, that subpopulation was considered to have gone extinct. Visual observation indicated that it usually meant that the local subpopulation from which the sample had been taken had indeed gone extinct above ground. These data are not very different from those for the other 'biennials' in terms of extinction frequency. Up to fifty local subpopulations became extinct in one year. Of all subpopulations, 80% became extinct at least once during the study period. Almost all of the subpopulations that never went extinct were found in areas of scrub, where they are buffered against the two main factors causing population reduction: drought, and defoliation by the cinnabar moth (*Tyria jacobaea*). These buffered areas act as refuges. After local extinction, repopulation may take place eventually through dormant seeds that germinate and through seed dispersal from the refuges. Each of these two processes acts in a density-dependent way because their effect increases at low densities. In an earlier paper (van der Meijden, de Jong & Klinkhamer 1985) we analysed the importance of the two mechanisms. Both led to repopulation of similar numbers of populations. Ragwort has only a very small seedbank and the number of seedlings, resulting from delayed germination is small, but it seems to be effective. On the other hand it is a rather successful wind disperser, if refuges are present.

In trying to understand the relative importance of delayed germination and dispersal for population growth, Klinkhamer *et al.* (1987) studied the dependence of optimal combinations of the two mechanisms on environmental conditions. This model, in which no local extinction is incorporated, deals with an environment consisting of a number of patches. Furthermore, it assumes homogeneous dispersal of seeds over all patches and accounts for additional mortality of dispersing seeds. As the criterion of success, the geometric growth average was used. It was found that delayed germination should be favoured if fluctuations in the reproductive success values of the populations are positively correlated. Dispersal on the other hand should be favoured if populations fluctuate

TABLE 7.4. Local extinction of subpopulations of biennials (based on van der Meijden *et al.* 1985 and unpublished observations)

Species	Number of sub-populations in 1980	Percentage extinct in			
		1983	1984	1985	1986
Verbascum densiflorum and *V. thapsus*	14	77	79	58	67
Cirsium vulgare	18	61	78	91	17
Echium vulgare	18	44	67	0	17
Arctium pubescens	11	27	36	60	50
Cynoglossum officinale	33	18	39	7	0
Carlina vulgaris	13	15	62	67	60
Inula conyza	11	9	46	100	71

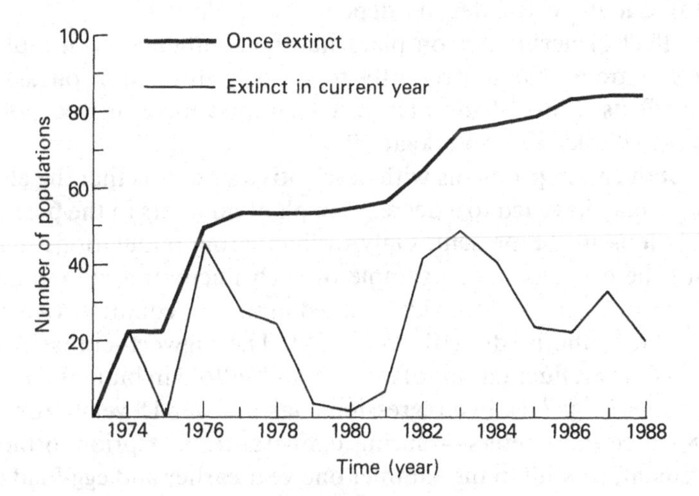

FIG 7.3 Above-ground extinction of local subpopulations of *Senecio jacobaea*, measured in 4 m^2 quadrats; once extinct means present, but extinct in at least 1 previous year during the study period.

in an uncorrelated manner. Because the two processes cancel out each other's effects, dormancy and dispersal should be negatively correlated in groups of species. That prediction was first made intuitively by Cook (1980). Several sets of data affirmed this prediction.

Local extinction is probably a more frequent event than is generally realized, and the importance of buffer mechanisms that allow recolonization can easily be underestimated.

WHAT CAUSES POPULATION FLUCTUATIONS?

Fluctuations may be caused by different levels of herbivory, pollination and disease, by changes in interspecific competition, by effects of weather and soil, and by their interactions.

All these processes are being studied by considerable numbers of scientists all over the world. It is known that pollinators do limit seed production in a number of cases, but it is equally well established that they do not influence it in other cases. The same is true for herbivory and disease. Herbivory or disease could lead to density-dependent regulation. Such density dependence should be operating on the scale of activity of the herbivore or pathogen population, which is usually much larger than the competition arena of the plants. As shown earlier, density-dependent regulation, except in monocultures, is far from common over larger areas, thus indicating that it is very doubtful whether herbivory and disease frequently cause density-dependent regulation.

The effect of herbivores on plant numbers, other than in biological control situations, has led recently to a lively discussion on assumed positive effects. Most of the literature examples have proved not fully convincing (Belsky 1986; Verkaar 1988).

One of the main problems with descriptive studies is that the effect of herbivores may have led to a decrease in plant numbers in the past which is not obvious in the present. Only in fluctuating populations may the effect still be obvious. One example of such fluctuating populations is that of ragwort, *Senecio jacobaea*, studied in several countries (Australia, Canada, the Netherlands, UK and USA). The ragwort–cinnabar moth system has shown fluctuations of over a hundredfold in biomass in Dutch coastal dunes (Fig. 7.4). In a regression analysis over 15 years from 1974 to 1988, three parameters — spring temperature just prior to biomass measurement, rainfall in the summer one year earlier and egg-load of the cinnabar moth one year earlier — accounted for most of the variance (Table 7.5). The last two years of this study were exceptional. There was an extremely late and cold spring followed by an extremely mild one with very different effects on spring growth. If we had not been able to compare these conditions with the other years, we should have concluded that spring temperature was the driving force of the system, whereas it is the least important parameter over the whole study period. This factual information was meant as an introduction to a paper by Weatherhead (1987) in which he asks 'How unusual are unusual events?' From a comparison of over 200 ecological studies he concluded that so-called unusual events are perhaps more usual than has been previously

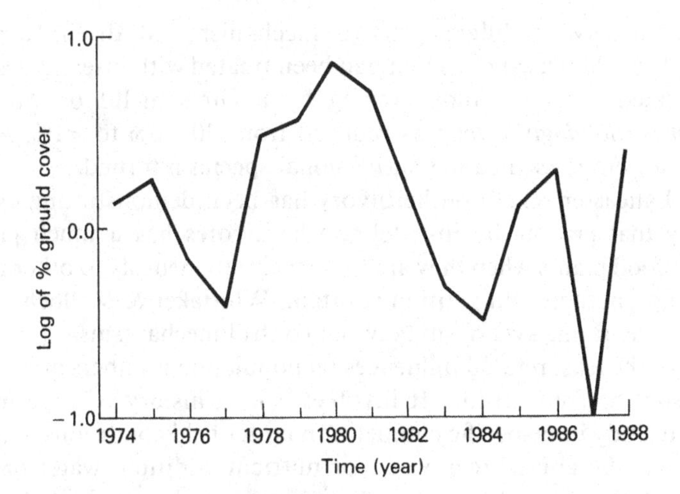

FIG 7.4 Fluctuations in the percentage ground cover of *Senecio jacobaea* on a Dutch sand dune system over a 15-year period.

recognized. He found one unusual event mentioned in each of ten studies. The proportion of studies with such an event increased with study duration from one to five years whereafter it levelled off. His interpretation is that short-term studies experience too many unusual events, and that these studies lack the benefit of the perspective provided by a longer study. If we had been studying the ragwort–cinnabar moth system only during the last two years it would have been impossible to get a correct idea about the importance of the driving forces.

Unusual events, however, certainly are not unimportant. They may change the relative abundances of species considerably, and they may change the relative importance of competition.

Plant ecology has not yet become an experimental science, although it is evolving in that direction. Experiments have elucidated some of the

TABLE 7.5. Multiple regression analysis on the percentage ground cover of *Senecio jacobaea* from 1974 to 1988

Variable	Partial correlation
Spring temperature in year n	0.35
Summer rainfall in year $n-1$	0.51
Egg load *Tyria jacobeae* in year $n-1$	−0.46

Multiple correlation = 0.82; F = 4.96; P = 0.018.

interactions between different control mechanisms. McBrien, Harmsen & Crowder (1983) used plots that had been treated with insecticides next to untreated plots to study grazing by a chrysomelid on *Solidago canadensis. Solidago* cover was reduced from 40–70% to <1%. At the same time, suppressed earlier successional species returned.

The Lancaster school on herbivory has been demonstrating experimentally that grazing by invertebrate herbivores has a much greater effect on food plants when they are subject simultaneously to other forms of pressure, notably competition (Cottam, Whittaker & Malloch 1986), thus demonstrating synergism between control mechanisms.

One of the least refuted influences on population numbers or biomass is interspecific competition. It involves a long history of experimentation. To study interspecific competition under field conditions, circumstances may be altered (e.g. water or nutrient addition, water or light reduction). Recently above-ground clipping experiments have become frequently used (Fowler 1986; Ellison 1987; Goldberg 1987). Because roots and root-fragments are left in the soil, there may be mineralization or even immobilization of nutrients, but it is probably the only practicable way to manipulate populations and achieve a lower density. Although we know much about species and community productivity in relation to light, water and nutrient availability (Grime 1979; Tilman 1982), still little is known about what individual plants are competing for in specific situations (Hubbell & Foster 1986). Goldberg (1987) found in her study on *Solidago* that per-unit weight competitive effects did not differ among neighbour-species. Are differences in size or abundance indeed more important for the outcome of competition than differences in resource use?

THE INDIVIDUAL IN THE POPULATION: THE IMPORTANCE OF GENETIC VARIATION

'One of the consequences of a population approach to the biology of plants is that it focuses attention at the level of individual behaviour; the population has no meaning except as the summed activities of its individuals and their interactions' (Harper 1977). Individual plants differ from each other and their differences may affect population control: 'the population dynamics is continually confounded in the population genetics' (Harper 1977). Similar remarks have been made by many others to argue for a closer fusion between population ecology and population genetics (Antonovics 1976; Roughgarden 1977; Hickman 1979; Solbrig 1980).

It is disappointing to see that this impulse had hardly produced any effect on studies of population control and related fields, although it is now forming a major component of life-history studies. In recent textbooks population genetics is hardly incorporated in considerations of the control of plant populations (Crawley 1986; Diamond & Case 1986). Hutchings (1986) wrote that 'Gottlieb (1977) could show no significant differences between large and small plants in natural populations of the annual herb *Stephanomeria*, thus apparently vindicating, at least for this species, the ecologist's tendency to ignore genotypes when studying fecundity differences between individual plants in populations'. Even in a final paper on the population biology of *Viola* spp. by Solbrig *et al.* (1988), genetic differences between individuals receive only minor attention. Differences in growth rate between plants in the field were found to be influenced more by the environment than by the genotype. Additive genetic variance for growth was very small.

Should it be concluded that population genetics are not important for population control? I strongly doubt it. There are too many situations in which it could be demonstrated that genetic variation significantly affects plant fitness under natural conditions to maintain the simplification of genetic uniformity in population control theory (see, for example, Clausen 1951; Haeck & Woldendorp 1985; Endler 1986).

The two extreme situations that can be envisaged in the dynamics of a surviving population are growth to extremely high densities and decline to very low densities. Two examples will illustrate the importance of genetic variation at both sides.

Kromer & Gross (1987) studied the effects of genotype upon seedling and adult performance in *Oenothera biennis* by using genetically identical seeds. Genotype effects became apparent with plant development. Reproductive output was strikingly different. When grown under very high-density regimes, genotypes with many, but light, seeds still flowered whereas genotypes with heavy seeds did not flower at all. So population density selectively affected different genotypes and had great consequences for individual fitness.

Ennos (1985) demonstrated under particular experimental conditions that seedlings of *Trifolium repens* which contained the hydrolysing enzyme linamarase (an essential requisite for cyanogenesis) had a 31% higher probability of survival than seedlings without this enzyme. The observed effect is probably a consequence of selective grazing by molluscs. Similar differences were found in many other studies, as in our own work on insect–plant relationships. Alkaloid concentration in *Senecio jacobaea* has a genetic basis, and genotypes with high levels of specific alkaloids are avoided by the cinnabar moth (A.M. van Zoelen & K. Vrieling, personal communication).

Not all *Trifolium* individuals and not all *Senecio* individuals are protected. There are indications that the production of secondary plant substances involves a considerable cost. Only under a high herbivore pressure will the advantage of investment in defence become obvious, and otherwise low densities of resistant genotypes may be expected. If herbivory or disease reduces population density, the composition of plant populations may change. Resistant genotypes may survive.

This field of research is extremely well suited for experimentation with identical plants reared by clonal propagation or other forms of multiplication. That possibility does not yet seem to be very popular in the ecological community.

DISCUSSION

The secret life of plants is being discovered. Ecological knowledge about widely different mechanisms has increased tremendously. One gets the impression, however, that generalizations and comprehensive theory are lagging far behind. Some of the modern textbooks are quite encyclopaedic in citing collections of unintegrated facts but fail to report on the development of an integrated science. Some of my problems with present-day plant population ecology are the following.

1 Theoretical studies and empirical studies do not exploit and reinforce each other exhaustively. When using models there are two complementary approaches: the formulation of simple conceptual models intended to study, but also to generate, general principles and the formulation of so-called realistic models intended to understand specific situations. Both approaches demand interaction between theoretician and experimenter. Judging from the rather strict separation of theoretical and empirical publications in ecological journals, I expect that a better integration would be very profitable. Theoreticians should perhaps say in a more detailed way what and how experimenters should measure, whereas experimenters should tell theoreticians what is relevant to incorporate in models.

2 Much of the factual information about population control indicates that density-independent factors are at least as important as density-dependent factors, and probably much more important. Theoreticians tend strongly to overestimate the importance of density dependence.

3 Plants differ greatly from one another in phenotype, especially in respect of resistance against herbivores and diseases. The importance of phenotypic differences has been demonstrated. Differences in individual characteristics are likely to be important in less obvious aspects of plant life too. It is against this background that experimental and theoretical analysis should proceed.

4 Many recent studies in plant ecology cover only a very short period and are not replicated. It is very likely that the literature is becoming enriched with information on many unique events that cannot properly be placed in perspective and will rather hamper formulation of unifying concepts.

5 The arena of interaction between individual plants has a very local scale, which is neglected in many empirical and model studies. Perhaps it is for this reason that the role of density dependence is still obscure. Probably too many ideas have been borrowed from our zoological colleagues. Because plants interfere only with their neighbours, clones of the Lotka–Volterra model are not suited for describing interactions between plants.

ACKNOWLEDGMENTS

I appreciated discussing the manuscript with Tom de Jong and Peter Klinkhamer, and I am grateful to two referees. Hans Haarlem, Ceciel Sassen, Erik Schravendeel and Karin van Wijk provided technical assistance.

REFERENCES

Ågren, G. I. & Fagerström, T. (1984). Limiting dissimilarity in plants: randomness prevents exclusion of species with similar competitive abilities. *Oikos,* **43**, 369–75.

Antonovics, J. (1976). The input from population genetics. *Systematic Botany,* **1**, 233–45.

Antonovics, J. & Levin, D. A. (1980). The ecological and genetic consequences of density-dependent regulation in plants. *Annual Review of Ecology and Systematics,* **11**, 411–52.

Belsky, A. J. (1986). Does herbivory benefit plants? A review of the evidence. *American Naturalist,* **127**, 870–92.

Clausen, J. (1951). *Stages in the Evolution of Plant Species.* Cornell University Press, Ithaca, New York.

Cook, R. (1980) The biology of seeds in the soil. *Demography and Evolution in Plant Populations* (Ed. by O. T. Solbrig), pp. 107–31. Blackwell Scientific Publications, Oxford.

Cottam, D. A., Whittaker, J. B. & Malloch, A. J. C. (1986). The effects of chrysomelid beetle grazing and plant competition on growth of *Rumex obtusifolius. Oecologia,* **70**, 452–6.

Crawley, M. J. (Ed.) **(1986).** *Plant Ecology.* Blackwell Scientific Publications, Oxford.

de Jong, T. J. & Klinkhamer, P. G. L. (1988). Population ecology of the biennials *Cirsium vulgare* and *Cynoglossum officinale* in a coastal sand-dune area. *Journal of Ecology,* **76**, 366–82.

den Boer, P. J. (1968). Spreading of risk and stabilization of animal numbers. *Acta Biotheoretica,* **18**, 165–94.

Diamond, J. & Case, T. J. (Eds) **(1986).** *Community Ecology.* Harper & Row, New York.

Ellison, A. M. (1987). Effects of competition, disturbance and herbivory on *Salicornia europaea. Ecology,* **68**, 576–86.

Endler, J. A. (1986). *Natural Selection in the Wild.* Princeton University Press.

Ennos, R. A. (1985). Measuring the effects of genetic variation upon plant fitness. *Structure and Functioning of Plant Populations.* Vol. 2. *Phenotypic and Genotypic Variation in Plant Populations* (Ed. by J. Haeck & J. W. Woldendorp), pp. 153–60. North-Holland, Amsterdam.

Fowler, N. L. (1986). Density-dependent population regulation in a Texan grassland. *Ecology,* **67**, 545–54.

Goldberg, D. E. (1987). Neighbourhood competition in an old-field plant community. *Ecology,* **68**, 1211–23.

Gottlieb, L. D. (1977). Genotypic similarity of large and small individuals in a natural population of the annual plant *Stephanomeria exigua* ssp. *conoraria* (Compositae). *Journal of Ecology,* **65**, 127–34.

Grime, J. P. (1979). *Plant Strategies and Vegetation Processes.* Wiley, Chichester.

Grubb, P. J. (1984). Some growth points in investigative plant ecology. *Trends in Ecological Research for the 1980s* (Ed. by J. H. Cooley & F. B. Golley), pp. 51–74. Plenum, New York.

Grubb, P. J. (1986). Problems posed by sparse and patchily distributed species in species-rich plant communities. *Community Ecology* (Ed. by J. Diamond & T. J. Case), pp. 207–25. Harper & Row, New York.

Grubb, P. J., Kelly, D. & Mitchley, J. (1982). The control of relative abundance in communities of herbaceous plants. *The Plant Community as a Working Mechanism* (Ed. by E. I. Newman), pp. 79–97. Special Publications of the British Ecological Society, 1. Blackwell Scientific Publications, Oxford.

Haeck, J. & Woldendorp, J. W. (Eds) **(1985).** *Structure and Functioning of Plant Populations.* Vol. 2. *Phenotypic and Genotypic Variation in Plant Populations.* North-Holland, Amsterdam.

Harper, J. L. (1977). *Population Biology of Plants.* Academic Press, London.

Hickman, J. C. (1979). The basic biology of plant numbers. *Topics in Plant Population Biology* (Ed. by O. T. Solbrig, J. Jain, G. B. Johnson & P. H. Raven), pp. 232–63. Macmillan, London.

Hubbell, S. P. & Foster, R. B. (1986). Biology, chance and history and the structure of tropical rain forest tree communities. *Community Ecology* (Ed. by J. Diamond & T. J. Case), pp. 314–29. Harper & Row, New York.

Hutchings, M. J. (1986). The structure of plant populations. *Plant Ecology* (Ed. by M. J. Crawley), pp. 97–136. Blackwell Scientific Publications, Oxford.

Kelly, D. (1982). *Demography, population control and stability in short-lived plants of chalk grassland.* Ph. D. dissertation, University of Cambridge.

Klinkhamer, P. G. L. & de Jong, T. J. (in press). A deterministic model to study the importance of density-dependence for regulation and natural selection in sparse plant populations. *Acta Botanica Neerlandica.*

Klinkhamer, P. G. L., de Jong, T. J., Metz, J. A. J & Val, J. (1987) Life history tactics of annual organisms: the joint effects of dispersal and delayed germination. *Theoretical Population Biology,* **32**, 127–56.

Kromer, M. & Gross, K. L. (1987). Seed mass, genotype, and density effects on growth and yield of *Oenothera biennis* L. *Oecologia,* **73**, 207–12.

MacArthur, R. H. & Wilson, E. O. (1967). *The Theory of Island Biogeography.* Princeton University Press.

McBrien, H., Harmsen, R. & Crowder, A. (1983). A case of insect grazing affecting plant succession. *Ecology,* **64**, 1035–9.

Pacala, S. W. & Silander, J. A. (1985). Neighborhood models of plant population dynamics I. Single-species models of annuals. *American Naturalist,* **131**, 385–411.

Phillips, P. L. & MacMahon, J. A. (1981). Competition and spacing in desert shrubs. *Journal of Ecology,* **69**, 97–115.

Pielou, E. C. (1962). The use of plant-to-neighbour distances for the detection of competition. *Journal of Ecology,* **50**, 357–67.

Roughgarden, J. (1977). Basic ideas in ecology. *Science,* **196**, 51.

Schat, H. (1982). *On the ecology of some dune slack plants.* Ph.D. dissertation, Free University, Amsterdam.

Schenkeveld, A. J. M. & Verkaar, H. J. P. A. (1984). *On the ecology of short-lived forbs in chalk grasslands.* Ph.D. dissertation, University of Utrecht.

Shaw, R. G. & Antonovics, J. (1986). Density dependence in *Salvia lyrata,* a herbaceous perennial: the effects of experimental alteration of seed densities. *Journal of Ecology,* **74**, 797–813.

Solbrig, O. T. (Ed.) **(1980).** *Demography and Evolution in Plant Populations.* Blackwell Scientific Publications, Oxford.

Solbrig, O., Curtis, W. F., Kincaid, D. T. & Newell, S. J. (1988). Studies on the population biology of the genus *Viola.* VI. the demography of *V. fimbriatula* and *V. lanceolata. Journal of Ecology,* **76**, 301–19.

Strong, D. R. (1986). Density vagueness: abiding the variance in the demography of real population. *Community Ecology* (Ed. by J. Diamond & T. J. Case), pp. 257–68. Harper & Row, New York.

Symonides, E., Silvertown, J. & Andreasen, V. (1986). Population cycles caused by overcompensating density-dependence in an annual plant. *Oecologia,* **71**, 156–8.

Tilman, D. (1982). *Resource, Competition and Community Structure.* Princeton University Press.

van der Meijden E., de Jong, T. J. & Klinkhamer, P. G. L. (1985). Temporal and spatial dynamics of biennial plants. *Structure and Functioning of Plant Populations.* Vol. 2. *Phenotypic and Genotypic Variation in Plant Populations* (Ed. by J. Haeck & J. W. Woldendorp), pp. 91–103. North-Holland, Amsterdam.

van der Meijden, E., van Bemmelenn, M., Kooi, R. & Post, B. J. (1984). Nutritional quality and chemical defence in the ragwort-cinnabar moth interaction. *Journal of Animal Ecology,* **53**, 443–53.

Verkaar, H. J. (1988). Are defoliators beneficial for their host plants in terrestrial ecosystems? A review. *Acta Botanica Neerlandica,* **37**, 137–52.

Weatherhead, P. J. (1987). How unusual are unusual events? *American Naturalist,* **128**, 150–4.

White, J. (1980). Demographic factors in populations of plants. *Demography and Evolution of Plant Populations* (Ed. by O. T. Solbrig), pp. 21–48. Blackwell Scientific Publications, Oxford.

White, J. (Ed.) **(1985).** *Studies on Plant Demography. A Festschrift for John L. Harper.* Academic Press, London.

Yeaton, R. I. & Cody, M. L. (1976). Competition and spacing in plant communities, the northern Mojave Desert, *Journal of Ecology,* **64**, 689–96.

Yoda, K., Kira, T., Ogawa, H. & Hozumi, K. (1963). Self thinning in overcrowded pure stands under cultivated and natural conditions. *Journal of Biology, Osaka City University,* **14**, 107–29.

IV. EVOLUTIONARY AND BEHAVIOURAL ECOLOGY

It is not surprising that most of the central tenets of evolutionary ecology were introduced by zoologists, because it is natural for them to think of such concepts as sexual selection and strategies of parental care. The 1980s have seen a considerable transfer of these ideas to the plant kingdom. Lloyd has spearheaded the adaptation of the concept of reproductive stategies to plants. He points to the need to link this new perspective with the classic studies of pollination and dispersal, as well as with morphological, taxonomic and general evolutionary aspects of reproduction. His approach is that of the modeller.

By contrast, Harvey and Pagel emphasize the fact that there is a wealth of information available in existing data bases which can be used to test the generality of new ideas about the basis of organic diversity. These ideas are emerging, in part, from specific experimental studies such as those described by Partridge, in which the central aim is to discover the costs of reproduction and their role in the evolution of life histories.

These three papers, taken together, illustrate well the interplay of theory, quantitative observation and experimentation which is also a theme of Sections III and V. In the present section, however, we also see the broadening of the ecological base of scholarship to overlap, if not embrace evolutionary and behavioural biology — an event which Darwin would have applauded as a return to his own style of thinking. As Harvey and Pagel point out, one discipline has been too inclined to view organisms as identical, while the other has perhaps been too ready to ignore environmental variation. It is a welcome development from which much progress is to be expected, particularly as techniques for field experimentation are refined.

8. THE REPRODUCTIVE ECOLOGY OF PLANTS AND EUSOCIAL ANIMALS

D. G. LLOYD

Department of Plant and Microbial Sciences, University of Canterbury, Christchurch 1, New Zealand

INTRODUCTION

Evolutionary biology was revolutionized in the mid-1960s by the development of a new paradigm, eventually labelled evolutionary ecology. This examines strategies for the 'optimal' deployment of adaptive mechanisms. Following the pioneering books by Williams (1966), MacArthur & Wilson (1967) and Levins (1968), diverse strategies have been explored, including those concerned with life-histories, phenology, sexual versus asexual reproduction, and mating patterns.

The present chapter primarily considers the evolutionary ecology of reproduction among seed plants. The reproductive ecology of plants has been revitalized by the concept of adaptive strategies. Most of the major innovations, such as Bateman's principle, sexual selection, sex allocations and parental care strategies, were developed first for animals and have been applied secondarily to the plant kingdom. The borrowings from zoology have been of enormous benefit to plant studies (Charnov 1982; Willson 1983). The coverage of strategy thinking in botanical studies is still very incomplete, however. Many aspects of plant reproduction have yet to be considered in depth from the new ecological perspective.

The central consideration of adaptive strategies is how individual organisms are selected to deploy their limited resources among various structures and behaviours. The diversity of evolutionarily stable strategies (ESSs) can be classified into two major superfamilies, allocation strategies and size–number strategies, which involve additive and multiplicative expenses respectively. Here I consider an example of each of the superfamilies, namely sex allocations and seed size, using phenotypic models of parental selection. The same methods of analysis are then extended to two topics that involve the death or sterilization (and zero fitness) of some progeny, self-incompatibility in plants and sterile castes in social animals. The two subjects present similar challenges to selection theory. It is argued that they can be explained in the same way as sex allocations and seed sizes, by the constrained optimization of parental fitness measured by the number of mature, fertile offspring in a parent's family.

ALLOCATION STRATEGIES

Allocation strategies deal with the deployment of an individual's resources among alternative expenses, where the total expenditure is the *sum* of the various allocations. This superfamily includes a wide range of plant and animal strategies, including those concerned with (i) life-histories (reproductive effort), (ii) phenology — the distribution of flowers, germinating seeds, etc., in the resource, time, (iii) dispersal — the distribution of diaspores or individual animals in space, (iv) breeding patterns, including sexual versus asexual reproduction, self- versus cross-fertilization, and male versus female investment, (v) habitat choice, (vi) optimal diets, and (vii) behavioural games. The relative expenditures on morphological structures in general can also be treated as a matter of ESS allocations.

All allocation strategies can be considered under a single theoretical framework (Lloyd 1987a), an extension of methods developed principally for sex allocations. Selection maximizes fitness subject to the constraint that the sum of expenditures is constant (assuming that all limiting resources are spent). The outcome of selection as a process of constrained optimization (Maynard Smith 1978) depends on two equations, the fitness and constraint equations. The fitness equation expresses the manner in which fitness varies as allocations and other variables change. If we consider the ability of a rare mutant of phenotype 2 to invade a resident population of phenotype 1, the fitness advantage of the mutant (cf. Hamilton 1967) is a function of the allocations by both phenotypes. That is, for any number of allocations a_i, b_i, c_i, . . . ,

$$w_2 - w_1 = f(a_2, b_2, c_2, \ldots a_1, b_1, c_1 \ldots). \quad (8.1)$$

The constraint equation may be written as

$$a_i + b_i + c_i + \ldots = 1. \quad (8.2)$$

This expresses the allocations as proportions, and assumes that all resources are spent and that each allocation can vary independently of the others (there is no additional constraint on expenditure).

For all such sets of allocations, the selected allocations can be obtained using a marginal advantage theorem (Lloyd 1988b). At the ESS, an increment in expenditure on any allocation brings the same rate of increase in the fitness advantage of a phenotype. That is, for three (or more) variables, the constrained ESS allocations $\hat{a}$, $\hat{b}$, $\hat{c}$, etc., obey the equation

$$\frac{\partial(w_2 - w_1)}{\partial \hat{a}_2} = \frac{\partial(w_2 - w_1)}{\partial \hat{b}_2} = \frac{\partial(w_2 - w_1)}{\partial \hat{c}_2} = \ldots \tag{8.3}$$

Many allocation problems, such as those dealing with sex ratios, consider only two alternative expenditures. Their proportions can be expressed in terms of one variable, as a and $1-a$. In this reduced case, the constraint equation is trivialized ($a + (1-a) = 1$, or $a = a$). The outcome of selection can be obtained by considering the fitness equation alone, on the assumption that the fitness advantage is maximized. That is,

$$\frac{\partial(w_2 - w_1)}{\partial \hat{a}_2} = 0. \tag{8.4}$$

If selection is treated as though it optimizes the fitness of a *single* phenotype rather than selects the winner(s) from a pool of competing phenotypes, eqn. (8.4) can be further simplified to

$$\frac{\partial w_2}{\partial \hat{a}_2} = 0, \tag{8.5}$$

the property that has been most frequently used to model allocation strategies, including sex allocations. Equations (8.4) and (8.5) give the same answer for non-invasibility of a monomorphic strategy, but a consideration of the fitness advantage is necessary to determine a polymorphic equilibrium and to portray selection realistically.

Reproductive allocations

Perhaps the most intensively studied allocations are those that partition parental investment into male and female components. The study of sex ratios began with the perceptive argument of Fisher (1930) that, in dioecious populations, parents are selected to spend equal resources on male and female offspring. This has served ever since as a null hypothesis for sex ratios. A considerable set of factors that can cause selection of unequal sex ratios is now known (Hamilton 1967; Charnov 1982) and is still being added to.

Sex allocation theory was first extended to hermaphrodites by Maynard Smith (1971). Through the efforts of Charnov (e.g. 1982), it is now widely recognized that sex ratios and maternal versus paternal expenditures of cosexes are two expressions of the same topic, sex allocation. This recognition is of paramount importance for botanists, because the large majority of plants have cosexual conditions (including hermaphroditism in the accepted, Linnaean sense) in which individuals act as both paternal and maternal parents. A number of phenomena first explored in the sex ratios of animals, such as local competition for mates or resources and fixed costs, have been successfully transferred to the study of sex allocations of cosexual plants (Charnov 1982; Lloyd 1984). Other phenomena of special interest in cosexes (e.g. self-fertilization, Charlesworth & Charlesworth 1981) or plants (e.g. pollen as a reward to pollinators as well as a type of 'progeny', Lloyd 1984 and below) have been added. The theoretical and empirical investigation of sex allocations in cosexual plants is now being energetically pursued.

Nevertheless, I offer the view here that the origin of studies of reproductive allocations of plants in the earlier studies of sex ratios of animals has misled botanists as well as benefited them. The reproductive allocations of cosexes involve much more than male versus female expenditures, and the topics of major interest in studies of sex ratios and cosexual allocations are only partly shared. The reason for this divergence is that in dioecious organisms, parents produce male and female offspring whose sexual activity does not begin until *after* parental investment ceases. In general the mating and parental care strategies of parents benefit sons and daughters equally, or almost so. In this situation, the strategies concerning the sex ratio of offspring are effectively decoupled from the mating and parental care strategies of the parents.

In cosexual species, on the other hand, reproductive allocations are concerned with the *current* paternal and maternal fitness of the parents themselves. Hence they must consider all factors that influence maternal and paternal fitness, including mating and parental care strategies in both gender roles. Many structures affect maternal and paternal fitness differently. Consequently, the relative allocations to the maternally and paternally derived progeny themselves are only a small part of the larger subject of reproductive allocations. In recent years, botanists have attempted to partition all reproductive expenditure into male and female categories. While this practice is adequate for theoretical considerations of the two-item subject of sex allocation, it is a gross simplification of the complex forces involved in the whole process of reproduction. In addition to the direct expenditure on the 'progeny', pollen and seeds, reproducing plants invest in a variety of 'accessory structures' (petals, sepals, fruit walls, flowering stems, etc.) which involve considerable costs. In a study of six cosexual species, my associates found that the dry weight of all accessory structures constituted between 42 and 90% of the reproductive costs (C. Aker, unpublished; R. Zyskowski, unpublished). The relative costs of the particular structures varied greatly among the species. The ratio of the weights of petals to those of supporting stems, for example, varied from 3.6 : 1 to 1 : 7.4, nearly a thirtyfold range.

We need to consider the contribution that each structure makes to maternal and paternal fitness. The marginal advantage theorem can be used to investigate theoretical expectations of expenditures on various reproductive structures. As an illustration, I will consider here expenditure on a single accessory structure (Lloyd 1988a) in addition to the investment in pollen and seeds. Suppose a plant can produce from a fixed reproductive investment either o pollen grains, n seeds, or floral stems of total size k. If the proportional allocations to these three items by phenotype i are a_i, b_i, and c_i respectively, the three expenditures are oa_i, nb_i and kc_i. If the stems are necessary to support flowers and subsequently fruit, paternal and maternal fitness may be assumed to be proportional to $(oa_i)^\alpha (kc_i)^\gamma$ and $(nb_i)^\beta (kc_i)^\kappa$ respectively. The exponents α, β, etc., describe the shapes of the fitness curves relating the fitness contribution from a structure to the allocation to that structure. Exponents less than one denote decelerating curves, and the size of such exponents describes the extent to which fitness continues to increase as allocations become large.

The fitness of an individual of either phenotype ($i = 1$ or 2) in a population in which a rare mutant of phenotype 2 is poised to invade a resident phenotype, 1, is the sum of its maternal and paternal fitness. If each plant has K mates and K is large,

$$w_i = (nb_i)^\beta (kc_i)^\kappa + K(nb_1)^\beta (kc_1)^\kappa \cdot \frac{(oa_i)^\alpha (kc_i)^\gamma}{K(oa_1)^\alpha (kc_1)^\gamma}.$$

The ESS allocations occur when the marginal advantages of increasing expenditure on each structure are equal. Calculating the fitness advantage, $w_2 - w_1$, and using equation (8.3) give ESS allocations in the ratio

$$\hat{a} : \hat{b} : \hat{c} = \alpha : \beta : \gamma + \kappa. \tag{8.6}$$

Each structure is allocated resources according to its own effects on parental fitness, and more particularly how much an increasing allocation continues to improve fitness. It is also evident that the two items of conventional sex allocations, a and b, refer only to the direct expenditure on pollen and seeds. The predicted ratio of expenditure on pollen and seeds, $\alpha : \beta$, is the same as that obtained by considering only those two items (Lloyd 1984). Hence if the various reproductive structures are selected independently, we can examine any subset of the total reproductive expenses. Even if the fitness curves for pollen and seeds have the same shape ($\alpha = \beta$) and the direct expenditures are equal, the total 'male' and 'female' expenditures need not be equal ($\hat{M} : \hat{F} = \alpha + \gamma : \beta + \kappa$; Lloyd 1988b).

If a structure has more than one effect, as stems do in the example above, it receives an allocation for each effect and its total allocation is based on the sum of its separate contributions. This may cause further deviations from an expectation of equal sex allocations. Suppose that seeds (but not pollen) perform two necessary functions, namely dispersal and establishment, which are achieved by components (seed coats and embryos) that receive allocations d_i and e_i. If maternal fitness is proportional to d_i^δ multiplied by e_i^ε, and pollen fitness is proportional to a_i^α, the marginal advantage theorem readily shows that, at the ESS,

$$\hat{a} : \hat{d} : \hat{e} = \alpha : \delta : \varepsilon. \tag{8.7}$$

If the exponents are equal, the ratio of direct male and female expenditure $[\hat{a} : (\hat{d} + \hat{e})]$ is now 1 : 2 rather than 1 : 1. Conversely, if we assume that pollen has two functions and serves as a reward for pollinators as well as providing gametes, its allocation will be correspondingly increased. Then, however, not even the direct expenditure on pollen and seeds can be strictly considered as 'male versus female' allocations. A similar situation occurs when some ovules are self-fertilized (Lloyd 1987b).

Further complications arise frequently in cosexual species. These include situations where paternal and maternal investments are limited by different types of resource (McGinley & Charnov 1988; D. G. Lloyd, unpublished), and the fact that paternal and maternal investment occur largely at different times and may therefore come from resource pools of different size (Charlesworth & Charlesworth 1987). We can conclude only that the numerous reproductive allocations of cosexes are routinely affected by a complex set of factors that juggle the allocations in response to all facets of reproductive ecology. Fisher's 'matching rule' (Lloyd 1987a), equal male and female expenditures for equal rewards, is only one of a multitude of factors routinely affecting reproductive allocations. Consequently, it plays a less valuable role as a null hypothesis for reproductive allocations in cosexes than it does for sex ratios. Moreover, in the next section we see that selection does not act on allocations for total seed or pollen expenditure as such. The variables that selection actually operates on are the components of the 'allocations', which are selected together with the components of other allocations in the competition for limited resources.

The method of analysis used here, the marginal advantage theorem, and the conclusion that each separate structure must be considered for its own summed effects are equally applicable to other sets of allocations, including those to any set of morphological structures.

SIZE–NUMBER STRATEGIES

The other major superfamily of strategies is concerned with subdivision of allocations into units whose size and number are *multiplied* together to give the expenditure to that allocation. Size–number compromises are most frequently considered for offspring or gametes (seeds and pollen in the case of seed plants). They are also involved in a variety of other contexts, such as the number of branches or leaves a plant bears, the number and duration of matings a male insect attempts in a given time (Parker 1978) and the number and duration of visits to habitat patches that a foraging animal makes (Charnov 1976).

Size–number strategies may involve only one level of components, as assumed in almost all models to date of offspring size versus number (e.g. Smith & Fretwell 1974; Brockelman 1975; Parker & Begon 1986; Temme 1986; Lloyd 1987c). The repetitive organization of plants, however, causes particular structures to be produced many times, providing a number of levels of subdivision. Pollen, for instance, is not produced in a single gonad or a pair, but in a number of sporophylls (stamens) that are organized into flowers that are in turn grouped into inflorescences whose structures may themselves be complex. Similarly, seeds are packaged into sporophylls (carpels) that are combined into fruits organized in infructescences. The total number of pollen grains or seeds can be varied by altering the number of units at any level. These multi-level size-versus-number strategies may be referred to as 'packaging strategies' (Lloyd 1987c).

The single- and multi-level size-versus-number strategies are subject respectively to the constraints,

$$s_i n_i = R, \tag{8.8a}$$

or

$$s_i p_i q_i \ldots = R, \tag{8.8b}$$

where the total amount of available resources is R, the amount of resources invested in each ultimate unit (its size) by an individual of phenotype i is s_i, and the number of units is n_i, or if these are packaged at several levels the number at each level is p_i, q_i, etc. Since each ultimate unit is a part of the package at every level, the fitness of the producing phenotype is the product of the fitness contributions from all levels. The fitness contribution of each level is a function of the number of components at that level. Thus for single-level size–number compromises,

$$w_i = f(s_i)g(n_i), \tag{8.9a}$$

and for multi-level strategies,

$$w_i = f(s_i)h(p_i)k(q_i)\ldots \tag{8.9b}$$

The expected size and the numbers at all packaging levels can be obtained simply with the appropriate form of the same marginal advantage theorem that was used above for allocation strategies. At the ESS for any number of packaging levels, increasing expenditure on the size of the ultimate unit or on any packaging component brings the same rate of increase in the fitness advantage of a phenotype (Lloyd 1988b). For single-level compromises, again considering the invasibility of a resident phenotype 1 by a rare mutant of phenotype 2,

$$\left[\frac{\partial(w_2 - w_1)}{\partial\hat{s}}\right]\left[\frac{1}{\hat{n}}\right] = \left[\frac{\partial(w_2 - w_1)}{\partial\hat{n}}\right]\left[\frac{1}{\hat{s}}\right]. \tag{8.10}$$

Substituting the general fitness functions of (8.9a) into (8.10) gives

$$\left[\frac{f'(\hat{s})}{f(\hat{s})}\right]\hat{s} = \left[\frac{g'(\hat{n})}{g(\hat{n})}\right]\hat{n}, \quad \text{or} \; \frac{f'(\hat{s})}{g'(\hat{n})} = \frac{f(\hat{s})}{\hat{s}} \bigg/ \frac{g(\hat{n})}{\hat{n}}. \tag{8.11}$$

Since $f(\hat{s})/\hat{s}$ and $g(\hat{n})/\hat{n}$ are the average returns per unit of expenditure, the ratio of the marginal returns from increasing size or number at the ESS is equal to the ratio of the average returns already attained. Equations that are equivalent to (8.10) and (8.11) can readily be written for multi-level strategies. In the case of single-level compromises (but not multi-level ones), the constraint specified in (8.8a) can be trivialized by writing the number of units as a function of their size and the available resources ($n = R/s$) and treating selection as a process of simple optimization of size. As for allocation strategies, most authors have adopted this technique and obtained the ESS by solving for $\partial(w_2 - w_1)/\partial\hat{s}_2 = 0$, or equivalently for rare mutants $\partial(w_2/\partial\hat{s}_2) = 0$.

Seed size-versus-number

As a specific example, I will examine the compromise between seed size and number. For simplicity, the topic will be treated as a single-level compromise, although this entails omitting the selective forces that are involved in packaging seeds successively into carpels, fruit and infructescences. The strategies will be considered first in terms of parental fitness (cf. Lloyd 1987c), and then from the viewpoint of the seed offspring.

It can readily be shown that, for an intermediate number of seeds (neither one nor infinity) to be the parental optimum, the fitness curve relating the success of individual seeds to their size must at first decrease slowly or not at all, and then increase rapidly before the rate of increase declines again (Fig. 8.1(a)). For a convenient approximation to this shape, assume that seeds have no chance of survival until they reach a minimum size, s_m. A minimum size might be imposed by the fixed cost of seed coats or a lack of competitive ability of smaller seeds, for example. The fitness of individual offspring is assumed to be proportional to their size beyond s_m raised to a power, x (Fig. 8.1(b)). Hence $f(s_i) = k_s(s_i - s_m)^x$. When the fitness of a seed parent is proportional to the number of offspring it produces, $f(n_i) = n_i$. Substituting in (8.10) or (8.11) for $i = 1$ and 2 and thus equating the marginal fitness advantages of increasing size and number gives a constrained optimum for the parents at

$$\hat{s} = \frac{s_m}{1-x}. \tag{8.12}$$

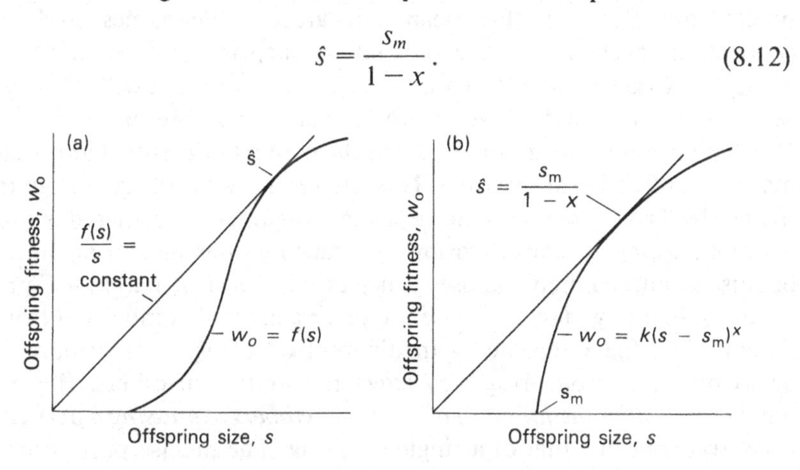

FIG 8.1 The selection of a single-level size–number compromise: (a) the general function, $w_0 = f(s)$ for the fitness curve relating the size of offspring at the end of parental investment, s, to their fitness w_0; (b) a specific model for the offspring fitness curve, $w_0 = k(s - s_m)^x$.

An intermediate seed size ($s_m < \hat{s} < R$) is selected if $x < 1$. The optimal size does not depend on the amount of parental resources, R. Hence variation in resources should be tracked by variation in offspring number rather than size, as both botanists and zoologists have often observed since the pioneering studies of Salisbury (1942) and Lack (e.g. 1954). The model also supports the hypothesis that the size of offspring at independence is related to the size at which they can 'procure food' (Itô 1980) or more generally become 'durable' in their natural environment (Lloyd 1987c). This hypothesis helps to explain correlations between the size of offspring at independence (seed maturation for plants) and various ecological parameters, such as K-strategies, parental size, physically demanding juvenile habitats, etc. The power of the current models to explain these general correlations is considerable (Lloyd 1987c). But as shown by Harper, Lovell & Moore (1970), the average seed size of seed plants varies over more than ten orders of magnitude! We have made little progress towards explaining exactly where on this scale the seedlings of each species become sufficiently durable to maximize parental fitness.

So far, seed strategies have been examined solely from the maternal parent's perspective. As long as the fitness curve relating the success of individual offspring to their size increases at levels of investment above the parental optimum, individual offspring will benefit from greater parental investment than the parent is selected to give. In this sense, there is a conflict between parents and *single* offspring, as Trivers (1974) pointed out. But does this mean that 'greedy' phenotypes (and their controlling genes) that are expressed in offspring and enhance their ability to extract resources beyond the parental optimum will actually be selected and cause a deviation from the parental optimum?

That question is examined here in the context of plants that produce many seed offspring at one time. This situation is very different from that treated by Trivers (1974), who in his key argument considered a single (caribou) young. A consideration of larger families has an advantage, because it immediately exposes a neglected aspect of parent–offspring conflicts. In a large family, a phenotype or gene that is inherited from a parent will appear repeatedly in different offspring. Furthermore, the behaviour of one offspring may affect the welfare of others. *We must therefore consider the fitness of all family members expressing a particular phenotype*, not just that of a single offspring. The precise proportion of family members expressing a phenotype will depend on the genotypes of the parents, the pattern of inheritance, whether the controlling alleles come from the mother or father, and in the latter case whether the family is derived from one or more than one father and the frequency of the alleles. Here the strategies of offspring are examined by considering the selection of part-families that share a phenotype.

Suppose that seed parents have been selected to nurture seeds to size s_1, as in the model above. Now assume that among a crop of seeds being nurtured by one parent, a fraction t of the seeds have a genetically determined phenotype, 3, that enables those seeds to extract more resources from their seed parent, i.e. $s_3 > s_1$. Assume too that there are no environmental differences in the position or timing of seeds affecting the ability of different seeds of the crop to extract resources and that seeds do not differ in the quality of their genotypes, which might affect the parental strategy (Temme 1986). Since all seeds of the same phenotype then reach the same size, we consider the fitness of members of those parts of the maternal family that reach sizes s_1 and s_3, having 'unselfish' and 'greedy' phenotypes respectively. The number of seeds that a parent can produce depends on the proportions of the two phenotypes in a crop. If all seeds produced by a family reach size s_1, the size–number constraint function is $R = ns_1$ or $n = R/s_1$. If a fraction t of seeds reaches size s_3, $n = R/[s_3t + s_1(1-t)]$. Assume that the curve relating the size and success of individual seeds increases monotonically and, as before, $f(s_i) = \mathrm{k}(s_i - s_m)^x$. We can now examine offspring strategies by comparing the fitnesses of part-families of phenotypes 1 and 3 that constitute the same fraction, t, of a seed crop, assuming the rest of the seeds have the resident strategy, $s_1 = s_m/(1-x)$. This examines the question of whether the members of a part-family should comply with the parent's strategy or extract more resources. The part-family fitnesses are

$$w_{p1} = \left[\frac{Rt}{s_1}\right]k(s_1 - s_m)^x$$

and

$$w_{p3} = \left[\frac{Rt}{s_3t + s_1(1-t)}\right]k(s_3 - s_m)^x .$$

There is no advantage to a part-family from increasing the rate of extraction of resources when $\partial(w_3 - w_1)/\partial s_3 = 0$. Then the value of s_3 that has the maximum advantage over other seed sizes for members of the part-family is

$$s^* = \frac{s_mt + xs_1(1-t)}{t(1-x)} \qquad (8.13a)$$

$$= \left[\frac{s_m}{1-x}\right]\left[1 + \left(\frac{1-t}{t}\right)\left(\frac{x}{1-x}\right)\right], \qquad (8.13b)$$

since $s_1 = s_m/(1-x)$. The (constrained) optimal strategy for members of a part-family that share a phenotype is indeed different from that of the seed parent, but it is also different from that of a single offspring treated as though it can vary by itself (for which, $s \rightarrow R$). The strategy for members of a part-family is not to grow larger indefinitely, unless the part-family comprises only one seed (when $t \rightarrow 0$). As the representation of a greedy phenotype in a family increases, greediness becomes progressively disfavoured because it increasingly reduces the number of seeds, including those of its own phenotype. When all seeds in a crop are identical ($t \rightarrow 1$), from (8.13b) the optimal offspring size, $s^* \rightarrow s_m/(1-x)$, the size favoured by the parent (8.12).

We now ask whether selection on an offspring phenotype can enforce a seed size, $s^* > \hat{s}$, contrary to the seed parent's interests. The answer is that it cannot because several factors prevent selection of s^*. First, the value of s^* depends on the fraction of a family expressing the over-extracting phenotype, t. If the allele(s) for s^* are not fixed in the population, the value of t will vary greatly among seed parents at any one time, depending particularly on the allele frequencies and whether the alleles in a family come from the seed parent or any number of the pollen parents. The flexible strategy required to reach s^* is unattainable, since no juvenile plant (or animal) could in practice adjust its resource–extraction strategy in response to its frequency in a family. Second, suppose though that such a conditional strategy were possible and the genes and phenotypes responsible invaded a population and increased in frequency. As they increased, their optimal strategy would converge on that of the seed parent. Third, consider the interests of a modifier gene that prevents greedy behaviour (reduces s_3 to s_1) in a family in which some seeds endeavoured to extract more resources than others ($0 < t < 1$). An allele of a modifier gene that is unlinked to the loci controlling the seed extraction phenotypes will appear equally frequently in seeds of different extraction phenotypes. Hence the optimal seed size for a modifier allele will coincide with that of the total family—that is, with the parent's strategy. A modifier allele that enforces the parental strategy will be favoured over one that does not. A modifier that prevents the greediness of an unconditionally greedy phenotype ($s_3 > s_1$ for any t) that has invaded a population would also be favoured by selection.

The conclusion from the parental and offspring models is that there may be a conflict between the optimal offspring size selected by genes expressed in parents and their offspring, depending on the proportion of offspring expressing a phenotype in a family. Such conflicts are essentially between family segments containing differing numbers of offspring, not between parents and their own offspring. In these conflicts, the investing parent's interests in the whole family will be upheld. The reasons offered here for supremacy of the whole-family parental strategy are quite different from the idea suggested earlier by Alexander (1974), that the parents' interests win because parents have control over their offspring.

The arguments presented above apply equally to families in which a single offspring per year is produced over a number of years, when parental resources 'overspent' in one year are subtracted from those available for reproduction in subsequent years. There too, the investment per young favoured by an investing parent will coincide with the collective interests of its whole family and those of any modifier genes and will therefore be evolutionarily stable. Similar arguments apply to families in which over-extracting offspring reduce the size rather than the number of the remainder of the family. In all these contexts, the offspring collectively *are* the parent's fitness, and the selection interests of investing parents and their whole families coincide.

Apparent instances of parent–offspring conflict, such as seed abortion or begging by young birds, may be best regarded as an exchange between two rules-of-thumb. Each progeny 'asks for as much as it can benefit from', and parents 'give as much as they can' according to their current resources and competing commitments. The outcome of such interplays maximizes, as nearly as possible, the lifetime fitness of a parent and the total fitness of its family in a situation where neither participant has all the information necessary for an optimal decision.

The principal reason that previous quantitative models of parent–offspring conflict have concluded that greedy offspring are selected against the parent's strategy appears to be that these models consider only the initial increase of greedy offspring, e.g. the models of Stamps, Metcalf & Krishnan (1978), Parker & MacNair (1978), and Law & Cannings (1984). Deductions of a parent–offspring conflict in that situation are not at variance with the present model (cf. equations (8.12) and (8.13) when $t < 1$), but they do not include the full effects of a number of greedy offspring. It is these that have led me to conclude that the parental and whole-family strategies coincide.

Whenever parents and offspring compete for reproductive contributions to the gene pool, however, a genuine parent–offspring conflict could be realized. Then the offspring are acting not as a source of fitness for their parents, but as a potential threat to the parent's contribution to the gene pool of subsequent generations. A kin selection model that adds the effects on parents and offspring is the appropriate means of analysing such situations, and could reveal when a parent's lifetime fitness is increased by actions to its own detriment. The reproductive values of the parent and its offspring would have to be taken into account. An intriguing case where parent–offspring competition may have caused selection of a shorter lifespan in a rosette plant has been described by Smith (1984).

SELF-INCOMPATIBILITY

Parental selection is also appropriate for the selection of self-incompatibility in angiosperms. In the simpler gametophytic systems, pollen that carries a particular S allele cannot grow down a style containing the same allele. Self-fertilization is therefore restricted, but as there are many alleles in a population, cross-fertilization is scarcely affected. The system generates a remarkable but rarely appreciated enigma (Charnov 1979; Beach & Kress 1980; Hamilton 1987) that challenges the theory of individual selection as severely as the occurrence of sterile castes in animals (Darwin 1859; Hamilton 1964; and below). Why should a pollen grain express an S allele whose only effect is to cause the suicide of its carrier when it is deposited on an incompatible stigma?

The solution to this paradox is already present in mathematical treatments of the selection of cross- and self-fertilization. A number of authors, starting from Fisher (1941), have considered the fitness of cosexual individuals in terms of the sum of fitness contributions gained from self-fertilized ovules, cross-fertilized ovules, and outcrossed pollen — three distinct families of the same individual. As an illustration, I will follow the formulation for 'prior selfing' in a previous publication (Lloyd 1979). In an individual of phenotype i, a proportion, b_i, of the n ovules are self-fertilized *before* a fraction e of the remaining ovules is cross-fertilized. Cross-fertilized ovules have a fitness of one, and selfed ovules have fitness $1-\delta$, where δ is the inbreeding depression (cf. Charlesworth & Charlesworth 1978). Assume that variation in the proportion of selfed ovules does not affect the amount of pollen, g, available to donate pollen in cross- fertilizations with K mates, where K is large. Again selection is examined as the outcome of competition between a resident phenotype (1) and a mutant phenotype (2). The fitness of phenotype i = 1 or 2 is then the sum of the contributions of the three families. That is,

$$w_i = n\left\{2b_i(1-\delta) + e\left[(1-b_i) + \frac{K(1-b_i)g}{Kg}\right]\right\}. \tag{8.14}$$

Putting $i = 1$ and 2 and subtracting gives the fitness advantage for the mutant,

$$w_2 - w_1 = n(b_2 - b_1)[2(1-\delta) - e]\,. \qquad (8.15)$$

In this simple situation, crossing is selected ($w_2 > w_1$ when $b_2 < b_1$) whenever $e > 2(1-\delta)$. The level of cross-pollination is then sufficient to outweigh the twofold 'automatic advantage of selfing' discounted by inbreeding depression. The selection models that have previously been offered to explain the occurrence of cross- and self-fertilization, e.g. those of Nagylaki (1976) and Uyenoyama (1988a), are actually multi-family models, although we have not regarded them in this way. In certain circumstances the loss to the (male and female) selfed family that self-incompatibility causes is compensated for by the gain to the outcrossed seed family. There is a conflict between the two families of the parent, and it is resolved in self- incompatible plants at the expense of self-pollen but to the benefit of the combined pollen and seed families. It is more accurate to interpret selection for self-incompatibility in terms of the sum of the effects on the outcrossed maternal family and the selfed family than to consider it as the outcome of a conflict between male and female parenthood as proposed by Willson (1983), or a conflict between sporophyte and gametophyte generations as suggested by Beach & Kress (1980), or solely as altruism on the part of pollen grains as put forward by Hamilton (1987), or solely as progeny choice by the female parent as indicated by Uyenoyama (1988a, b). It is simply parental selection.

STERILE CASTES IN EUSOCIAL ANIMALS

Parental selection can also be applied to the occurrence and characteristics of sterile castes in eusocial animals. Darwin (1859, p. 242) considered these castes to be 'by far the most serious special difficulty, which my theory [of natural selection] has encountered'. It is difficult to see how selection can generate sterile castes when it favours individuals with the highest reproductive success. In one of his most ingenious arguments, Darwin proposed that 'This difficulty . . . disappears, when it is remembered that selection may be applied to the family, as well as to the individual' (1859, p.237). Accordingly, Darwin briefly outlined how the occurrence of sterile castes, their differences from the fertile sexes, and above all differences among distinct sterile castes could be explained as advantageous to whole colonies.

Since the development of the concept of kin selection by Hamilton (1963, 1964), the occurrence of sterile castes, particularly in the haplo-diploid Hymenoptera, has been explained by a kin selection hypothesis. In haplo-diploid species, a worker is more closely related to its sisters ($r = 3/4$) than to its own daughters ($r = 1/2$), and hence increases its inclusive fitness by producing sisters rather than daughters. The kin selection hypothesis has been widely accepted, and indeed the accompanying explanation of the prevalence of haplo-diploids among eusocial animals (Wilson 1971; Trivers 1985) has been considered to be the major triumph of kin selection theory. Family selection, Darwin's alternative hypothesis, has been ignored or considered equivalent to the kin selection hypothesis (Wilson 1975; Maynard Smith 1986). Here I re-examine the two hypotheses with the aid of a quantitative model, and conclude that parental (= family) selection offers a valid explanation and that the kin selection hypothesis is not applicable when workers become common.

The kin selection argument that workers in haplo-diploid species favour assistance to sisters because they are more closely related to them than to their own daughters cannot be accepted in its (usual) simplest form, because taken at face value it leads to the conclusion that all diploid daughters in a colony should be sterile. The advantage of being a worker (and the advantage to a colony from producing additional workers) should decrease as the proportion of workers rises, for two reasons. The benefit of the further aid that each additional worker gives to each sib is likely to diminish, at least at higher frequencies of workers when juveniles are well cared for. Moreover, the number of the fertile offspring that provide the female fitness of a colony decreases as the number of workers increases. Consequently, the combined altruistic effects of numerous workers on their joint inclusive fitness are not additive. In such situations, Hamilton's rule does not give an accurate prediction of the outcome of selection (see Charlesworth 1980). The rule is accurate for single workers and is therefore applicable to the initial origins of eusociality, which Hamilton (1964, 1977, personal communication 1988) emphasized. But we also wish to address such matters as the optimal proportions of castes and their distinctive properties in advanced eusocial species. Then we must consider the *collective* fitnesses of the various part-families and the parental fitness, as in the cases of parental care and self-incompatibility considered above.

Parental fitness will be examined first in a simple formulation of Darwin's hypothesis. If we assume that the frequency of male progeny in a colony and the ratio of fertile females to sterile workers are selected independently of each other, we can examine a subset of the allocations, the relative proportions of females and workers (see the discussion of eqn

(8.6) above). Consider a population of social animals in which females mate only once and a single female establishes a colony in which the non-male progeny consist of fertile females and sterile workers in the proportions f_i and s_i, respectively. Assume that a single generation of females and workers grow up together, and the females then mate and found new colonies. Hence the fitness of parents (male or female) obtained by producing females is calculated over only one generation of offspring. A family can mature n females and/or workers. If workers are produced, parental fitness, w_i, measured in terms of parental genes transmitted to the mature female offspring, is a product of four items — family size, the relatedness of parents to their offspring, r_{po}, some function of the frequency of females, $g(f_i)$, and the effect of worker assistance on the survival or fecundity of females, which is a function of the frequency of workers, $h(s_i)$. That is,

$$w_i = nr_{po}g(f_i)h(s_i)\,.$$

To find the proportions of females and workers that maximize parental fitness, the fitness of a resident phenotype (1) that produces a proportion f_1 of females is compared with the fitness of a mutant of phenotype 2 that produces a slightly higher proportion, f_2, of females. The fitness advantage of the mutant over the resident phenotype is

$$w_2 - w_1 = nr_{po}[g(f_2)\,h(s_2) - g(f_1)h(s_1)]\,.$$

At the parental optimum when the marginal advantages of producing more females and workers are equal, from (8.3),

$$\frac{g'(\hat{f})}{g(\hat{f})} = \frac{h'(\hat{s})}{h(\hat{s})}\,. \tag{8.16}$$

The proportionate gains from increasing workers and females are then equal. The nature of the optimum can be seen more clearly by substituting particular functions in (8.16). Suppose parental fitness is directly proportional to the frequency of female progeny, that is $g(f_i) = f_i$, and workers cause an increase in female survival rate that is proportional to the frequency of workers. Let $h(s_i) = 1 + ks_i$. Substituting in (8.16) gives

$$\hat{f} = \frac{1+k,}{2k} \quad \text{or } \hat{s} = \frac{k-1}{2k}\,. \tag{8.17}$$

Parental fitness is not increased by the presence of workers unless $k > 1$, when workers increase the survival of females by an amount that compensates for their own sterility.

The relatedness of the workers to the fertile females, r_{sf}, does not appear in the equation for parental fitness or those for the optimal proportions of female and worker progeny, since parental fitness considers only the transmission of genes from parents to fertile offspring. This result contrasts with the usual conclusion obtained from consideration of the inclusive fitness of individual workers, thus demonstrating that Darwin's hypothesis and Hamilton's kin selection hypothesis are not alternative versions of the same argument. Even the relatedness of parents to their offspring does not affect the outcome of parental selection, because parents are assumed to be equally related to their fertile and sterile offspring. The model is applicable to diploid and haplo-diploid species.

A parental selection hypothesis can explain the existence of sterile individuals in a population and it can predict the ESS proportions of sterile castes as a function of the assistance they give to the welfare of fertile colony members. Moreover, the differences between sterile castes and the fertile sexes, or between different sterile castes, can readily be explained in terms of benefits that accrue to parents through their families (or to genes that are passed on by fertile members from one family to another in successive generations). The evolution of sterile castes is analogous to the differentiation of the various structures of one individual, as Darwin (1859) recognized when he wrote of the 'division of labour' in insect colonies.

The model developed here has been described hitherto as one of parental selection, assuming that parents control the reproduction of their offspring and that the genes determining the strategy act in the parents. But the parental strategy describes the maximization of the fitness of the whole family. Consequently the strategy of the offspring collectively coincides with that of the parents. Hence the strategy derived above will also be obtained by genes acting in the offspring. In contrast to the situation with seed size considered previously, there is only one fertile part-family under consideration and there is no selection conflict between the fertile members of the family and their parents.

According to the model of parental (family) selection, the relatedness of workers and females does not influence the selection of a sterile caste. Thus the occurrence of sterile castes in diploid termites and naked mole rats and in asexual aphids (see Trivers 1985), as well as haplo-diploid Hymenoptera, is easily explained by a parental selection hypothesis. This hypothesis suggests that the distribution of eusociality should be sought in ecological factors rather than the genetic system (cf. Crozier 1984;

Stubblefield & Charnov 1986). These include the tendencies of various groups to evolve co-operative care of the young or to form colonies with multiple overlapping generations, or the effectiveness of care that workers give to the offspring of others compared with the care provided by fertile females (embodied in k above) or predation (Strassman, Queller & Hughes 1988).

There are further analogies between caste structure of the social insects and the reproductive biology of plants. The endosperm of angiosperms is a sterile tissue (in genetical terms, an individual) and therefore it is reproductively a dead-end, like a sterile caste. Several authors have recently postulated that the properties of endosperm can be understood in terms of kin selection arguments, e.g. Charnov (1979), Westoby & Rice (1982), Law & Cannings (1984), Queller (1984) and Haig & Westoby (1988). But, in this case too, the effect of the sterile individual can only be considered via the transmission of the genes of the male and female parents, which are fully considered in models of parental selection. And again, as in the cases of self–incompatibility and sterile workers, there are alternative hypotheses available that can explain the existence of double fertilization and the nature of endosperm in terms of parental selection. One evident advantage of double fertilization is that it causes expenditure on the endosperm and embryo to be laid out only when fertilization occurs and there is an offspring present that benefits from the expenditure. This does not happen in the less efficient system of gymnosperms (Westoby & Rice 1982). The expenditure trigger benefits both male and female fitness, as it must do because both paternal and maternal nuclei contribute to endosperm.

DISCUSSION

The models presented above have been predicated on the assumption that natural selection is a process of constrained optimization. The only constraint explicitly included in the calculations is the virtually universal one, a finite amount of resources. A more complete consideration of the selection of the reproductive features of plants would include other structural constraints where appropriate. On occasions, these may be of a historical, genetical, developmental or behavioural nature (Maynard Smith *et al.* 1985). Whenever these structural constraints can be formulated algebraically, they can be included in quantitative models of evolutionarily stable strategies, as was done above for the resource constraint. The realization that explanations of biological characters as the product of natural selection can incorporate structural constraints whenever they apply, enlightens the controversy concerning adaptationist versus structuralist explanations, as put forward by Gould & Lewontin (1979). It is artificial to make a dichotomy between separate adaptationist and structuralist 'programmes'. Satisfactory explanations of the evolution of biological properties should involve a synthesis of structuralist and adaptationist elements, not a choice between them (cf. Maynard Smith 1978; Krebs & McCleery 1984). The concept of selection as constrained optimization regards selection as providing not the best of all possible worlds, but only the best of those attainable.

The major instrument employed here to examine the constrained optimization of reproductive characters is the marginal advantage theorem, the principle that at the ESS increments in any component of any allocation bring the same rate of increase in fitness advantage. In a biological context, this has been proved formally only for the forms of allocation and size–number constraints used in equations (8.2) and (8.8) and their combination (Lloyd 1988b). Intuitive proofs have been provided in the past for theorems that are equivalent to the marginal advantage theorem, in both economics (e.g. Jevons 1879) and biology (for the ideal free distribution for habitat choice, Fretwell & Lucas 1970). These run along the lines that if allocations (or their size–number components) do not bring the same marginal rates of return, selection or the constrained maximization of utility will change them until they do so—or an item is excluded. It will be shown elsewhere that the marginal advantage theorem holds for any single continuous constraint function, but it does not always apply when the number of constraints equals or exceeds the number of variables under selection (D. G. Lloyd & D. L. Venable, unpublished).

The particular approach used here to examine the selection of reproductive allocations, size–number compromises, self-incompatibility and sterile castes is the consideration in each case of parental fitness. This involves simply summing the contributions that each fertile offspring makes to parental fitness. It is appropriate to do this because the fitness of a parent comes from the contributions its fertile descendants make to the gene pool of subsequent generations. The parental fitness of an individual is simply the summation of its fertile progeny, weighted where appropriate by their reproductive values. The summation of parental fitness can be made over a single one-generation family as in the models of seed size–number strategies, over a number of families considered for a single generation as in reproductive allocations and self–incompatibility, or over a multi-generation family as in more realistic models of colonies of social animals. Models of parental selection (= family selection) are routinely used for analysing a wide variety of reproductive strategies. Models of sexual versus asexual reproduction, sex ratios, and the evolution of dioecy and heterostyly, to name just a few topics, are models of family selection — a form of individual selection, not group selection. The models presented above may have reached some unorthodox conclusions, particularly for offspring size and number and sterile castes, but their formulation in terms of parental selection conforms to a well-established approach.

Reproductive topics such as parent–offspring conflict and sterile individuals in both plant and animal kingdoms have previously been explained by kin selection, using Hamilton's (1963, 1964) rule. In these behaviours, however, a number of family members are doing the same things, whether they are begging juveniles, sterile workers or dying pollen grains. The behaviour of each individual affects the consequences of altruism or selfishness by other family members. Kin selection arguments conventionally deal with decisions of individuals considered separately and make no allowance for interaction between simultaneous actors. They are therefore inappropriate for situations where a number of family members are acting simultaneously. Parental selection considers events in the way that genes are passed on, from parents to their fertile descendants, and incorporates family interactions. It is therefore the simplest and most illuminating procedure for analysing the selection of events that involve a number of family members acting simultaneously.

ACKNOWLEDGMENTS

I thank Peter Grubb, Bill Hamilton, Curt Lively and an anonymous referee for their helpful comments on a draft of the manuscript.

REFERENCES

Alexander, R. D. (1974). The evolution of social behavior. *Annual Review of Ecology and Systematics*, **4**, 325–83.

Beach, J. H. & Kress, W. J. (1980). Sporophyte versus gametophyte: a note on the origin of self-incompatibility in flowering plants. *Systematic Botany*, **5** , 1–5.

Brockelman, W. Y. (1975). Competition, the fitness of offspring, and optimal clutch size. *American Naturalist*, **109**, 677–99.

Charlesworth, B. (1980). Models of kin selection. *Evolution of Social Behavior: Hypotheses and Empirical Tests* (Ed. by H. Markl), pp. 11–26. Verlag Chemie GmbH, Weinheim.

Charlesworth, B. & Charlesworth, D. (1978). A model for the evolution of dioecy and gynodioecy. *American Naturalist*, **112**, 975–97.

Charlesworth, D. & Charlesworth, B. (1981). Allocation of resources to male and female functions in hermaphrodites. *Biological Journal of the Linnean Society*, **15**, 57–74.

Charlesworth, D. & Charlesworth, B. (1987). The effect of investment in attractive structures on allocation to male and female functions in plants. *Evolution*, **41**, 948–68.

Charnov, E. L. (1976). Optimal foraging, the marginal value theorem. *Theoretical Population Biology*, **9**, 129–36.

Charnov, E. L. (1979). Simultaneous hermaphroditism and sexual selection. *Proceedings of the National Academy of Sciences*, **76**, 2480–4.

Charnov, E. L. (1982). *The Theory of Sex Allocation.* Princeton University Press.

Crozier, R. H. (1984). On insects and insects: twists and turns in our understanding of the evolution of eusociality. *The Biology of Social Insects* (Ed. by M. D. Breed, C. D. Michener & H. E. Evans), pp. 4–9. Westview Press, Boulder, Colorado.

Darwin, C. (1859). *The Origin of Species.* Facsimile edition. Harvard University Press, Cambridge, Massachusetts (1964).

Fisher, R. A. (1930). *The Genetical Theory of Natural Selection.* Clarendon Press, Oxford.

Fisher, R. A. (1941). Average excess and average effect of a gene substitution. *Annals of Eugenics*, **11**, 53–63.

Fretwell, S. D. & Lucas, H. L. (1970). On territorial behaviour and other factors influencing habitat distribution in birds. I. Theoretical development. *Acta Biotheoretica*, **19**, 16–36.

Gould, S. J. & Lewontin, R. C. (1979). The spandrels of San Marco and the Panglossian paradigm: a critique of the adaptationist programme. *Proceedings of the Royal Society*, **B**, **205**, 581–98.

Haig, D. & Westoby, M. (1988). Inclusive fitness, seed resources, and maternal care. *Plant Reproductive Ecology: Patterns and Strategies* (Ed. by Lovett Doust & L. Lovett Doust), pp. 60–79. Oxford University Press.

Hamilton, W. D. (1963). The evolution of altruistic behavior. *American Naturalist*, **97**, 354–6.

Hamilton, W. D. (1964). The genetical evolution of social behaviour. I & II. *Journal of Theoretical Biology*, **7**, 1–16 & 17–51.

Hamilton, W. D. (1967). Extraordinary sex ratios. *Science*, **156**, 477–88.

Hamilton, W. D. (1977). Review of E. O. Wilson (1975): *Sociobiology: the New Synthesis. Journal of Animal Ecology*, **46**, 975–83.

Hamilton, W. D. (1987). Discriminating nepotism: expectable, common, overlooked. *Kin Recognition in Animals* (Ed. by D. J. C. Fletcher & C. D. Michener), pp. 417–37. Wiley, New York.

Harper, J. L., Lovell, D. H. & Moore, K. G. (1970). The shapes and sizes of seeds. *Annual Review of Ecology and Systematics*, **1**, 327–56.

Itô, Y. (1980). *Comparative Ecology.* Cambridge University Press.

Jevons, W. S. (1879). *The Theory of Political Economy*, 2nd edn. Macmillan, London.

Krebs, J. R. & McCleery, R. H. (1984). Optimization in behavioural ecology. *Behavioural Ecology: an Evolutionary Approach*, 2nd edn. (Ed. by J. R. Krebs & N. B. Davies), pp. 91–121. Blackwell Scientific Publications, Oxford.

Lack, D. (1954). *The Natural Regulation of Animal Numbers.* Oxford University Press.

Law, R. & Cannings, C. (1984). Genetic analysis of conflicts arising during development of seeds in the Angiospermophyta. *Proceedings of the Royal Society*, **B, 221**, 53–70.

Levins, R. (1968). *Evolution in Changing Environments.* Princeton University Press.

Lloyd, D. G. (1979). Some reproductive factors affecting the selection of self-fertilization in plants. *American Naturalist*, **113**, 67–79.

Lloyd, D. G. (1984). Gender allocations in outcrossing cosexual plants. *Perspectives on Plant Population Ecology* (Ed. by R. Dirzo & J. Sarukhan), pp. 277–300. Sinauer, Sunderland, Massachusetts.

Lloyd, D. G. (1987a). Parallels between sexual strategies and other allocation strategies. *The Evolution of Sex and Its Consequences* (Ed. by S. C. Stearns), pp. 263–81. Birkhäuser, Basel.

Lloyd, D. G. (1987b). Allocations to pollen, seeds and pollination mechanisms in self-fertilizing plants. *Functional Ecology*, **1**, 83–9.

Lloyd, D. G. (1987c). Selection of offspring size at independence and other size-versus-number strategies. *American Naturalist*, **129**, 800–17.

Lloyd, D. G. (1988a). Benefits and costs of biparental and uniparental reproduction in plants. *The Evolution of Sex* (Ed. by R. E. Michod & B. R. Levin), pp. 233–52. Sinauer, Sunderland, Massachusetts.

Lloyd, D. G. (1988b). A general principle for the allocation of limited resources. *Evolutionary Ecology*, **2**, 175–87.

MacArthur, R. H. & Wilson, E. O. (1967). *The Theory of Island Biogeography*. Princeton University Press.

McGinley, M. A. & Charnov, E. L. (1988). Multiple resources and the optimal balance between size and number of offspring. *Evolutionary Ecology*, **2**, 77–84.

Maynard Smith, J. (1971). The origin and maintenance of sex. *Group Selection* (Ed. by G. C. Williams), pp. 163–75. Aldine-Atherton, Chicago.

Maynard Smith, J. (1978). Optimization theory in evolution. *Annual Review of Ecology and Systematics*, **9**, 31–56.

Maynard Smith, J. (1986). *The Problems of Biology*. Oxford University Press.

Maynard Smith, J., Burian, R., Kauffman, S., Alberch, P., Campbell J., Goodwin, B., Lande, R., Raup, D. & Wolpert, L. (1985). Developmental constraints and evolution. *Quarterly Review of Biology*, **60**, 265–87.

Nagylaki, T. (1976). A model for the evolution of self-fertilization and vegetative reproduction. *Journal of Theoretical Biology*, **58**, 55–8.

Parker, G. A. (1978). Searching for mates. *Behavioural Ecology: an Evolutionary Approach* (Ed. by J. R. Krebs & N. B. Davies), pp. 214–44. Blackwell Scientific Publications, Oxford.

Parker, G. A. & Begon, M. (1986). Optimal egg size and clutch size: effects of environment and maternal phenotype. *American Naturalist*, **128**, 573–92.

Parker, G. A. & MacNair, M. R. (1978). Models of parent–offspring conflict. I. Monogamy. *Animal Behaviour*, **26**, 97–110.

Queller, D. C. (1984). Models of kin selection on seed provisioning. *Heredity*, **53**, 151–65.

Salisbury, E. J. (1942). *The Reproductive Capacity of Plants*. Bell, London.

Smith, A. P. (1984). Postdispersal parent–offspring conflict in plants: antecedent and hypothesis from the Andes. *American Naturalist*, **123**, 354–70.

Smith, C. C. & Fretwell, S. D. (1974). The optimal balance between size and number of offspring. *American Naturalist*, **108**, 499–506.

Stamps, J. A., Metcalf, R. A. & Krishnan, V. V. (1978). A genetic analysis of parent–offspring conflict. *Behavioural and Ecological Sociobiology*, **3**, 369–92.

Strassman, J. E., Queller, D. C. & Hughes, C. R. (1988). Predation and the evolution of sociality in the paper wasp, *Polistes bellicosus*. *Ecology*, **69**, 1497–505.

Stubblefield, J. W. & Charnov, E. L. (1986). Some conceptual issues in the origin of eusociality. *Heredity*, **57**, 181–7.

Temme, D. H. (1986). Seed size variability: a consequence of variable genetic quality among offspring? *Evolution*, **40**, 414–17.

Trivers, R. L. (1974). Parent–offspring conflict. *American Zoologist*, **14**, 249–64.

Trivers, R. L. (1985). *Social Evolution*. Benjamin/Cummings, Menlo Park, California.

Uyenoyama, M. K. (1988a). On the evolution of genetic incompatibility systems: incompatibility as a mechanism for the regulation of outcrossing distance. *The Evolution of Sex* (Ed. by R. E. Michod & B. R. Levin), pp. 212–232. Sinauer, Sunderland, Massachusetts.

Uyenoyama, M. K., (1988b). On the evolution of genetic incompatibility systems. III.

Introduction of weak gametophytic self-incompatibility under partial inbreeding. *Theoretical Population Biology*, **34**, 47–91.

Westoby, M. & Rice, B. (1982). Evolution of the seed plants and inclusive fitness of plant tissues. *Evolution*. **36**, 713–24.

Williams, G. C. (1966). *Adaptation and Natural Selection.* Princeton University Press.

Willson, M. F. (1983). *Plant Reproductive Ecology.* Wiley, New York.

Wilson, E. O. (1971). *The Insect Societies.* Belknap Press, Cambridge, Massachusetts.

Wilson, E. O. (1975). *Sociobiology: the New Synthesis.* Belknap Press, Cambridge, Massachusetts.

9. COMPARATIVE STUDIES IN EVOLUTIONARY ECOLOGY: USING THE DATA BASE

P. H. HARVEY AND M. D. PAGEL
Department of Zoology, University of Oxford, Oxford OX1 3PS, UK

INTRODUCTION

In the 75 years since the birth of the British Ecological Society there has been a relentless accumulation of data. Imaginative use of those data enables us to tackle new and varied questions about the reasons for organic diversity. Experimental, observational and comparative studies are all needed if we are to make sense of the scheme of things. At the extremes, experiments under controlled conditions allow us to test carefully formulated explanations for the occurrence of particular patterns of behaviour or of morphological features, while comparative studies help us to generate explanations and are necessary to test their generality. For example, the theory of optimal foraging has resulted in a number of elegant experimental investigations showing why a handful of animal species feed in the ways that they do (Stephens & Krebs 1986). Species can be classified by the optimality criterion used in selecting food: they may be energy maximizers, central load foragers, and so on. At the same time, the increasing data base on diet, body size, energy needs, habitats occupied and home range or territory size for literally thousands of species allows some more general tests of those explanations (Harvey & Mace 1983). The expected relationship between foraging area and energy requirements across species varies according to the optimization criterion used (Schoener 1983).

Optimal foraging theory is one of the few areas where we can move from theory through experiments to comparative tests. More usually the comparative data base allows us to narrow down the types of theory that are most likely to explain the generation and maintenance of organic diversity. But that is not a bad start. For example, years of experiments and observations have failed to explain the evolutionary functions of sleep, but we do now know how patterns of sleep in some taxa vary with habitat and the occurrence of natural enemies. Analysis of the comparative data (Elgar, Pagel & Harvey 1988) shows that quiet sleep patterns do not correlate with metabolic turnover in the way expected by one theory if sleep has an energy-conserving function. At the same time, correlations of active sleep patterns in a number of taxa support the idea that the upper limit on the amount of active sleep is set by the need for efficient thermoregulation. Such comparative studies have the potential to shift emphases of research from testing one set of ideas to focusing on others. Similarly, life-history theory is a 'can of worms' when it comes to experimental studies (Partridge & Harvey 1985, 1988; Reznick 1985), but the comparative data reveal associations between life-history variables across species that point to particular processes (such as mortality patterns in natural populations) as being important while other factors (such as metabolic turnover or brain size) can be relegated to the scrap heap of history (Read & Harvey, in press).

With each passing decade it has become possible to tackle new sets of questions using comparative data from ecological studies. An example comes from population structure in birds. Through the 1930s there was debate about whether young birds breed near to their birthplace or whether they select at random a suitable nesting site anywhere within the species range (Lincoln 1934; Chapman 1935; Kluijver 1935; Kendeigh & Baldwin 1937; Nice 1937). By the 1950s enough long-term ringing and recapture data had become available to give us a rough feel for the extent to which various species of birds disperse from their natal area to breed, and how those differences relate to habitat use (Kluijver 1935, 1951; Kendeigh & Baldwin 1937; Nice 1937, 1943; Kendeigh 1941; Austin 1949, 1951). By the 1970s, the data were sufficient to detect fairly fine-grained but near-universal patterns of sex differences in dispersal behaviour with the implication that intrasexual selection and inbreeding avoidance were somehow involved (Greenwood 1980; Greenwood & Harvey 1982). Now, at the end of the 1980s, we have a fairly general picture of rates of parent–offspring and sib–sib matings in natural populations of birds, and it is clear that small, spatially constrained populations have higher rates of close inbreeding (Ralls, Harvey & Lyles 1986). We can look forward to a time when the right data are available to tell us whether, as a general rule, inbreeding is actively avoided — at the moment we have only tantalizing glimpses (McGregor & Krebs 1982; Koenig, Mumme & Pitelka 1984; van Noordwijk, van Tienderen & de Jong 1985). And, with the application of DNA fingerprinting to populations of birds (Burke & Bruford 1987; Wetton *et al.* 1987), we shall obtain more precise information for a reasonable sample of species on extra-pair copulation and egg dumping (Harvey 1985). We shall know who is raising whose offspring.

Our increased understanding of the population structure of birds has come through the accumulation of enormous amounts of hard-won data that have had to be carefully compiled and painstakingly analysed. This chapter is about the accumulation and analysis of such data sets. We do not attempt comprehensive coverage of this type of work. Instead, through a series of descriptive vignettes we hope to illustrate where the major problems lie and how they should be faced. Our chapter is divided into three main parts: (i) the accumulation and availability of data, (ii) the analysis of the data, and (iii) checking that the conclusions are correct.

MAKING SURE IT IS THE BEST DATA BASE

Collecting the right data

The data used in comparative studies, like those on the population structure of birds, were often collected with a different purpose in mind. For example, many bird-ringing studies were performed with a view to estimating mortality and migration parameters. It is increasingly true that opportunistic comparative studies can provide us with answers to questions that we might not have expected to be able to answer. But it is also true that many data have been lost simply because they were not considered worth recording at the time. The ultimate source of many of our best data bases is the amateur naturalist whose records are accumulated on national data bases or in the more specialist journals. For example, the British Trust for Ornithology (BTO) data base is enormous, based on the continuing records of some eight thousand field naturalists. Those records have been used in hundreds of publications. Dr. J. J. D. Greenwood, the present Director of the BTO, has provided us with an appendix to this chapter which details the tasks and problems faced by the Trust. Of the many points he makes, we wish to highlight two.

The first point is that data are accumulating faster than they can be computerized. This is not a trivial issue. Many research workers spend many hours copying and computerizing raw data before they can be used. Another Greenwood, Dr. P. J. Greenwood, and one of us spent several weeks at BTO headquarters in the mid-1970s copying out the ringing return data for a number of species before computerizing them. When we came to publish our analyses, it was necessary to publish the raw data in case subsequent workers needed to use them (e.g. Greenwood & Harvey, 1976). In fact Taylor & Taylor (1977) reanalysed the data for different reasons and, presumably, recomputerized them. Had those data been available on a national data base in computer-readable form, months of work would have been saved. As it turned out large portions of those same data had been copied by several workers previously and the process has often been repeated since. An even more extreme version of the same story can be told of 30 years' worth of data from the Wytham Wood great tit project: most of the data were computerized at Sussex, and then a more comprehensive version was independently produced at Oxford. The duplication of effort entailed several hundred hours' work. Fortunately, this general problem is decreasing in magnitude as computers become more widely used. Nevertheless, additional funds are necessary if this process is to be completed, even on such nationally important data bases as those housed by the BTO. At an international level, the problem is actually more acute. As May (1988) points out, there is not a single collection (let alone a computerized collection) of the records of known species on earth. Many species of animals and plants were described in monographs, the last copies of which were lost many decades ago. The scientific community does not even know which species have been described!

The second of Dr Greenwood's points which we focus on is the need for rapport between those needing to use the data, and those collecting it. Feedback from the scientific community on the sorts of data that are likely to be most useful can help at the planning stage as record schemes are developed. Very careful planning must go into integrating the efforts of eight thousand data-gatherers, but the occasional suggestion from those who use the data is generally welcomed and can have important consequences.

Publishing the data

We have already pointed to the need to publish (or otherwise make available) raw data sets. This issue is far more important than is generally realized for at least four reasons: specialist interest, contradictory results, alternative analyses, and wider-ranging analyses.

Specialist interest

Comparative analyses are performed by research workers who do not have specialist knowledge of all the species being studied. Accordingly, aberrant data can be overlooked and correct data misused. If comparative studies are to achieve the respect they deserve, it is important that experts are able to verify the data as being reasonable. For example

Stearn's (1983) comparative study of life-history variation in reptiles was correctly criticized on these and other grounds by Vitt & Seigel (1985), and reanalysis of the data by Dunham & Miles (1985) produced different conclusions. Another example comes from work co-authored by one of us on testes size in primates (Harcourt *et al.* 1981; Harvey & Harcourt 1984). Primate species in which females often mate with more than one male during a given oestrus (multi-male) have larger testes for their body sizes than species in which females mate with a single male (single-male). Sperm competition in the multi-male species was thought to select for larger testes. However, the proboscis monkey (*Nasalis larvatus*) stood out as an exception: it has small testes for its body size but was recorded in the literature as being multi-male. Fortunately, a specialist referee was able to point us to two continuing field studies which demonstrated that the apparently multi-male proboscis monkey was actually single-male.

Even the best comparative relationships reveal genuine biological exceptions which are worth studying because the taxa are testifying to unusual selective pressures or unusual evolutionary responses. For example, many pinnipeds have unusually early ages at weaning associated with the production of particularly rich milk, presumably in response to strong predator pressure selecting for short periods on the pupping ground before leaving for the safer aquatic environment. In contrast, the massive eggs of kiwis (*Apteryx* spp.) in relation to the mother's body size may result from selection to produce large young (Calder 1984) or could be the result of some sort of phylogenetic inertia — for example, secondary body size reduction may not have been matched by reduction in egg size (Gould 1986). We find this latter explanation unappealing because we see no reason why egg size cannot decrease in just the same way that it can increase over evolutionary time.

Contradictory results

Many sources of data are now available for a variety of characters and it is possible for different workers to reach quite different conclusions because they have been analysing different data sets. For example, Millar (1977) mentioned the absence of a correlation between age at weaning and body size in mammals, whereas other reports found highly significant correlations (see Clutton-Brock & Harvey 1984). Fortunately, Millar published his raw data which, it turned out, were taxonomically biased, were subject to considerable error variance, and contained crucial inaccuracies (Clutton-Brock & Harvey 1984).

Alternative analyses

There are two main reasons why data often need to be reanalysed. First, many different evolutionary scenarios can produce the same biological result, and we may need to model the effects of a scenario different from that envisaged by the original analysis. Second, different statistical tests make different assumptions about the data, and we may wish to reanalyse them under a different and perhaps more realistic set of statistical assumptions. These two reasons together make for a diversity of comparative methods, indeed it sometimes seems there are almost as many comparative methods as there are data sets that have been analysed. The methods that are currently available have been reviewed elsewhere (Pagel & Harvey 1988a). Here we give particular examples of the two reasons outlined above for reanalysing data sets.

Alternative evolutionary scenarios. Different evolutionary scenarios are not uncommon. The constancy of a character in a particular monophyletic taxon might result from the constituent species sharing common environments and similar selective pressures, or we might suppose that inadequate genetic variance has been available to allow the generation of species differences. Similarly, it has often been claimed that allometric relationships found across species (that is, power functions relating the size of a character to body size) result from 'common growth mechanisms' (Huxley 1932: see Clutton-Brock & Harvey 1979) so that adaptive explanations for species differences in organ sizes, which often scale allometrically, are not 'necessary' (Lewontin 1979, p. 13). Alternatively, there may be very good adaptive reasons why organs are selected to scale with body size in the ways that they do (Harvey & Clutton-Brock 1983).

If we cannot distinguish between alternative explanations for patterns (or lack of patterns) in cross-species data sets, we may often be forced to perform separate analyses that make different assumptions. For example, the intercept on the body size axis of the power function relating organ size to body size may differ if allometric relationships are fitted separately for species belonging to different taxonomic families. Similarly, intercepts may differ for species grouped according to differences in ecology. If we wanted to assume that common growth mechanisms were responsible for allometric relationships within families, but that adaptive differences were responsible for the different intercepts among families, we might estimate separate lines. But, if we wanted to assume that each species was independently adapted, we might fit separate lines for groups of species with different ecologies rather than taxonomic affinities (Harvey & Clutton-Brock 1983).

Ideally, of course, we should look for separate effects of ecology and taxonomy, using appropriate methods, but in many cases it may not be possible to distinguish between the two. A case study comes from McNab's (1986) suggestion that metabolic rate is correlated with diet and habitat utilization independently of body size across mammals. McNab is correct but, as Elgar & Harvey (1987) point out, metabolic rate differs among taxa when body size effects are taken into account. Other than on a special case basis, which can be used to argue either way (Harvey & Elgar 1987; McNab 1987), it is not currently possible to distinguish ecological from taxonomic correlates of metabolic rate differences among species. This is not to argue that, in some special sense, taxonomy influences metabolic rate. Ecology may be important, but there are many behavioural, morphological and ecological correlates of taxonomy in mammals, and to single out diet or habitat utilization seems unjustified on present evidence.

Different statistical assumptions. Although evolutionary and other biological assumptions must be incorporated into statistical comparative tests, so must assumptions about the form of the data. For example, are relationships among variables linear, and if so what data transformations are necessary? What statistical distributions do the data best fit? Are the data measured with error, or are there sources of biological error which must be accounted for? Model 1 linear regression analysis is a common procedure in comparative tests, but the statistical error variance on the independent variable (x) under that model is assumed to be zero. Alternative line-fitting procedures, such as major axis and reduced major axis analysis, make different assumptions about error of the y and x variables. If we are interested in the slope of the line relating the y and x variables, as we often are when attempting to interpret comparative relationships in terms of mechanical scaling principles, it is important that we make the correct assumptions about error variance.

In fact, error variance may be the cause of one particular set of statistical artefacts which have been given biological interpretation. When regression lines are fitted across logarithmically transformed measures of brain (y variable) and body size (x variable) for mammal species, it is frequently found that the slope increases as more higher-level taxa are incorporated into the analysis (Gould 1966, 1975; Lande 1979). For example, lines fitted across species within genera may lie in the 0.2–0.4 region (Lande 1979), while the best-fit line fitted across a wide range of mammal species from different orders has a slope of 0.75. But, as Martin & Harvey (1985) pointed out, as a wider range of species' body-weights is incorporated into the data set, so the effects of measurement error are reduced and slopes increase under model 1 regression. On reanalysing the data, Martin & Harvey (1985) found an increase in slope with taxonomic level of analysis under nested analysis of covariance models using both regression analysis and major axis analysis, although the effect was less under the latter model. However, Martin & Harvey had not gone far enough: major axis and reduced major axis models make their own particular assumptions which are violated by the data, and produce artefactual increases in slope with taxonomic level (Pagel & Harvey, 1988b). When a more appropriate 'structural relations model' (Rayner 1985) that incorporates estimates of error variances is applied to the data, there is no general increase in slope with taxonomic level (Pagel & Harvey, 1989). Indeed, taxonomic differences in encephalization (brain size differences with body size effects removed) are associated with differences in occupancy of ecological niches. In these various analyses, many of the same data were subjected to increasingly refined analysis and, because raw data had sometimes been included in the original publications, estimates of statistical error variance often proved possible.

Wider-ranging analyses

Comparative biologists may rightly be proud of the nature and generality of their findings. But they cannot expect investigations to stop there. Eventually their analyses will be extended both taxonomically and critically. It is important that the original data can be incorporated into subsequent extensions of the original analysis.

We have already mentioned the allometric analysis that related testes size to breeding system in primates (Harcourt *et al.* 1981). It was only natural that mammalogists interested in other orders should examine the generality of this finding. Clutton-Brock, Guinness & Albon (1982) found the same relationship in cervids, although they did not compare relative testes size between cervids and primates. And, more recently, Kenagy & Trombulak (1986) have extended the study to a wider range of mammalian orders.

Not only is it possible to generalize findings using comparative data, but it is also possible to examine further predictions as new data become available. For example, Møller (1988a,b) argued that, if Harcourt *et al.* (1981) were correct about the reasons for the relationship between relative testes size and breeding system in primates, other predictions should hold true concerning sperm motility, sperm production rates, and sperm reserves in males. Møller's predictions were confirmed but, although there were data on sperm motility for a number of primates species (Møller 1988a), he needed to extend his net to mammals as a whole in order to get data on sperm production and sperm reserves (Møller 1988b).

Other explanations of comparative trends have fared less well when additional data have been incorporated. For example, Martin (1981) suggested that hatchling brain size in birds and neonatal brain size in mammals are evolutionarily linked to maternal metabolic turnover. More direct tests of his explanation for both birds and mammals (Bennett & Harvey 1985; Pagel & Harvey 1988c) than Martin performed in his original papers have failed to produce predicted relationships. For example, Martin had argued that the 0.75 exponent linking maternal body-weight to neonatal brain size resulted from a direct link between maternal basal metabolic rates and neonatal brain size (he argued that brains are energetically costly to produce). In fact mammals with high metabolic rates for their body size do not produce young with larger brains than those with low metabolic rates for their body size or, in statistical terms, there is not a significant partial correlation between adult metabolic rate and neonatal brain size when the effects of adult body size are held constant (Pagel & Harvey 1988c).

ASKING THE RIGHT QUESTIONS

Tukey (1962) wrote 'Far better an approximate answer to the right question, which is often vague, than an exact answer to the wrong question, which can always be made precise.' That, perhaps, is by far the most important message concerning the best use of the data base in evolutionary ecology. However good the data, it is absolutely essential that we are clear about both the questions we ask of the data, and what the results of any particular analysis mean.

Optimality models

Optimality models force us to be explicit about the constraints and the selective forces that are implicit in many verbal explanations for the adaptive significance of particular traits (see Maynard Smith 1978). In the introduction to this paper, we mentioned the link between optimal foraging theory and territory size. A literature developed in the 1960s and 1970s on the ways species' territory sizes change with body size and diet (e.g. McNab 1963; Schoener 1968; Milton & May 1976; Clutton-Brock & Harvey 1977). Perhaps a reasonable question to ask at the time was whether larger animals would need larger territories in order to satisfy their larger metabolic needs. As predicted, territory size did increase with body size and, furthermore, species living on more sparsely distributed food resources were also found to have larger territories.

The next step was to ask whether territory size increased with body size in a quantitatively sensible way. Since metabolic rate (energy needed per unit time) increases with body-weight raised roughly to the 0.75 power, presumably the minimum-sized, continuously productive territory (energy produced per unit time) for animals with similar diets living in similar habitats would also be expected to increase with the 0.75 power of body-weight (*contra* Lindstedt, Miller & Buskirk 1986). The data did not accord with that expectation: territory size in a variety of taxa increased with body size with an exponent appreciably greater than 0.75 (e.g. Harvey & Clutton-Brock 1981; Gittleman & Harvey 1982; Mace & Harvey 1983; Lindstedt, Miller & Buskirk 1986). Two possible causes for the discrepancy are likely. First, suitable habitat is not continuous, and larger species must take in a disproportionate area of unsuitable vegetation. Second, the acceptable food spectrum might change with territory size (Schoener 1983). For example, single-prey loaders supplying food to a nest in the middle of their territory might have evolved an optimal foraging strategy resulting in the selection of a smaller-size range of only the larger food items at an increased foraging distance from the nest. The major point to be made with this example is the importance of defining and refining the optimality model that underlies the comparative test.

It seems that comparative analyses in evolutionary ecology would often profit from the explicit use of either kinematic population genetic or optimality models. Perhaps the best example of the need for the use of such models comes from that section of the comparative literature which deals with the allometry of life-history variation. For many years, biologists have realized that body-weight has far-reaching implications for the optimal construction of organisms (Huxley 1932; Thompson 1942). For example, bones of heavy animals need to be relatively wider (McMahon & Bonner 1983). Body-weight is an easily measured and widely recorded trait, and comparative biologists following the morphological tradition noted that it is highly correlated with life-history differences among species. Correlation does not necessarily imply causation, yet the comparative literature seems fairly unified on the importance of body size as a *determinant* of life-history variation (McMahon & Bonner 1983; Calder 1984; Schmidt-Nielsen 1984). When the reasons for the correlations are discussed (and often they are not), authors either suggest that (i) larger organisms take longer to reach adult size, or (ii) that common growth mechanisms (presumably hormones) determine both size and life-histories (for a review, see Read & Harvey in press). In either case, size is seen as the target for selection, and life-history variation as the consequence. Occasionally, it has been argued that other (confounding) variables, such as brain size or metabolic rate, that are correlated with body size, are responsible for the relationship between body size and life-history (e.g. Sacher & Staffeldt 1974; McNab 1980, 1983, 1986). In such a scenario, some variable other than body size is viewed as the target of selection.

The view of life-history evolution with life-history patterns evolving as a consequence of selection on some target variable is given by a number of authors. For example 'Gestation time, postnatal growth rate and age at maturity [can be viewed] as a single growth continuum related to adult weight, which alone might be the target for selection' (Western & Ssemakula 1982, p. 287) or 'One need not speculate that evolution has selectively dealt with any single biological cycle' (Lindstedt & Calder 1981, p. 4). These statements are at odds with the data because species of very similar body sizes can have very different life-history patterns, and because life-history variables tend to be correlated with each other independently of body size (the data in support of these and other arguments are discussed by Read & Harvey in press). The allometric tradition of life-history evolution has developed independently of a rich and relevant literature on life-history theory (reviewed in Charlesworth 1980, and more recently by Partridge & Harvey 1988).

Life-history theory recognizes that, since reproduction is costly, fecundity at all ages in the lifespan cannot simultaneously be maximized. When a potentially long-lived individual reproduces and there is a cost for reproduction, its chances of surviving to breed again (and hence its fecundity at later ages) are reduced. The important factors in life-history evolution are trade-offs between different components of fitness. In the above example, fecundity is traded off against survival. How then does body size enter the picture? Body size might be viewed as a factor that can influence both survival and fecundity. Under some circumstances, larger individuals may have higher chances of survival and increased fecundity, but energy put into growth cannot simultaneously be put into reproduction. Individuals that delay reproduction can put more resources into growth, can achieve a larger body size and may, subsequently, have a higher rate of offspring production (benefits of delayed reproduction). However, if mortality rates are high early in life, individuals that delay reproduction may die before attempting to reproduce (a cost of delayed reproduction). The optimal body size is the one which results in that life-history that maximizes inclusive fitness or, to a reasonable approximation, the Malthusian parameter (Charlesworth 1980). Species occupying different niches have different optimal body sizes, depending upon the ways trade-off curves are influenced by body size. For example, the optimal size for a weasel feeding on small rodents may be small so that they can chase their prey down holes, while the optimum might be larger for rabbit-eating weasels so they can overpower their prey, produce larger offspring, and have higher chances of overwinter survival on stored food reserves (Ralls & Harvey 1985).

Actually, the relationships of life-history variables to each other, even when body size effects have been factored out, hold hope for fairly general explanations of life-history diversity. For example, among mammals and among birds, different species or higher-level taxa can be placed on a fast–slow continuum: those species with short gestation lengths or incubation periods for their body sizes, have relatively early ages at weaning or fledging, mature when relatively young, produce litters or clutches at more frequent intervals, and have shorter maximum recorded lifespans (Bennett & Harvey 1988; Harvey & Read 1988; Saether 1988; Read & Harvey in press). The shorter lifespans seem part of the key to understanding the covariation of life-history traits. On the whole, fecundity must be balanced by mortality in natural populations (Sutherland, Grafen & Harvey 1986). When mortality rates are high, some components of fecundity must also be high, though why all components change together is not yet understood. It is important to emphasize here that we are not discounting the importance of body size in life-history evolution, but we simply view it as one of many variables that influence trade-off curves.

GETTING THE RIGHT ANSWERS

As we tackle more general questions in evolutionary ecology, we are ever more likely to produce results that relate to and may appear to contradict findings in other fields of biological research, such as physiology or molecular genetics (e.g. exchanges between Greenough & Harvey 1987, 1988 and Dover 1988a, b, following Hudson, Kreitman & Aguadè 1987). The links between ecology and evolution are sufficiently fragile for internal consistency even to be an issue between these two closely related fields. Fortunately, the resolution of contradictory findings often leads to interesting advances in one or both fields. Comparative studies can often provide insight into how an inconsistency might be resolved. We give an example below.

It is well known that animals' energy needs scale with approximately the 0.75 power of body-weight (Kleiber 1961; Nagy 1987; McNab 1988) while population density seems to scale to the −0.75 to −1.0 power of body-weight (Fig. 9.1; Damuth 1981, 1987). Population density is often the reciprocal of home-range or territory size, which scales near to the 1.0 power of body-weight (Lindstedt, Miller & Buskirk 1986). Evolutionary optimization explanations have been suggested for both relationships (McMahon 1973 for metabolic needs, and see p.218 for home-range/population density). If the two exponents are roughly 0.75 and −0.75, each species occupying an area would be using about the same amount of energy and probably other resources, irrespective of body-weight (Damuth 1981, 1987). However, it is well known that species-abundance curves in natural communities are often canonically log-normally distributed, and one interpretation of the community data would predict an interspecific, canonically log-normal, resource utilization function (Sugihara 1980). There is a contradiction between the two conclusions (Harvey & Godfray 1987). From the ecological perspective, Sugihara (1989) suggests that within communities the cross-species variance of species energy needs and body size are the same on logarithmically scaled axes, a suggestion which seems to contradict the facts (Kleiber's law). Sugihara also suggests that variance in population density and body size are the same on logarithmically scaled axes, a suggestion which may be true.

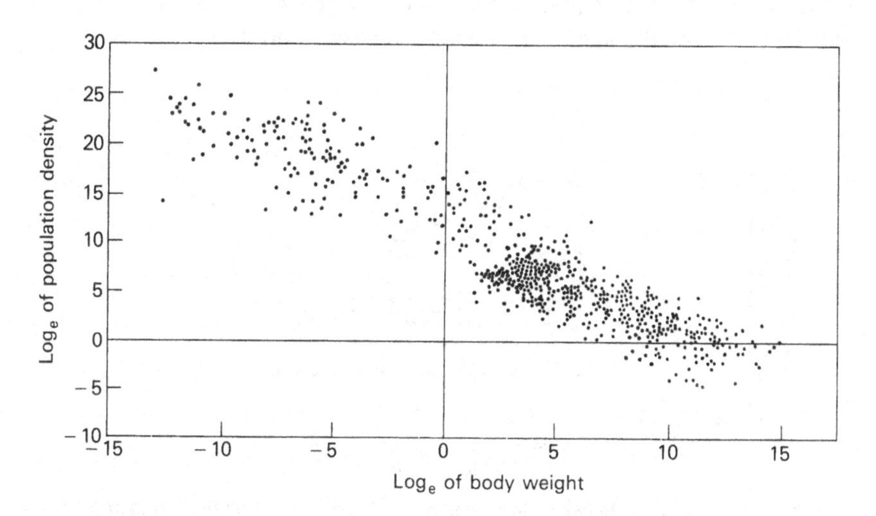

FIG 9.1 The relationship between body-weight (g) and population density (number per km^2) for animals (after Damuth 1987).

Closer examination of the available data suggests that Damuth's finding may represent an ecological limit rather than a norm of community structure. Species' population densities, like their territory or home-range sizes, tend to be recorded in areas where it is worth studying the species in question, that is where population density is high. To a reasonable approximation, Damuth's results may describe maximum population densities for the species concerned. In support of this assertion, the few studies that attempt to record population densities of a number of species in natural communities reveal a quite different pattern of population density in relation to body size (Fig. 9.2; Brown & Maurer 1987; Morse, Stork & Lawton 1988). The biological significance of these community structure findings is not yet clear, since, if there was *no* relationship between population density and body size, but population densities differed among species, larger ranges of population densities would be found for body sizes represented by more species. This effect needs to be removed in any correct analysis of the data. A further problem with the comparative data from within single communities is the small range of body-weights of the constituent species, which would be expected to result in much lower correlation coefficients between body-weight and population density than found in Damuth's sample.

In a previous section, we discussed how comparative data often need to be reanalysed as more information becomes available. As in any area of science, conclusions must be ever open to scrutiny. New tests of established theory must always be pursued when that theory seems inconsistent with other areas of scientific endeavour. Darwin's conflict with Kelvin over the age of the earth is a case in point.

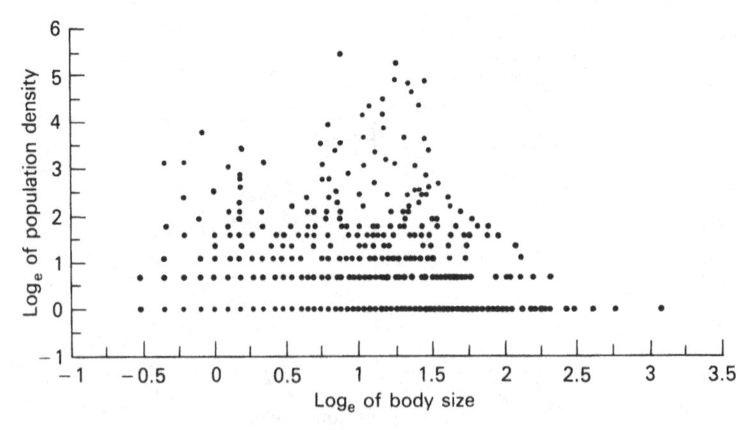

FIG 9.2 The relationship between body length (mm) and population density (sample size) for species of beetles from a single Bornean lowland rain forest community (after Morse, Stork & Lawton 1988).

EVOLUTIONARY ECOLOGY AS A FIELD

The relationship between the disciplines of ecology and evolution has never been an easy one. For example, and quite understandably, the classic models of population dynamics view the members of a species as identical without phenotypic variation for intrinsic growth rates, carrying capacities, coefficients of competition and so on (Maynard Smith 1974). On the other hand, evolutionary biologists often seem obsessed by the genetic forces that can influence or constrain evolutionary change without paying due regard to environmental variance. An example here is the attempts a few years ago to explain the evolution of epigamic traits through Fisher's runaway process. The process is of course important but, in the case of coloration among birds, environmental variation measured as incidence of blood parasites might be expected to correlate with species differences in coloration (Hamilton & Zuk 1981). A careful analysis of the then available data seemed to provide the necessary evidence (Read 1987). Similarly, just as it appeared that life-history variation was unrelated to ecological differences once the effects of body size had been removed among primates (Harvey & Clutton-Brock 1985), the picture was dramatically reversed when Ross (1987) showed size-independent associations, of the sort that would be predicted from life-history theory, between variation in life-history and ecology.

The data base in evolutionary ecology is growing at a most satisfying rate. It has enormous power to tackle questions that have long remained unanswered and we should be harnessing that power right now.

REFERENCES

Austin, O. L. (1949). Site tenacity, a behavior trait of the common tern. *Bird-Banding,* **20**, 1–39.

Austin O. L. (1951). Group adherence in the common tern. *Bird-Banding,* **22**, 1–15.

Bennett, P. M. & Harvey, P. H. (1985). Brain size, development and metabolism in birds and mammals. *Journal of Zoology, London,* **207**, 491–509.

Bennett, P. M. & Harvey, P. H. (1988). How fecundity balances mortality in birds. *Nature,* **333**, 216.

Brown, J. H. & Maurer, B. A. (1987). Evolution of species assemblages: effects of energetic constraints and species dynamics on the diversification of the North American avifauna. *American Naturalist,* **130**, 1–17.

Burke, T. & Bruford, M. W. (1987). DNA fingerprinting in birds. *Nature,* **327**, 149–52.

Calder, W. A. (1984). *Size, Function and Life History.* Harvard University Press, Cambridge, Massachusetts.

Chapman, L. B. (1935). Studies of a tree swallow colony. *Bird-Banding,* **6**, 45–57.

Charlesworth, B. (1980). *Evolution in Age-structured Populations.* Cambridge University Press.

Clutton-Brock, T. H., Guinness, F. E. & Albon, S. D. (1982). *Red Deer: Behavior and Ecology of Two Sexes.* Chicago University Press.

Clutton-Brock, T. H. & Harvey, P. H. (1977). Primate ecology and social organisation. *Journal of Zoology, London,* **183**, 1–39.

Clutton-Brock, T. H. & Harvey, P. H. (1979). Comparison and adaptation. *Proceedings of the Royal Society of London,* **B, 205**, 547–65.

Clutton-Brock, T. H & Harvey, P. H. (1984). Comparative approaches to investigating adaptation. *Behavioural Ecology: an Evolutionary Approach (*Ed. by J. R. Krebs & N. B. Davies), pp. 1–29. Blackwell Scientific Publications, Oxford.

Damuth, J. (1981). Population density and body size in mammals. *Nature,* **290**, 699–700.

Damuth, J. (1987). Interspecific allometry of population density in mammals and other animals: the independence of body mass and population energy use. *Biological Journal of the Linnean Society,* **31**, 193–246.

Dover, G. A. (1988a). Three into two won't go. *Nature,* **331**, 121.

Dover, G. A. (1988b). Evolution of the third kind. *Nature,* **332**, 402.

Dunham, A. E. & Miles, D. B. (1985). Patterns of covariation in life history traits of squamate reptiles: the effects of size and phylogeny reconsidered. *American Naturalist,* **126**, 231–57.

Elgar, M. A. & Harvey, P. H. (1987). Basal metabolic rates in mammals: allometry, phylogeny and ecology. *Functional Ecology,* **1**, 25–36.

Elgar, M. A., Pagel, M. D. & Harvey, P. H. (1988). Sleep in mammals. *Animal Behaviour,* **36,** 1407–19.

Gittleman, J. L. & Harvey, P. H. (1982). Carnivore home-range size, metabolic needs and ecology. *Behavioural Ecology and Sociobiology,* **10,** 57–64.

Gould, S. J. (1966). Allometry and size in ontogeny and phylogeny. *Biological Reviews,* **41,** 587–640.

Gould, S. J. (1975). Allometry in primates with emphasis on scaling and the evolution of the brain. *Contributions to Primatology,* **5**, 244–92.

Gould, S. J. (1986). Kiwi eggs and the liberty bell. *Natural History,* **11**, 20–9.

Greenough, J. A. & Harvey, P. H. (1987). Too much neutral polymorphism. *Nature,* **329**, 585–6.

Greenough, J. A. & Harvey, P. H. (1988). Greenough and Harvey reply. *Nature,* **331**, 121–2.

Greenwood, P. J. (1980). Mating systems, philopatry and dispersal in birds and mammals. *Animal Behaviour,* **28**, 1140–62.

Greenwood, P. J. & Harvey, P. H. (1976). The adaptive significance of breeding area fidelity of the blackbird (*Turdus merula* L.). *Journal of Animal Ecology,* **45**, 887–98.

Greenwood, P. J. & Harvey, P. H. (1982). The natal and breeding dispersal of birds. *Annual Review of Ecology and Systematics,* **13**, 1–21.

Hamilton, W. D. & Zuk, M. (1981). Heritable true fitness and bright birds: a role for parasites? *Science,* **218**, 384–7.

Harcourt, A. H., Harvey, P. H., Larson, S. G. & Short, R. V. (1981). Testis weight, body weight and breeding system in primates. *Nature,* **293**, 55–7.

Harvey, P. H. (1985). Raising the wrong children. *Nature,* **313**, 95–6.

Harvey, P. H. & Clutton-Brock, T. H. (1981). Primate home-range size and metabolic needs. *Behavioural Ecology and Sociobiology,* **8**, 151–5.

Harvey, P. H. & Clutton-Brock, T. H. (1983). Allometry and evolution: the survival of the theory. *New Scientist,* **98**, 312–15.

Harvey, P. H. & Clutton-Brock, T. H. (1985). Life history variation in primates. *Evolution,* **39**, 559–81.

Harvey, P. H. & Elgar, M. A. (1987). In defence of the comparative method. *Functional Ecology,* **1**, 160–1.

Harvey, P. H. & Godfray, H. C. J. (1987). How species divide resources. *American Naturalist,* **129**, 318–20.

Harvey, P. H. & Harcourt, A. H. (1984). Sperm competition, testis size and breeding system in primates. *Sperm Competition and the Evolution of Animal Mating Systems* (Ed. by R. L. Smith), pp. 589–600. Academic Press, New York.

Harvey, P. H. & Mace, G. M. (1983). Foraging models and territory size. *Nature*, **305**, 14–15.

Harvey, P. H. & Read, A. F. (1988). How and why do mammalian life histories vary? *Evolution of Life Histories of Mammals: Theory and Pattern* (Ed. by M. S. Boyce), pp. 213–32. Yale University Press, New Haven.

Hudson, R. R., Kreitman, M. & Aguadè, M. (1987). Properties of a neutral allele model with intragenic recombination. *Genetics*, **116**, 153–9.

Huxley, J. S. (1932). *Problems of Relative Growth.* MacVeagh, London.

Kenagy, J. G. & Trombulak, S. C. (1986). Size and function of mammalian testes in relation to body size. *Journal of Mammalogy*, **67**, 1–22.

Kendeigh, S. C. (1941). Territorial and mating behaviour of the house wren. *Illinois Biological Monographs*, **18**, 1–120.

Kendeigh, S. C. & Baldwin, S. P. (1937). Factors affecting yearly abundances of passerine birds. *Ecological Monographs*, **7**, 91–123.

Kleiber, M. (1961). *The Fire of Life.* Wiley, New York.

Kluijver, H. N. (1935). Ergebnisse eines Versuches über das Heimfindevermögen von Staren. *Ardea*, **24**, 227–39.

Kluijver, H. N. (1951). The population ecology of the great tit, *Parus m. major* L. *Ardea*, **39**, 1–135.

Koenig, W. D., Mumme, R. L. & Pitelka, F. A. (1984). The breeding system of the acorn woodpecker in central coast California. *Zeitschrift für Tierpsychologie*, **65**, 289–308.

Lande, R. (1979). Quantitative genetic analysis of multivariate evolution applied to brain : body size allometry. *Evolution*, **33**, 402–16.

Lewontin, R. C. (1979). Sociobiology as an adaptationist program. *Behavioral Science*, **24**, 5–14.

Lincoln, F. C. (1934). The operation of homing instinct. *Bird-Banding*, **5**, 24–30.

Lindstedt, S. L. & Calder, W. A. (1981). Body size, physiological time and longevity of homeothermic animals. *Quarterly Review of Biology*, **56**, 1–16.

Lindstedt, S. L., Miller, B. J. & Buskirk, S. W. (1986). Home range, time and body size in mammals. *Ecology*, **67**, 413–18.

Mace, G. M. & Harvey, P. H. (1983). Energetic constraints on home-range size. *American Naturalist*, **121**, 120–32.

McGregor, P. K & Krebs, J. R. (1982). Mating and song types in the great tit. *Nature*, **297**, 60–1.

McMahon, T. A. (1973). Size and shape in biology. *Science*, **179**, 1201–4.

McMahon, T. A. & Bonner, J. T. (1983). *On Size and Life.* Scientific American, New York.

McNab, B. K. (1963). Bioenergetics and the determination of home range size. *American Naturalist*, **97**, 133–40.

McNab, B. K. (1980). Food habits, energetics and population biology of mammals. *American Naturalist*, **106**, 116–24.

McNab, B. K. (1983). Energetics, body size and the limits to endothermy. *Journal of Zoology, London*, **199**, 1–29.

McNab, B. K. (1986). The influence of food habits on the energetics of eutherian mammals. *Ecological Monographs*, **56**, 1–19.

McNab, B. K. (1987). Basal rate and phylogeny. *Functional Ecology*, **1**, 159–60.

McNab, B. K. (1988). Complications in scaling the basal rate of metabolism in mammals. *Quarterly Review of Biology*, **63**, 25–54.

Martin, R. D. (1981). Relative brain size and basal metabolic rate in terrestrial vertebrates.

Nature, **293**, 57–60.

Martin, R. D. & Harvey, P. H. (1985). Brain size allometry: ontogeny and allometry. *Size and Scaling in Primate Biology* (Ed. by W. L. Jungers), pp. 147–73. Plenum, New York.

May, R. M. (1988). How many species are there on earth? *Science*, **241**, 1441–9.

Maynard Smith, J. (1974). *Models in Ecology*. Cambridge University Press.

Maynard Smith, J. (1978). Optimization theory in evolution. *Annual Review of Ecology and Systematics*, **9**, 31–56.

Millar, J. S. (1977). Adaptive features of mammalian reproduction. *Evolution*, **31**, 370–86.

Milton, K. & May, M. L. (1976). Body weight, diet and home range area in primates. *Nature*, **259**, 459–62.

Møller, A. P. (1988a). Ejaculate quality, testes size and sperm competition in primates. *Journal of Human Evolution*, **17**, 479–88.

Møller, A. P. (1988b). Ejaculate quality, testes size and sperm production in mammals. *Functional Ecology*, **3**, 91–6.

Morse, D. R., Stork, N. E. & Lawton, J. H. (1988). Species number, species abundance and body length relationships or aboreal beetles in Bornean lowland rain forest trees. *Ecological Entomology*, **13**, 25–37.

Nagy, K. A. (1987). Field metabolic rate and food requirement scaling in mammals and birds. *Ecological Monographs*, **57**, 111–28.

Nice, M. M. (1937). Studies of the life history of the song sparrow. Part 1. *Transactions of the Linnean Society of New York*, **4**, 1–247.

Nice, M. M. (1943). Studies of the life history of the song sparrow. Part 2. *Transactions of the Linnean Society of New York*, **6**, 1–328.

Pagel, M. D. & Harvey, P. H. (1988a). Recent developments in the analysis of comparative data. *Quarterly Review of Biology*, **63**, 413–40.

Pagel, M. D. & Harvey, P. H. (1988b). The taxon level problem in mammalian brain size evolution: facts and artifacts. *American Naturalist*, **132**, 344–59.

Pagel, M. D. & Harvey, P. H. (1988c). How mammals produce large brained offspring. *Evolution*, **42**, 948–57.

Pagel, M. D. & Harvey, P. H. (in press). Taxonomic differences in the scaling of brain on body weight in mammals. *Science*.

Partridge, L. & Harvey, P. H. (1985). The costs of reproduction. *Nature*, **316**, 20.

Partridge, L. & Harvey, P. H. (1988). The ecological context of life history evolution. *Science*, **241**, 1449–55.

Ralls, K. & Harvey, P. H. (1985). Geographical variation in size and sexual dimorphism of North American weasels. *Biological Journal of the Linnean Society*, **25**, 119–67.

Ralls, K., Harvey, P. H. & Lyles, A. M. (1986). Inbreeding in natural populations of birds and mammals. *Conservation Biology: the Science of Scarcity and Diversity* (Ed. by M. E. Soulé), pp. 35–56, Sinauer, Sunderland, Massachusetts.

Rayner, J. M. V. (1985). Linear relations in biomechanics: the statistics of scaling functions. *Journal of Zoology, London*, **206A**, 415–39.

Read, A. F. (1987). Comparative evidence supports the Hamilton and Zuk hypothesis on parasites and sexual selection. *Nature*, **327**, 68–70.

Read, A. F. & Harvey, P. H. (in press). Life history differences among the eutherian radiations. *Journal of Zoology, London*.

Reznick, D. (1985). Costs of reproduction: an evaluation of the empirical evidence. *Oikos*, **44**, 257–67.

Ross, C. R. (1987). The intrinsic rate of natural increase and reproductive effort in primates. *Journal of Zoology*, **214**, 199–200.

Sacher, G. A. & Staffeldt, E. F. (1974). Relationship of gestation time to brain weight for placental mammals. *American Naturalist*, **108**, 593–616.

Saether, B. E. (1988). Pattern of covariation between life-history traits of European birds. *Nature*, **331**, 616–17.

Schmidt-Nielsen, K. (1984). *Scaling. Why is Animal Size so Important?* Cambridge University Press.

Schoener, T. W. (1968). The sizes of feeding territories among birds. *Ecology*, **49**, 123–41.

Schoener, T. W. (1983). Simple models of optimal feeding territory size: a reconcilation. *American Naturalist*, **121**, 608–29.

Stearns, S. C. (1983). The effects of size and phylogeny on patterns of covariation in the life-history traits of lizards and snakes. *American Naturalist*, **123**, 56–72.

Stephens, D. W. & Krebs, J. R. (1986). *Foraging Theory*. Princeton University Press.

Sugihara, G. (1980). Minimal community structure: an explanation of species abundance patterns. *American Naturalist*, **116**, 770–87.

Sugihara, G. (in press). How *do* species divide resources? *American Naturalist*.

Sutherland, W. J., Grafen, A. & Harvey, P. H. (1986). Life history correlations and demography. *Nature*, **320**, 88.

Taylor, L. R. & Taylor, R. A. J. (1977). Aggregation, migration and population mechanics. *Nature*, **265**, 415–21.

Thompson, D. W. (1942). *Growth and Form*. Macmillan, New York.

Tukey, J. W. (1962). The future of data analysis. *Annals of Mathematical Statistics*, **33**, 1–67.

van Noordwijk, A. J., van Tienderen, P. H. & de Jong, T.J. (1985). Genealogical evidence for random mating in a natural population of the great tit (*Parus major* L.) *Naturwissenschaften*, **72S**, 104–5.

Vitt, L. J. & Seigel, R. A. (1985). Life history traits of lizards and snakes. *American Naturalist*, **125**, 480–4.

Western, D. & Ssemakula, J. (1982). Life-history patterns in birds and mammals and their evolutionary interpretation. *Oecologia*, **54**, 281–90.

Wetton, J. H., Carter, R. E., Parkin, D. T. & Walters, D. (1987). Demographic study of a wild house sparrow population by DNA fingerprinting. *Nature*, **327**, 147–9.

APPENDIX: DATA BANKS OF THE BRITISH TRUST FOR ORNITHOLOGY

J. J. D. GREENWOOD

BTO, Beech Grove, Tring, Hertfordshire HP23 5NR

The BTO conducts research through a network of over 8000 amateurs and its data banks extend over several decades. Here I describe briefly those schemes that provide data most relevant to evolutionary ecology.

Over 19 million birds had been ringed (banded) by the end of 1987, of which 2.2% have been subsequently recovered. In addition, all ringers supply numbers of certain species ringed during the summer months, broken down into birds hatched that spring and older birds, from which we can model survival rates. Ringing also generates population numbers at 'constant effort' sites from which production and survival rates can be estimated.

We hold records of habitat, site and contents for over 700 000 nests, providing information on the breeding biology of many species. Full computerization of these records would allow us to answer questions that would be inaccessible to most research projects, such as regional and annual differences in breeding success over long periods of time.

The 'Common Bird Census' has been running for about 25 years and the 'Waterways Birds Census' for 15. They produce indices of the variation in numbers of the commoner species on farmland, in woodland, and on waterways. Changes can be related to environmental changes and the data are sufficient to illuminate, with caution, more general questions, such as long-term patterns of change in population levels, and changes in habitat use.

In a similar way, the 'Birds of Estuaries Enquiry' provides information on population changes, seasonal patterns of use, and the ways in which these are related to weather and to characteristics of the estuaries.

Other surveys and censuses of selected species include the annual census of herons (*Ardea cinerea*) started in 1928. We can integrate diverse information in single studies, such as that of changing population levels, breeding success and nesting season of stock doves (*Columba cenas*) in relation to changing crops and pesticide usage. In future, individual schemes will be better integrated, to provide assessments of key population characteristics and how they change from year to year.

The primary criterion for data collection is whether there is an important scientific question to be answered, followed by whether we can answer it. This involves greater co-operation between amateurs and professionals. It is important that professionals acknowledge the amateur contribution to the raw data when they publish in the technical literature and that they write accounts of their work that are seen by amateurs, such as in the Trust's newsletter.

10. AN EXPERIMENTALIST'S APPROACH TO THE ROLE OF COSTS OF REPRODUCTION IN THE EVOLUTION OF LIFE-HISTORIES

L. PARTRIDGE
Department of Zoology, University of Edinburgh, West Mains Road, Edinburgh EH9 3JT, UK

INTRODUCTION

The use of experiments varies greatly between different areas of science. The role of experiment is to control confounding variables and to manipulate those of interest, so that their effects can be deduced. An experimental approach is therefore likely to be of value whenever the mechanisms underlying the behaviour of a complex system are examined. Use of experiments is obviously limited in the first instance by practical considerations; astronomers cannot at present perturb planetary motion and palaeontologists cannot do breeding experiments with fossils. In ecology too, the ultimate limits are set by physical constraints on what can be manipulated (Kareiva & Anderson 1988). However, other factors such as economics and the perceived value of an experimental approach are also likely to be important in practice (May 1989).

A survey of the recent papers published in two long-established non-applied British journals (*Journal of Animal Ecology* and *Journal of Ecology*) and an American equivalent (*Ecology*) shows some differences in habits between the communities on the two sides of the Atlantic (Table 10.1). The proportion of experimental papers does not differ significantly between the two British journals (χ^2 = 1.1, d.f. = 1, N. S.), and is a minority in both cases. The majority of the papers in the American journal contain some experimental work, and the difference in proportion is significantly different from the two British journals combined (χ^2 = 24.8, d.f. = 1, $P < 0.001$). A possible explanation is that the nature of the studies conducted in the two communities does not differ, but that there is a difference in editorial policy between these American and British journals. One might then expect American observational studies to be sent to the British journals and the British experimental work to the American one. However, a much higher proportion of the papers in the

TABLE 10.1. An analysis of the descriptive and experimental papers published in three ecological journals in the 2 years up to and including July 1988. Short notes, symposia, reviews and invited papers were omitted from the analysis. Papers were classified as 'descriptive' if they contained no standardization or manipulation of variables; pure measurement, however sophisticated, was therefore included in this category. Contributions were classified as 'experimental' if they contained any experimental work, however little. Theoretical papers (which were few) were assigned on the basis of the type of data with which they were designed to be used

	Numbers of papers		Percentage experimental
	Descriptive	Experimental	
Journal of Animal Ecology	85	58	41
Journal of Ecology	103	53	34
Ecology	150	199	57

American journal are by American authors. It is possible that the 'missing' studies are sent to other journals. It seems much more likely that the perceived value of experiments differs, and is perhaps associated with differences in funding policy for ecological research.

The aim of the present paper is to explore the value of experiments in the study of costs of reproduction, which play a central role in current life-history theory. It is argued that an experimental approach is frequently indispensable to an advance in understanding in this area, and that in order to make progress we need a great deal more information about the exact mechanisms involved in mediating reproductive costs.

NATURAL SELECTION AND LIFE-HISTORY

The life-history of an organism is the combination of age-specific survival rates and fertilities seen in a typical individual. The life-history leading to the greatest fitness is the one that maximizes the Malthusian parameter r, or intrinsic rate of population increase of the genotype specifying the trait, under the prevailing ecological conditions. The parameter r is determined by age-specific survival and reproductive rates, and is defined by the characteristic equation:

$$1 = \int_0^w l_x m_x e^{-rx} dx,$$

where x is age, l_x is survival probability to age x, m_x is fertility at age x and w is the age of last breeding. The main assumptions involved are frequency-independent weak selection, stable age distribution and a constant environment (Charlesworth 1980; Caswell 1989).

An evolutionarily ideal organism would therefore commence breeding at birth at the maximum possible rate and would continue to do so throughout its infinite lifespan. The great diversity seen in age at first breeding, reproductive rate and lifespan would make no evolutionary sense if this were possible, and most theoretical models of life-history evolution assume that reproduction incurs costs in terms of survival and future fertility, so that only certain constrained combinations are possible in practice. The life-history favoured by selection therefore depends upon the extrinsic and intrinsic constraints on age-specific fertility and survival rates.

Reproductive costs have important implications for the timing and extent of reproduction. For instance, it may pay to postpone reproduction completely if growth, maturation or learning leads to an increase in fertility with age, because the loss of present offspring may be more than compensated by the gain in survival of the parent to later ages or by an increase in reproductive rate at later ages (Gadgil & Bossert 1970; Pianka & Parker 1975; Charlesworth 1980). It may also be advantageous to reproduce at less than the maximum possible rate, if the gain in future reproductive potential is sufficiently high.

Much of the theoretical and empirical work on optimal reproductive rate has been devoted to understanding the evolution of avian clutch size. Lack (1954) suggested that nidicolous birds would be selected to lay the most productive clutch size, defined as that clutch size resulting in the largest number of young surviving to breed, and he thought that the ability of the parents to care for the hatched young would be the major limitation on reproductive success. Subsequent empirical studies (reviewed by Lessells 1986) made it clear that birds frequently lay a clutch smaller than the most productive.

Several explanations for the discrepancy have been suggested. Theoretical studies by Williams (1966) showed that one possible reason could be a deleterious effect on the survival or future fertility of adults laying large clutches. An explicit model of this suggestion for birds formulated by Charnov & Krebs (1973), is illustrated in Fig. 10.1. A concave down relationship between the probability of nestling survival to breed leads to a domed relationship between total clutch productivity and clutch size, with the most productive brood size b_o at the peak (Fig. 10.1(a)). If a linear increase in parental mortality before the next breeding season is superimposed on this picture (Fig. 10.1(b)), then the parent is selected to maximize the difference between gain of progeny and loss of self as breeders in the next season (the model assumes parthenogenesis and breeding by young in the season following their birth, so that they become reproductively equivalent to their parents at that time), which leads to a

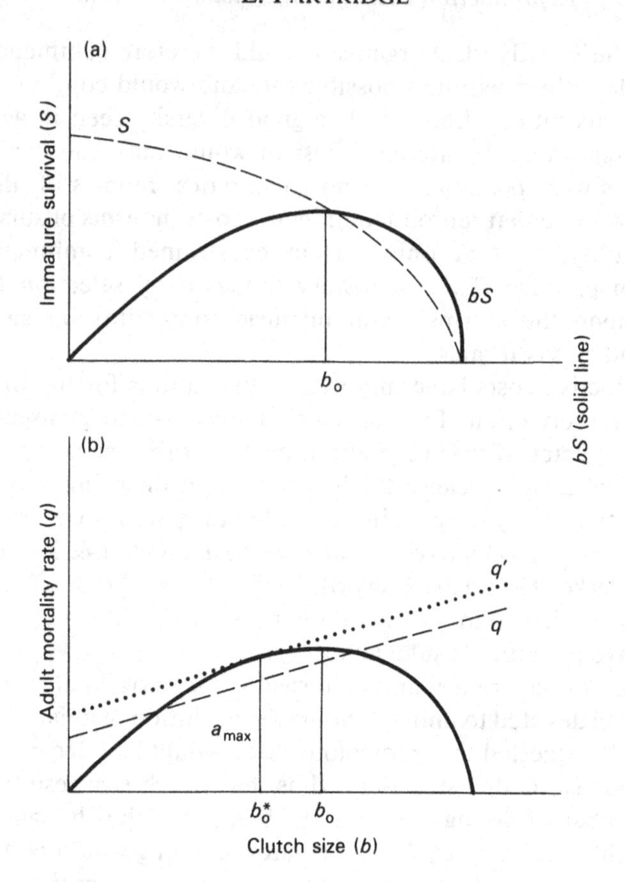

FIG 10.1 The model of optimal clutch size for birds formulated by Charnov & Krebs (1974). (a) Immature survival (S) declines with increasing clutch size (b) leading to a domed relationship between total clutch productivity (bS) and clutch size, with the most productive brood size at b_o. (b) Adult mortality rate (q) shows a linear increase with clutch size, and the optimal clutch size b_o^* maximizes the difference (α) between the loss of the parent and gain of offspring, leading to an optimum at a lower clutch size than b_o.

new lower optimal brood size b_o. Testing of this kind of model therefore requires measurement of both production of recruited offspring and costs of reproduction to parents in relation to different clutch sizes.

MEASURING COSTS OF REPRODUCTION

Despite the central role of costs of reproduction in life-history theory, there is a continuing dispute in the literature about their existence. Much of this controversy has stemmed from differences in the techniques used to detect costs, some of which fail to control major sources of confounding variation (Partridge & Harvey 1988). The problem arises in purely correlational studies, where no experimental manipulation of reproductive costs is made. Individuals are allowed to reproduce at their chosen rate, in field or laboratory, and their subsequent survival and/or fertility monitored. This approach can suffer from four major difficulties (Table 10.2). Perhaps the most serious of these is individual variation. For instance, in birds, evidence is starting to accumulate for individual differences in ability to rear broods of different sizes, with each individual tending to produce its own optimal clutch (Gustaffson & Sutherland 1988; Pettifor, Perrins & McCleery 1988). If the 'best' birds lay the largest clutches and survive well, then a spurious positive correlation between reproductive rate and survival could be generated. This and the other confounding variables listed in Table 10.2 mean that, to measure costs, it is essential that comparable individuals be randomly assigned to experimental treatments and environments. A survey of laboratory and field studies of plants and animals (Reznick 1985) and one of avian field studies (Partridge, in press) showed the widespread occurrence of reproductive costs, with negative results and cases of apparent negative costs of reproduction occurring mainly in correlational studies. Those correlational studies that did show evidence of costs may therefore have underestimated their magnitude.

The experimental studies covered by these surveys were mainly phenotypic; environmental manipulations were used to vary reproductive rate. However, a second area of dispute here concerns the relative merits of genetic and phenotypic manipulations of reproductive rate. Genetic correlations can be measured in appropriately designed breeding experiments and by artificial selection, and the few such studies of reproductive costs have in general revealed a negative effect of reproductive rate on subsequent survival and fertility (Reznick 1985). There are strong practical motives for avoiding work on genetic correlations, such as the relatively small changes in reproductive rate that can be effected, the large errors generally associated with the measurements and the extreme difficulty of measuring genetic correlations adequately under field conditions. However, phenotypic manipulations are only of any evolutionary relevance if the response to them is the same as that to a genetic manipulation of the same magnitude. It is at present not clear if this is always or ever the case. Phenotypic studies have come to have a bad name amongst geneticists because the correlations between reproductive rate and subsequent survival or fertility revealed by *correlational* phenotypic studies often differ in sign from the genetic correlations.

TABLE 10.2. Potential confounding variables in tests of costs of reproduction

---

1 *Environmental variation*
The individuals being compared inhabit different environments that vary in suitability
Potentially problematic where naturally occurring correlations between reproductive rate and subsequent survival or fertility are measured in the field or in inadequately controlled laboratory conditions, so that individuals experience environments of varying degrees of severity (Bell & Koufopanou 1985; Partridge & Harvey 1985; Reznick 1985; van Noordwijk & de Jong 1986)

2 *Gene/environment interaction*
Individuals are placed in an identical non-natural environment to which they are not equally suited
Potentially problematic where naturally occurring correlations are measured on individuals of different genotype (species, clones or individuals) placed in identical controlled field or laboratory conditions (Reznick 1985; Service & Rose 1985; Reznick, Perry & Travis 1986)

3 *Phenotypic variation*
The individuals compared differ in phenotype in a way that affects their ability both to survive and to reproduce
Potentially problematic if naturally occurring correlations are measured on individuals varying in genotype or in developmental or more recent history, so as to produce variation in total reproductive potential (Partridge & Farquhar 1983; Rose 1984a; Bell & Koufapanou 1985; Partridge & Harvey 1985; Reznick 1985; van Noordwijdk & de Jong 1986)

4 *Genetic correlation by linkage disequilibrium*
The individuals compared differ in life-history traits with an independent genetic basis
Potentially problematic if individuals from previously at least partially reproductively isolated populations are compared, so that some linkage disequilibrium has arisen in genes affecting survival or fertility

---

However, there have been very few cases where the results of genetic and *experimental* phenotypic studies have been compared, and the available results do give some cause for optimism. For example in female *Drosophila melanogaster* the genetic correlations between early fertility and subsequent survival and fertility are negative (Rose & Charlesworth 1981a,b; Rose 1984a; Luckinbill *et al.* 1984), while the phenotypic correlations are positive (Partridge 1988). However, an experimental increase in the rate of egg-production or of exposure to males produces a negative effect on female survival (Partridge, Green & Fowler 1987). More studies of this kind are sorely needed, and meanwhile the main task for phenotypic experiments is to identify the nature of reproductive costs and hence design the appropriate experiments to measure their magnitude.

HAVE COSTS OF REPRODUCTION AFFECTED THE EVOLUTION OF AVIAN CLUTCH SIZE?

Brood size manipulation experiments have clearly revealed costs of reproduction in several bird species. However, few have attempted any quantitative assessment of their possible impact on the evolution of clutch size, and in one case at least the costs were insufficient to account for the difference between the most productive clutch size and the clutch size generally laid (Reid 1987). However, there are several reasons for suspending judgement on this question.

First, it is likely that manipulation of brood size misses several costly aspects of reproduction. For the purposes of understanding the evolution of reproductive rate, we are interested only in those costs which change with the number of offspring produced, not with one-off costs like nest building or the acquisition of a territory. Costs of egg-production and of incubation are known to be high in at least some species (e.g. Winkler 1985; Tinbergen *et al.* 1987), and there is a need for techniques to manipulate these activities. In addition, the results of some studies have shown clearly that costs of reproduction can last into the ensuing breeding season in the form of a drop in fertility (Roskaft 1985; Tinbergen 1987; Gustaffson & Sutherland 1988; Nur 1988), and few studies of reproductive costs have attempted to cover both survival and fertility over a long period. In addition, brood size manipulations may not always have the effect of manipulating parental work rate. If a bird has in some sense 'decided' on its work rate for the brood before the manipulation is done, then the manipulation will be informative about the effect of parental work rate on the offspring, but not about the effects on the parent. In one study at least (Korpimaki 1988), parental effort appeared to be little affected by an increase in the brood, although in other studies (e.g. that of Dijkstra 1988) there was clearly some effect. More measurements of behavioural effects on parents would be valuable.

Second, it is important that the expected evolutionary effects of costs are correctly entered into optimality models. If individual birds differ in their breeding potential, then the exact design of experiments needs careful thought; the effect of an increase of three in the clutch size on offspring recruited and future parental reproduction will probably not be the same for a great tit female in poor condition as for one in good condition, and the effects on overall population optimal clutch size may not even be additive. It may therefore be necessary to take some explicit account of individual variation.

Third, it will be important to bear in mind that other forces, particularly year-to-year variation in conditions for breeding, may select for a clutch size below the one that is on average the most productive (Boyce & Perrins 1987).

Lastly, there is a real problem with sample sizes. The error variances in brood size manipulation studies have typically been very large, but quite small reproductive costs can shift the optimum clutch size by several eggs (Nur, in press). Large-scale experiments are therefore needed.

These considerations raise a general difficulty with the testing of optimality models, which is that, if the real world appears not to comply with the model, it is tempting to fiddle with the assumptions or with the type of data used in testing until a fit appears (Maynard Smith 1978; Gould & Lewontin 1979). Clearly such a procedure is unacceptable, and the measurement of costs of reproduction in relation to reproductive rate can be best applied to the construction of realistic optimality models, which can then provide an explicit basis for comparative tests of the general importance of reproductive costs in life-history evolution. We need to know which aspects of reproduction are most costly in different ztaxa, the timing of their impact and the shapes of the curves relating them to reproductive rate. Birds and plants are likely to be suitable material for field work on this problem.

MECHANISMS OF REPRODUCTIVE COSTS

There are two broad reasons for wishing to disentangle the mechanisms producing costs of reproduction. One has been mentioned, and is that only by understanding how costs work can we design appropriate experiments to measure them. The second is that they raise fundamental questions about physiology and constraints on biological function; we do not at present understand why high rates of reproduction lead to an elevated risk of death, or why high growth trajectories can have the same effect (Clutton-Brock, Albon & Guiness 1985). Nor is it known what, if any, aspects of somatic repair are nutritionally costly (Kirkwood 1981). Identification of the structures or processes that set limits to longevity, growth and fertility is of fundamental and applied importance.

A useful distinction has been drawn between ecological and physiological costs (Calow 1979). The former occur when reproduction increases the level of exposure to external hazards such as disease, predation or accidents (Tuttle & Ryan 1981). Reproduction could cause physiological costs if it, for instance, removed resources from other functions such as growth, somatic repair or maintenance. Although competition for nutrients is usually assumed to be responsible for physiological costs, there is little direct evidence that this is the case (Bryant 1988; this volume).

Laboratory experiments with *Drosophila* females have gone some way to unravelling the nature of physiological costs of reproduction, and the results do not at present strongly implicate nutrient limitation as an important mechanism. X-irradiation of wild-type females and short exposures to high temperatures have the paradoxical effect of extending the lifespan. These effects are not seen in mutant *ovaryless* females, which have about the same longevity as virgin wild-type females, given one of the above sterilizing treatments (Maynard Smith 1958; Lamb 1964). These results implicate ovarian activity as a cause of death, and could mean either that some aspect of oogenesis or vitellogenesis is responsible, or that mating is dangerous, since females that produce a lot of eggs also re-mate more frequently (Trevitt, Fowler & Partridge 1988). Both processes seem to be involved. If mating is standardized, and egg-production rate manipulated by variation in nutrition or by availability of oviposition sites, longevity is reduced by egg production. This result occurs with the same magnitude irrespective of the nutritional status of the females, which suggests that competition for nutrients may not be at the basis of the effect. In addition, exposure to males *per se* reduces lifespan in females whose egg production and egg fertility are experimentally standardized (Partridge *et al.* 1986; Partridge, Green & Fowler 1987).

Exposure to males could be costly as a result of some consequence of mating itself, such as mechanical injury, infection or damaging effects of sperm or accessory fluid. There could also be an effect of events prior to mating such as increased activity, competition, harassment or hormonal changes consequent on courtship. To test for post-mating effects, the survival rates of females continuously exposed to two intact males was compared with that of females exposed to intact males on 1 day in 3, and to males with their external genitalia microcauterized on the remaining 2 days. The microcauterized males courted normally, but were unable to mate with the female, and they served as a control for any effects of pre-mating exposure on female survival. The females continuously exposed to intact males showed significantly lower survival rates (Fig. 10.2), showing that the survival cost of exposure to males is in part a consequence of mating (Fowler & Partridge 1989).

Pre-mating effects were investigated in an experiment where the survival rates of females kept alone were compared with those of females kept with males, but prevented from mating, by exposure either to two microcauterized males, or to two mutant fruitless males. Males of the fruitless type will court but not mate with females. Their behaviour is not the same as wild-type males, because they court other males as readily as they court females and are also subject to unusually high levels of

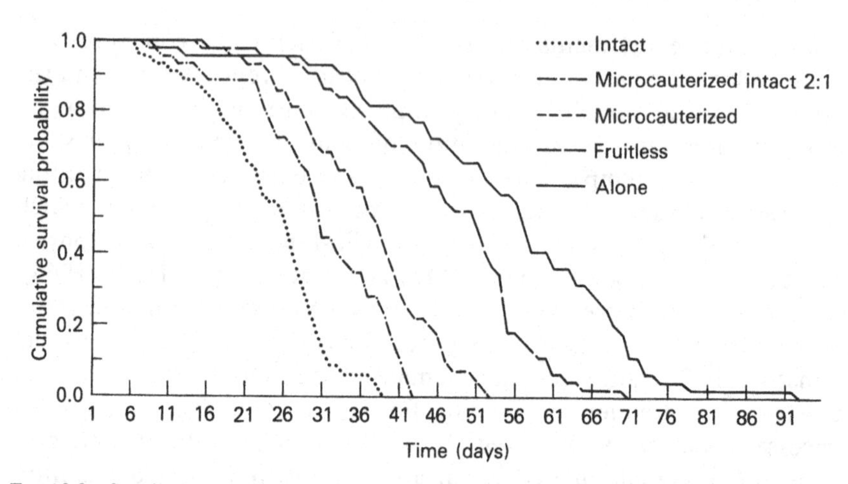

FIG 10.2 Survival curves for females of *Drosophila melanogaster* in pre- and post-mating experiments. Pre-mating costs are evident because females exposed to microcauterized or fruitless males survived less well than females kept alone. This cannot have been entirely an effect of competition, because females exposed to fruitless males survived better than those exposed to microcauterized males. Post-mating costs are evident because continuous exposure to intact males leads to a higher death-rate than that experienced by females intermittently exposed to microcauterized males (Fowler & Partridge, 1989).

courtship by other males (Hall 1978; personal observation). Both groups of females exposed to males had lower longevity than females kept alone, with females exposed to microcauterized males being significantly more affected than those exposed to fruitless males (Fig. 10.2). This second finding suggests that there must be a pre-mating cost of exposure to males other than competition for food, because otherwise the two kinds of males should not differ in their effects on the female (K. Fowler & L. Partridge, unpublished results).

There is clearly some way to go in understanding reproductive costs in even this relatively simple system; we do not know how the costly processes exert their effects, nor have we yet identified the physiological targets of the costly activities. In addition, the effects of reproduction on subsequent fertility are virtually unexplored. What is clear from the results is that several costly processes are involved.

DOES REPRODUCTION ACCELERATE AGEING?

One suggested cost of reproduction is the acceleration of ageing, which is the inherent decline in biological performance with increasing age in the adult lifespan; in evolutionary terms this is manifested as a drop in life expectancy and fertility with advancing age. This decline is especially obvious in animals in zoos or laboratories, where they are protected from the normal hazards of the wild, but is also evident in natural populations. For instance field studies of birds have shown a decline in fertility later in the lifespan in arctic terns (Coulson & Horobin 1976), great tits (Perrins 1979) and sparrowhawks (Newton, Marquiss & Moss 1981), while a decrease in survival rate has been reported in great tits (Webber 1975), fulmars (Dunnet & Ollason 1978) and black-capped chickadees (Loery *et al.* 1987).

There are some difficulties with detecting and measuring the effects of ageing. One is that the individuals that survive to later ages may well not be a random sample of the cohort from which they are drawn. If older individuals tend to be ones which were of better quality to start with, then the effects of senescence will be underestimated. For fertility, this problem can be circumvented by making repeated measurements on the same individuals. A second problem is that senescence is not necessarily the only age-related change in performance. Especially in animals with indeterminate growth, fertility is likely to show an increase with age, which could mask the effects of ageing. Even where growth is determinate, under some circumstances theory predicts an increase in reproductive effort with age. For instance, if senescence itself lowers life expectancy and this is detectable to the animal, then, since future reproductive potential is lowered, current reproduction becomes relatively more valuable. This may explain the increase in parental effort and reproductive success found in elderly female red deer by Clutton-Brock (1984). Longitudinal data on both survival rates and the fertility of known individuals are likely to yield the most convincing results.

Ageing could occur as a result of the inevitable effects of wear and tear and accumulation of irreparable errors. If this idea were correct, then ageing would require no evolutionary explanation. However, such a view cannot explain the very different potential lifespans and rates of senescence in otherwise biologically similar organisms (Comfort 1956; Williams 1957), or the association between the rate of ageing and mode of reproduction (Bell 1984), both of which imply variation in the extent to which the effects of damage are combated. One theory of the evolution of ageing, the 'pleiotropy theory', suggests that senescence is a deleterious side-effect of genes causing high fertility early in the lifespan (Williams 1957). The reason that such genes are expected to survive in natural populations is that the force of natural selection declines on cohorts of increasing age, because externally imposed mortality from disease, predation and accidents means that there will be fewer survivors to later ages, irrespective of any effects of ageing itself. A gene that affects the survival probability or reproductive success of older adults will therefore be expressed in a lower proportion of its carriers, and will hence be subject to less intense selection than a gene affecting younger age classes (Medawar 1952; Charlesworth 1980). If a gene affects performance at more than one age, then selection will be more intense on the earlier effects, so that a gene that increases survival or fertility early in the lifespan at the expense of decreasing it later can spread by natural selection (Hamilton 1966; Charlesworth 1980).

Physiological costs of reproduction provide a possible basis for gene action of the kind postulated in this theory; if the allocation of nutrients to reproduction denies them to growth, maintenance or repair, then future fertility and survival probability could be irreversibly compromised, so that ageing would be accelerated as a consequence of reproduction. The different rates of senescence seen in different species could then be interpreted as a consequence of different allocation decisions, and the species with the highest potential survival rates and future fertility would be those with the highest levels of resource input into growth and somatic repair and maintenance (Kirkwood 1977, 1981; Kirkwood & Holliday 1979).

There have been very few direct tests of the pleiotropy theory, and none in the field. It relies on the idea that reproduction has deleterious effects which are persistent and increase in impact with age. One experiment with male *Drosophila* has not supported this idea. Exposure to virgin females shortens male lifespan, but this appears to be because of an elevated risk of death at the time of exposure, not afterwards. If the exposure to females is switched part way through the male lifespan, males of the same age but with a different history of exposure to virgin females do not differ in their life expectancy when their current exposure to females is standardized (Partridge & Andrews 1985). A similar experiment with females revealed a reversible effect of exposure to males, but this time there was a lag period during which reproductive history did affect life expectancy (Partridge *et al.* 1986). The results of environmental manipulations of reproductive rate therefore do not at present support the idea that reproduction accelerates senescence.

A second approach to the problem has exploited genetic variation in reproductive rate in female *Drosophila*, mainly by using artificial selection on age at breeding (e.g. Rose & Charlesworth 1981 a, b; Luckinbill *et al*, 1984; Rose 1984b). 'Old' lines were propagated by breeding from old adults while in the 'young' lines the breeders were newly eclosed adults. The 'old' lines were therefore selected for high longevity and high fertility late in the lifespan. The results have in general revealed a negative genetic correlation of high fertility early in life with longevity and fertility later in the lifespan, apparently supporting the pleiotropy theory. However, the results of the environmental manipulations described above make it clear that reproduction can reduce longevity by a short-term risk effect, rather than by an acceleration of senescence, so that the negative genetic correlation with longevity could occur by a similar mechanism (Partridge 1987). Furthermore, the increased late fertility found in 'old' lines in artificial selection experiments could be a direct response to selection for high fertility late in life, rather than a secondary consequence of lower reproductive rate earlier in the lifespan. Further work on these lines is required, to determine exactly how the increase in longevity and late fertility in the 'old' lines has been achieved, and in particular what role is played by reproductive rate earlier in the lifespan.

CONCLUSIONS

Clearly there is a need for more work on the role of reproductive costs in ageing and in the evolution of life-histories generally. For progress to be made, we need to know much more about the mechanisms by which reproductive costs occur. An experimental approach is clearly going to be required, with a judicious use of environmental and genetic manipulations and of laboratory and field studies. Ultimately, the best tests of the evolutionary theories may be comparative, but an understanding of mechanisms is vital for the construction of realistic explicit optimality models to provide the theoretical underpinning of the comparative work.

REFERENCES

Bell, G. (1984). Evolutionary and nonevolutionary theories of senescence. *American Naturalist,* **124**, 600–3.

Bell, G. & Koufapanou, V. (1985). The cost of reproduction. *Oxford Surveys in Evolutionary Biology,* Vol. 3 (Ed. by R. Dawkins & M. Ridley), pp. 83–131. Oxford University Press.

Boyce, M. S. & Perrins, C. M. (1987). Optimizing great tit clutch size in a fluctuating environment. *Ecology,* **68**, 142–53.

Bryant, D. M. (1988). Energy expenditure and body mass changes as measures of reproductive costs in birds. *Functional Ecology,* **2**, 23–34.

Calow, P. (1979). The cost of reproduction—a physiological approach. *Biological Reviews.* **54**, 23–40.

Caswell, H. (1989). Life-history strategies. *Ecological Concepts* (Ed. by J. M. Cherrett), pp. 285–307. *Symposia of the British Ecological Society,* 29. Blackwell Scientific Publications, Oxford.

Charlesworth, B. (1980). *Evolution in Age-structured Populations.* Cambridge University Press.

Charnov, E. L. & Krebs, J. R. (1973). On clutch size and fitness, *Ibis,* **116**, 217–19.

Clutton-Brock, T. H. (1984). Reproductive effort and terminal investment in iteroparous animals. *American Naturalist,* **123**, 212–29.

Clutton-Brock, T. H., Albon, S. D. & Guiness, F. E. (1985). Parental investment and sex differences in juvenile mortality in birds and mammals. *Nature,* **313**, 131–3.

Comfort, A. (1979). *The Biology of Senescence.* Elsevier, New York.

Coulson, J. C. & Horobin, J. (1976). The influence of age on the breeding biology and survival of the Arctic tern *Sterna paradisea. Journal of Zoology, London,* **178**, 247–60.

Dijkstra, C. (1988). *Reproductive tactics in the kestrel.* Ph.D. thesis, University of Groningen.

Dunnet, G. M. & Ollason, J. C. (1978). The estimation of survival rate in the fulmar. *Journal of Animal Ecology,* **47**, 507–20.

Fowler, K. & Partridge, L. (1989). A cost of mating in female fruit-flies. *Nature,* **388**, 760–1.

Gadgil, M. & Bossert, W. (1970). Life historical consequences of natural selection. *American Naturalist,* **104**, 1–24.

Gould, S. J. & Lewontin, R. C. (1979). The spandrels of San Marco and the Panglossian paradigm: a critique of the adaptationist programme. *Proceedings of the Royal Society of London,* **B, 205**, 581–98.

Gustaffson, L. & Sutherland, W. J. (1988). The cost of reproduction in the collared flycatcher *Ficedula albicollis. Nature,* **335**, 813–15.

Hall, J. C. (1978). Courtship among males due to a male-sterile mutation in *Drosophila melanogaster. Behavior Genetics,* **8**, 125–41.

Hamilton, W. D. (1966). The molding of senescence by natural selection. *Journal of Theoretical Biology,* **12**, 12–45.

Kareiva, P. & Anderson, M. (1988). Spatial aspects of species interactions: the wedding of models and experiments. *Theoretical Community Ecology* (Ed. by A. Hastings). American Mathematical Society, Providence, Rhode Island.

Kirkwood. T. B. L. (1977). Evolution of ageing. *Nature,* **270**, 301–4.

Kirkwood, T. B. L. (1981). Repair and its evolution: survival versus reproduction. *Physiological Ecology.* (Ed. by C. R. Townsend & P. Calow), pp. 165–89. Blackwell Scientific Publications, Oxford.

Kirkwood, T. B. L. & Holliday, R. (1979). The evolution of ageing and longevity. *Proceedings of the Royal Society of London,* **B, 205,** 531–46.

Korpimaki, E. (1988). Costs of reproduction and success of manipulated broods under varying food conditions in Tengmalm's owl. *Journal of Animal Ecology,* **57,** 1027–39.

Lack, D. (1954). *The Natural Regulation of Animal Numbers.* Clarendon Press, Oxford.

Lamb, M. J. (1964). The effects of radiation on the longevity of female *Drosophila subobscura. Journal of Insect Physiology,* **10,** 487–97.

Lessells, C. M. (1986). Brood size in Canada geese: a manipulation experiment. *Journal of Animal Ecology,* **55,** 669–89.

Loery, G., Pollock, K. H., Nichols, J. D. & Hines, J. E. (1987). Age-specificity of black-capped chickadee survival rates: analysis of capture-recapture data. *Ecology,* **68,** 1038–44.

Luckinbill, L. S., Arking, R., Clare, M. J., Cirocco, W. & Buck, S. A. (1984). Selection for delayed senescence in *Drosophila melanogaster. Evolution,* **38,** 996–1003.

May, R. (1989). Levels of organization in ecology. *Ecological Concepts.* (Ed. by J. M. Cherrett), pp. 339–60. Symposia of the British Ecological Society, 29. Blackwell Scientific Publications, Oxford.

Maynard Smith, J. (1958). The effects of temperature and of egg-laying on the longevity of female *Drosophila subobscura. Journal of Experimental Biology,* **35,** 832–42.

Maynard Smith, J. (1978). Optimization theory in evolution. *Annual Review of Ecology and*

Systematics, **9**, 31–56.

Medawar, P. B. (1952). *An Unsolved Problem of Biology*. Lewis, London.

Newton, I., Marquiss, M. & Moss, D. (1981). Age and breeding in sparrowhawks. *Journal of Animal Ecology*, **50**, 839–53.

Nur, N. (1988). The consequences of brood size for breeding blue tits. III. Measuring the cost of reproduction: survival, future fecundity, and differential dispersal. *Evolution*, **42**, 351–62.

Nur, N. (in press). The cost of reproduction in birds: an examination of the evidence. *Ardea*.

Partridge, L. (1987). Is accelerated senescence a cost of reproduction? *Functional Ecology*, **1**, 317–20.

Partridge, L. (1988). Lifetime reproductive success in *Drosophila*. *Lifetime Reproductive Success* (Ed. by T. H. Clutton-Brock), pp. 11–23. Chicago University Press.

Partridge, L. (in press). Lifetime reproductive success and life history evolution. *Lifetime Reproductive Success in Birds* (Ed. by I. Newton), Academic Press, London.

Partridge, L. & Andrews, R. (1985). The effect of reproductive activity on the longevity of male *Drosophila melanogaster* is not caused by an acceleration of ageing. *Journal of Insect Physiology*, **31**, 393–5.

Partridge, L. & Farquhar, M. (1983). Lifetime mating success of male fruitflies (*Drosophila melanogaster)* is related to their size. *Animal Behaviour*, **31**, 871–7.

Partridge, L., Fowler, K., Trevitt, S. & Sharp, W. (1986). An examination of the effects of males on the survival and egg-production rates of female *Drosophila melanogaster*. *Journal of Insect Physiology*, **32**, 925–9.

Partridge, L., Green, A. & Fowler, K. (1987). Effects of egg-production and of exposure to males on female survival in *Drosophila melanogaster*. *Journal of Insect Physiology*, **33**, 745–9.

Partridge, L. & Harvey, P. H. (1985). Costs of reproduction. *Nature*, **316**, 20–1.

Partridge, L. & Harvey, P. H. (1988) . The ecological context of life history evolution. *Science*, **241**, 1449–54.

Perrins, C. M. (1979). *British Tits*. Collins, London.

Pettifor, R. A., Perrins, C. M. & McCleery, R. H. (1988). Individual optimization of clutch size in great tits. *Nature*, **336**, 160–2.

Pianka, E. R. & Parker, W. S. (1975). Age-specific reproductive tactics. *American Naturalist*, **109**, 453–64.

Reid, W. V. (1987). The cost of reproduction in the glaucous-winged gull. *Oecologia*, **74**, 458–67.

Reznick, D. (1985). Costs of reproduction: an evaluation of the empirical evidence. *Oikos*, **44**, 257–67.

Reznick, D. N., Perry, E. & Travis, J. (1986). Measuring the cost of reproduction: a comment on papers by Bell. *Evolution*, **40**, 1338–44.

Rose, M. R. (1984a). Genetic covariation in *Drosophila* life-history: untangling the data. *American Naturalist*, **123**, 565–9.

Rose, M. R. (1984b). Laboratory evolution of postponed senescence in *Drosophilia melanogaster*. *Evolution*, **38**, 1004–9.

Rose, M. R. & Charlesworth, B. (1981a). Genetics of life history in *Drosophila melanogaster*. I. Sib analysis of adult females. *Genetics*, **97**, 173–86.

Rose, M. R. & Charlesworth, B. (1981b) Genetics of life history in *Drosophila melanogaster*. II. Exploratory selection experiments. *Genetics*, **97**, 187–96.

Roskaft, E. (1985). The effect of enlarged brood-size on the future reproductive potential of the rook. *Journal of Animal Ecology*, **54**, 255–60.

Service, P. M. & Rose, M. R. (1985). Genetic covariation among life-history components: the effect of novel environments. *Evolution*, **39**, 943–5.

Tinbergen, J. M. (1987). Costs of reproduction in the great tit: interseasonal costs associated with brood size. *Ardea,* **75,** 111–22.

Tinbergen, J. M., van Balen. J. H., Drent, P. J., Cave, A. J., Mertens, J. A. L. & den Boer-Hazewinkel, J. (1987). Population dynamics and cost-benefit analysis. *Netherlands Journal of Zoology,* **37,** 180–213.

Trevitt, S., Fowler, K. & Partridge, L. (1988). An effect of egg-deposition on the subsequent fertility and remating frequency of female *Drosophila melanogaster. Journal of Insect Physiology,* **34,** 821–8.

Tuttle, M. D. & Ryan, M. J. (1981). Bat predation and the evolution of frog vocalizations in the neotropics. *Science,* **214,** 677–8.

Van Noordwijk, A. J. & de Jong, G. (1986). Acquisition and allocation of resources: their influence on variation in life history tactics. *American Naturalist,* **128,** 137–42.

Webber, M. I. (1975). *Some aspects of the non-breeding population dynamics of the great tit* (*Parus major*). D. Phil. thesis, University of Oxford.

Williams, G. C. (1957). Pleiotropy, natural selection and the evolution of senescence. *Evolution,* **11,** 398–411.

Williams, G. C. (1966). Natural selection, the costs of reproduction, and a refinement of Lack's principle. *American Naturalist,* **100,** 687–90.

Winkler, D. W. (1985). Factors determining a clutch size reduction of California gulls (*Larus californicus*): a multi-hypothesis approach. *Evolution,* **39,** 667–77.

V. INTERRELATIONSHIPS BETWEEN ORGANISMS

Interactions between organisms, both within and between species, form a significant part of the subject matter of ecology. They have been touched on in Section III, but here the mechanisms involved in three kinds of interrelationships are examined in some detail.

Because of their size and longevity, forest trees are not normally considered good subjects for the study of dynamic ecology, but Shugart shows how modern techniques of computer simulation may be used to provide very large-scale and very long-term analyses of such systems. Of course, like all simulation models, they are only as good as the available data and assumptions permit, but Shugart and Urban's studies are based on vast data sets. This approach promises to lead to new insights into the long-term dynamics of forests.

The two other chapters in this section examine different aspects of interactions between dissimilar organisms (insects and plants) and a special but until recently neglected topic, mutualism (between ants and butterflies in this case). Edwards takes a somewhat iconoclastic look at the much debated concept of plant defence, and argues the need for a reappraisal of current thinking. Pierce shows by way of an intricately worked example how progress is being made by the combination of traditional field techniques, sophisticated field manipulations and laboratory chemistry uniting to relate ecology to evolution and behaviour to biochemistry.

11. FACTORS AFFECTING THE RELATIVE ABUNDANCES OF FOREST TREE SPECIES

H. H. SHUGART AND D. L. URBAN
Environmental Sciences Department, University of Virginia, Charlottesville, VA 22903, USA

INTRODUCTION

It has long been appreciated that the relative abundances of tree species in a forest vary along environmental gradients, and change at a given site-type as a result of gap-creation. The aim of this paper is to show how mathematical models can be used to describe more precisely the processes controlling relative abundances of trees, to predict the effects of changing conditions, and to test the completeness of our understanding of forest dynamics.

We begin by reviewing a conceptual model of the forest as a dynamic mosaic. We then proceed to consider the demographic mechanisms that underlie forest dynamics: tree establishment, growth and mortality. We will argue that these are qualitatively different phenomena that operate on disparate scales of space and time. This scale-incompatibility effectively decouples these mechanisms empirically, in that they cannot be resolved simultaneously in the same set of stand-level observations. Yet it is the very couplings among these mechanisms that generate forest pattern.

In order to examine the factors that affect the relative abundances of forest trees we need some means of integrating demographic mechanisms at a common scale. In the following sections, we present an overview of two approaches that we have found useful. The first approach is to use a set of forest simulation models based on the fates of individual trees. These are hierarchically structured simulators that integrate the demographics of individual trees to generate forest dynamics. This step allows us to explore the long-term implications of tree-to-tree as well as tree–environment interactions.

The second approach involves the use of a functional classification of tree species, in which species roles are defined according to the coupling of mortality and regeneration. These couplings imply a finite set of life-history 'strategies', which in turn imply simple rules that dictate patterns of forest dynamics. We will argue that this approach can serve as a powerful framework for interpreting forest pattern.

Processes of two kinds influence the relative abundances of forest trees. The first is environmental filtering: the sorting of trees by environmental constraints such as available light, temperature, soil moisture and mineral nutrient supply. The second ordering force is competition among trees, a phenomenon which can assume several different guises in forests. We shall explore both sets of factors, using empirical patterns as well as model-generated examples to illustrate what we believe are general patterns common to many forests.

A CONCEPTUAL MODEL OF FORESTS

A conceptual model of the forest as a dynamic mosaic serves as the framework for much of the following discussion. Whitmore (1982) reviewed several of the unifying features of gap dynamics of forests, including the allied features of tree species biology. The essential concept is that a forest can be abstracted as a mosaic of patches, each patch being the area locally dominated by a canopy tree. With the death of the dominant tree, the forest micro-environment is changed radically: more light reaches the forest floor and other abiotic factors change accordingly (notably temperature, soil moisture content and mineralization rate). A wave of seedlings may be established, and previously established but suppressed trees are released. These trees compete, and in the simplest case one comes eventually to dominate the canopy. When this tree dies, the cycle repeats.

The gap–mosaic concept is a venerable concept in ecology, with major contributions from Watt (1925, 1947) and Aubréville (1933, 1938). One of the expectations of the gap–mosaic view of forest dynamics is that mature or 'climax' forest should comprise a steady-state mixture of mosaic elements in all stages of development (Whittaker 1953; Bormann & Likens 1979).

DEMOGRAPHIC MECHANISMS OF FOREST DYNAMICS

The essential processes of forest dynamics include tree establishment, subsuming all events from germination to successful establishment of a sapling with a fair chance of survival; growth, emphasizing especially the relative success of different trees; and mortality, including natural age-related death as well as 'premature' mortality induced by environmental events.

Tree establishment

Germinating seeds and seedlings occupy sites of a scale of 0.0001 to 0.01 m^2; germination occurs over periods of weeks, and establishment over a few years. Trees typically produce very large numbers of seeds, and, despite the intervention of herbivores which may consume >90% of a seed crop, very numerous seedlings are produced by many species. The very high fecundity of the adults must be balanced by correspondingly high mortality rates of seedlings, and one may need to study many individuals — perhaps thousands — to estimate death–rates accurately (Hett & Loucks 1968). The patchy distribution of 'safe sites' *sensu* Harper *et al.* (1961) in forests, together with the logistical unlikelihood of tracking individuals to a point where their relative success can be determined, tends to decouple the fine-scale details of tree establishment from the dynamics of the forest *per se*; we may understand the mechanisms of germination, environmental filtering *sensu* Harper (1977) and establishment, but these details are logistically intractable at the level of the forest stand.

Tree growth

Tree growth, suppression and stand thinning take place within the zone of influence of canopy dominants (about 0.01–0.1 ha), and are resolved on time scales of several years to decades. At these scales the behaviour of individuals can be tracked empirically, so that the logistical problems are not as pronounced as those inherent in studies of seedling establishment. Further, the factors affecting tree growth are at this scale intuitively straightforward. Growth is the result of gross photosynthetic assimilation minus maintenance costs, and the effects of single environmental factors (biotic as well as abiotic) influencing photosynthesis and respiration have been well documented at the tissue level (Farquhar & von Caemmerer 1982; Jarvis & Leverenz 1983), although little attention has been paid to the interactions of relevant factors on tree growth. We conclude optimistically that we actually understand, in a deterministic sense, how the relevant mechanisms influence the growth rates of trees in different environs. But forest dynamics at this stage are subject to confounding factors that do not plague studies of demographic phenomena at other scales. The scaling-up of what is known of the physiology of tree tissues to the level of the whole plant continues to be a daunting problem. At present, it is difficult to derive mechanistically either annual or decadal tree growth expected under a particular environmental condition (Landsberg 1986), even for important commercial tree species growing in plantations (presumably the most straightforward and most studied case). One problem here is that the abiotic factors influencing net photosynthesis (light, temperature, moisture, mineral nutrients) covary with one another; further, these are influenced by the trees themselves. Thus, it is difficult to partial out these effects empirically.

Tree mortality

Trees have the potential to live for very long periods of time (centuries to millennia). Thus, natural mortality has a temporal scale of the order of hundreds of years, corresponding to annual rates of the order of 0.01 (Harcombe 1987); the spatial domain of mortality varies from that of the individual tree to large groups of trees which are, for example, destroyed by fire or blown down together (0.1 to 100 ha). At this scale, mortality is essentially a stochastic process: one may know what kinds of factors might make mortality more likely (Waring 1987), but to estimate mortality rates one must generally make inferences from a very large number of individuals, in effect substituting a large area for the long time span over which we would like to observe. Thus, the death of an individual tree is a chance event, and mortality schedules are stand-level approximations that apply to individuals only in a probabilistic sense.

Of course, a 'disturbance' — here defined, following Pickett & White (1985) as 'any relatively discrete event in time that disrupts ecosystem, community or population structure and changes resources, substrate availability, or the physical environment' — may alter patterns in mortality in a profound sense and, in doing so, alter the relative abundances of tree species (White 1979; Noble & Slatyer 1980; Mooney & Godron 1983; Pickett & White 1985). Relative to the death of an individual tree, most disturbances tend to increase the spatial scale and decrease the temporal scale of mortality. Some disturbances may also be more or less regular and predictable, and so seemingly less stochastic.

Lorimer (1980) reviewed temperate-zone studies of mature ('climax') forests and reported a substantial proportion of them to have uneven size or age distributions. Such distributions imply episodic mortality and/or regeneration. Episodic mortality favours species with adequate regeneration under the specific environmental conditions following the episode. Irregular episodes of mortality can maintain species richness by favouring the regeneration of one species in one incident (say if the annual climate was wet for a few years following the dieback of several canopy trees) and favouring another species in another episode (dry conditions following the dieback). Because of the longevity of trees, these events may be reflected in stand composition or structure for centuries.

INTERPRETATIVE TOOLS

It should be emphasized that there are two related aspects of these demographic mechanisms that make forests difficult subjects to study. First, as has already been shown, there is the incompatibility of the scales of the key events, which makes it logistically impractical to deal simultaneously with any two demographic process on an empirical basis. Secondly, there are the couplings and feedbacks between these processes that drive forest dynamics. Hence we must reconcile these processes in order to understand forests. Hierarchically structured, individual-based simulation models provide a useful means to this end.

Individual-based forest simulation models

Computer models that simulate the dynamics of a forest by following the fates of each individual tree were developed initially in the mid-1960s and have become increasingly used as the computer power available to ecologists has increased (Munro 1974; Shugart & West 1980). Huston, DeAngelis & Post (1988) point out that one advantage of individual-based models is that two implicit assumptions associated with more traditional modelling approaches are not necessary. These are the assumptions that (i) the unique features of individuals are sufficiently unimportant for individuals to be treated as identical, and (ii) the population is 'perfectly mixed' so that there are no local spatial interactions of any important magnitude. These assumptions seem particularly inappropriate for trees, which are sessile, vary greatly in size over their lifespan, and influence (and are influenced by) neighbouring plants in a size-dependent manner.

Most individual-based models of forests are parsimonious in that they are based on simple rules for interactions among individuals, e.g. shading, competition for limiting resources, in conjunction with simple rules for birth, death and growth of individuals. Further, there is a considerable body of information on the performance of individual trees (growth rates, establishment requirements, height/diameter relations) that can be used directly in estimating the parameters of such models. Admittedly the generality of such models is commonly limited by the site-specific nature of the information on growth rates etc.

The individual-tree models that will be used in the following discussion are based on the FORET model and its descendants, in particular a spatially explicit individual tree model call ZELIG (Smith & Urban 1988). Shugart (1984) provides more detailed discussion of these models, their derivation and testing.

The FORET model was originally derived from the JABOWA model of Botkin, Janak & Wallis (1972). Both models simulate the fate of individual trees, computing the birth, changes in diameter and death of each tree on a small simulated plot (about 0.01 to 0.1 ha). The diameter of each tree is used in conjunction with empirically derived allometric equations to compute height, leaf area, and above-ground biomass for each tree. These variables can then be summed for entire plots; a larger set of simulated plots can be aggregated to estimate stand-level responses.

Establishment of trees is a stochastic process. The species and number of seedlings 'planted' in a given year are probabilistic functions of the simulated environment at the forest floor, species regeneration characteristics and random factors. Tree growth is based on species-specific functions that predict, according to a tree's current diameter, the expected diameter increment for the tree under optimal environmental conditions (Botkin, Janak & Wallis 1972). This optimal increment is reduced to reflect sub-optimal conditions (shading or moisture- or nutrient-shortage). The simulated plots self-thin as the trees get larger. Mortality of trees is constant for trees under optimal conditions and has a value that is characteristic of the species. Trees that are growing sub-optimally, as a result of shading, are subjected to elevated rates of mortality.

These models have been tested using independent data sets from a variety of forests in different parts of the world (Shugart 1984, chapter 4). Tests of the models have involved the prediction of species composition changes over time spans as long as 18 000 years, along various environmental gradients, in response to different disturbance regimes, and in response to 'stress' and disease. The models seem to reproduce adequately realistic patterns in forest structure and composition at this coarse level of resolution. As one illustration, let us return to the notion of scale-dependent pattern in forests.

Scale and resolution of forest pattern: model results

There are several factors that confound the interpretation of diameter distribution data. Important among these are stand age, disturbance history and various environmental factors, as well as the diameter increments used and quadrat size. There is an expectation that tree-to-tree interactions are likely to be most easily interpreted using diameter distributions drawn from surveys at smaller spatial scales (Goff & West 1975). But sample variation (particularly for large trees) is greater at these same scales.

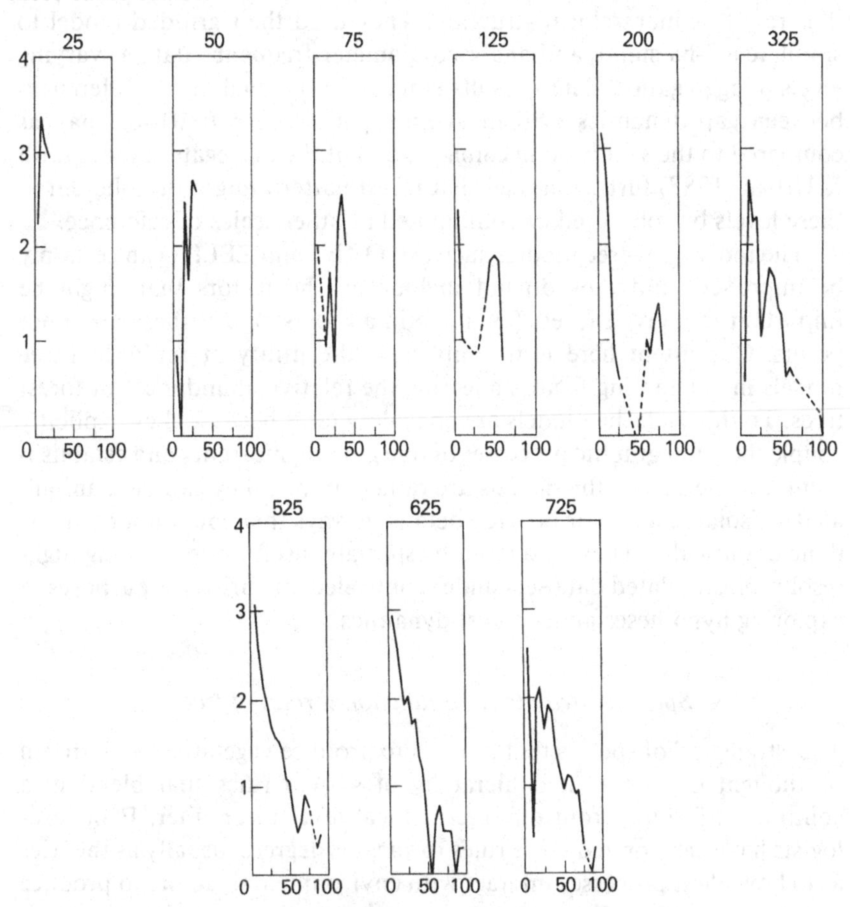

FIG 11.1 Successional changes in diameter distribution in a simulation of a 1 ha forest sample, made using the ZELIG model: the years indicated refer to the time after the simulation was initiated with a non-forest plot but with an adequate seed supply for all species.

Simulation models can control these confounding factors to some extent, and offer additional insights into the mechanics of diameter distribution. For example, Fig. 11.1 illustrates the temporal trend in the diameter distribution of a simulated 1-ha stand of forest subject only to light limitation (all other environmental factors being held at constant, non-limiting levels). The pattern shown in the forest 725 years old compares well with that found empirically, e.g. in the study of Goff & West (1975).

Simulation results similarly argue for a two-level interpretation of forest pattern. Smith & Urban (1988) have used the ZELIG model to illustrate this hierarchical structure. They used their gridded model to simulate a 9-ha stand, and analysed diameter-frequency data at varying levels of aggregation. Their results emphasize the qualitative differences between gap dynamics evident at fine spatial scales (0.01–0.1 ha), as compared to the steady-state configuration of the aggregate stand. Smith & Urban (1988) further argued that forest pattern might be coherent at these levels but obscured or confounded at other scales of reference.

The individual-tree models such as FORET and ZELIG can certainly be improved, and they do not include all the factors that might be important in a given forest (let alone in all forests). But these are moot points. Our intent here is to emphasize the utility of individual-tree models in elucidating factors affecting the relative abundances of forest trees. To this end, the models are appealing tools because they explicitly couple the demographic processes of tree growth, mortality and establishment. Further, since the models are rather simple, they can be manipulated to isolate factors or provide details in ways that could not easily be done empirically. These models are especially useful in producing high-resolution simulated data sets under controlled scenarios for purposes of exploring hypotheses about forest dynamics.

Species attributes and functional roles of trees

The 'strategies' of species that interact to produce vegetation pattern can be thought of as a nested hierarchy of system rules that blend in a continuous fashion from one hierarchical level to another. Plant ecologists have categorized these rules to various degrees, usually as theories as to how plant processes interact with environmental factors to produce vegetation pattern. These abstractions are usually constrained to particular space and time scales — either explicitly, or implicitly by the domain of their successful applications. One theory can work in producing successful predictions at one scale, while another, with different fundamental assumptions, can also work, albeit at a different scale.

For example, the triangle of plant strategies proposed by Grime (1977, 1979) involves a set of rules that can be used to predict a pattern of occurrence of strategies (and associated life-forms) to be expected under a particular environmental regime. There are several higher-level ordering rules that could be used to pattern the fundamental properties of the vegetation that depend on the climatic conditions, e.g. those of Holdridge (1967), Box (1981) and Woodward (1987).

In this section, we will discuss a lower-level set of ordering rules that might be applied under climate conditions in which one would expect trees, and further, in terms of Grime's triangle, in conditions in which one would expect the interactions between trees to be controlled by competition for space. This is the domain of gap dynamics in forests.

This case has been treated by Shugart (1984, 1987) in the context of minimal categories of gap competition in trees. His scheme categorizes trees according to two dichotomies that couple the demographic processes of mortality and regeneration: (i) does the species require a canopy gap for successful regeneration — yes or no? or (ii) does the species typically generate a gap with the death of a mature individual of typical size—yes or no?

In this pair of dichotomies, there are implied four resultant tree strategies which were categorized as 'species roles' (Fig. 11.2). Life-history traits associated with each role are straightforward and intuitive: gap-forming trees are large, while species that do not create gaps are typically smaller; gap-regenerating species are shade-intolerant, while species that do not require gaps are shade-tolerant.

This classification is intentionally simplistic, and other schemes could be devised by subdividing these roles or considering resources other than light. For example, it is common to divide trees into about five categories of shade-tolerance (Baker 1949; Ellenberg 1963). Grubb (1977), in his review of the importance of the 'regeneration niche' in separating species, emphasized the many different variables in a gap beside size, and the great importance of variation through time in the availability of seed. The latter issue has been taken up in mathematical models under the term 'variable recruitment' (Chesson & Warner 1981; Chesson 1986). Wherever careful studies are made, evidence is found for marked differences between years in the local seed crop; a recent study showed this for twenty-two species of Lauraceae over 6 years in a neotropical rain forest (Wheelwright 1986). It is also true that differences exist in the requirements of different tree species at different ages

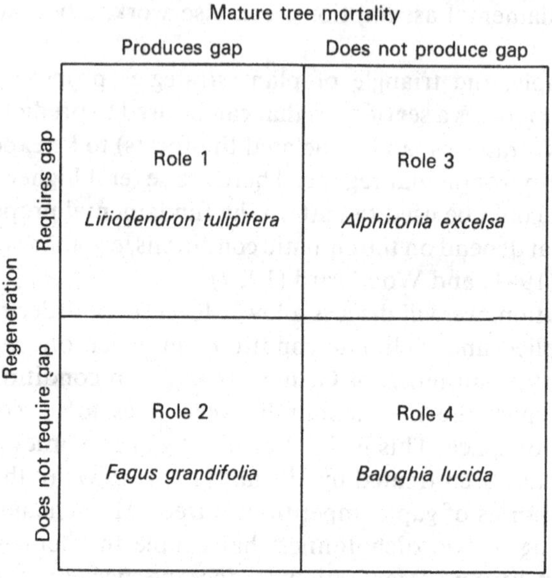

Regeneration	Mature tree mortality: Produces gap	Mature tree mortality: Does not produce gap
Requires gap	Role 1 *Liriodendron tulipifera*	Role 3 *Alphitonia excelsa*
Does not require gap	Role 2 *Fagus grandifolia*	Role 4 *Baloghia lucida*

FIG 11.2 A categorization of four roles for forest trees, dependent upon whether or not they produce a gap when they die, and whether or not they require a gap for regeneration (based on Shugart 1984). Information for the examples named is used in the simulations shown in Fig. 11.5.

(Franklyn & Dyrness 1969), and species with similar tolerances of deep shade may differ widely in their response when a gap is created (Kohyama 1987). Most of the studies referred to have been done in temperate forests, but studies on 'gap-partitioning' have also been made in tropical rain forests, for example by Denslow (1980) and Brokaw (1987). An alternative approach to modelling relative abundances in extremely species-rich tropical forest, based on the premise that many species have essentially the same requirements for regeneration, has been offered by Hubbell & Foster (1986).

Granted that many important axes of differentiation exist, it is still true that a basic 2 × 2 categorization of the type shown in Fig. 11.2, and discussed in more detail by Shugart (1984, 1987), can produce a rich array of forest dynamics. As shown in Fig. 11.3, two of the four roles are self-reinforcing, Role-1 species can create the gaps they need for regeneration, while role-4 species that can regenerate in the shade do not drastically open the canopy when they die. Species within the other two roles tend to give away the space they occupy when they die: role-2

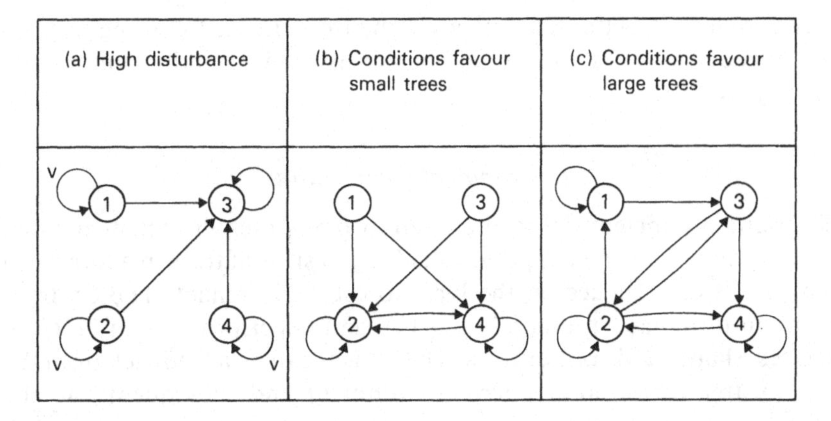

FIG 11.3 The effects of the conditions under which a forest may develop on the patterns of interrelationships found among trees with the four roles defined in Fig. 11.2 (based on Shugart 1987). Arrows show the most likely transfers of space following the death of a tree.

species let in role-1 and role-3 species when they die, while role-3 species cannot create the conditions they need for regeneration, but instead their death favours role-2 and role-4 species. Of course, if the trees playing roles 1 and 2 die at a relatively small size, they behave as role-3 and role-4 species respectively. Some combinations of roles can lead to reciprocal replacement: a role-1 species can take over a gap created by a role-2 species and vice versa.

The simple notion of functional roles of trees has clear implications in interpreting forest dynamics. The basic premise is that life-history traits effectively couple demographic processes of mortality, establishment and growth. Morphological and physiological considerations impose a correlation structure among life-history traits, such that only certain combinations of traits are commonly realized in any single species. For example, long-lived trees tend to grow more slowly than short-lived trees (Wells 1976), and shade-tolerant trees tend to be intolerant of water shortage (Huston & Smith 1987).

FACTORS AFFECTING THE RELATIVE ABUNDANCES OF FOREST TREES

To be successful within a particular forest a tree must not merely be able to survive and grow in the gaps that are created, i.e. under the local abiotic environmental constraints (the seasonal regime of radiation, temperature, soil moisture and nutrient supply), but it must be able to do this as well as or better than other nearby trees. In the following sections we consider first environmental constraints and secondly competition among the trees.

Environmental constraints

Trees attain sufficient size to alter their own micro-environment and that of subordinate trees. The species, shades and sizes of trees in a forest can have a direct influence on the local forest environment. The environment, in turn, has a profound influence on the performance of different species, shapes and sizes of trees. Thus, there can be a feedback from the canopy tree to the local micro-environment and subsequently to the seedlings and saplings from which the next canopy trees may come.

The alteration of the local environment by a canopy tree is perhaps most readily appreciated in terms of the radiation environment on the forest floor. The nature of the leaf area profile and the canopy geometry are dominant factors in the amount and nature of the radiation reaching the forest floor (Anderson 1964; Cowan 1968, 1971; Kira, Shinozaki & Hozumi 1969). Different tree species can also have specific effects on the pattern and quality of throughfall (Zinke 1962; Helvey & Patric 1965), soil moisture (Shear & Stewart 1934; Swift, Heal & Anderson 1979), soil nutrient availability (Zinke 1962; Challinor 1968), and a myriad of other factors (Penridge & Walker 1986; Sterner, Ribis & Schatz 1986).

Pastor and Post (1986) used an individual-tree simulation model to inspect the effects of microclimate, soil moisture and nitrogen availability on sites with different soils (Fig. 11.4). They tallied the factor (temperature, light, soil moisture or nitrogen availability) that was most limiting to growth on a tree-by-tree basis over successional time. Their results illustrate one way that simulation models can provide information that could not easily be obtained empirically. They found that the most limiting factor varied as a function of forest successional stage and as a function of the position of a tree in the canopy. During the first 50 years of succession when tree heights were below 10 m, nitrogen tended to be the most limiting factor. For stands in years 100 to 150 of the succession, light was the most growth-limiting factor for trees less than 10 m in height, but for taller trees (10–20 m in height) nitrogen was most limiting. By year 200–250 of the simulation, the demand for nitrogen by the tallest trees (20–30 m tall), in conjunction with the amount of nitrogen sequestered in living biomass, combined to make nitrogen the most limiting factor for most trees, regardless of height. Also, in years 200–250, soil moisture was an important constraint for many trees, while light was limiting only to the smallest trees.

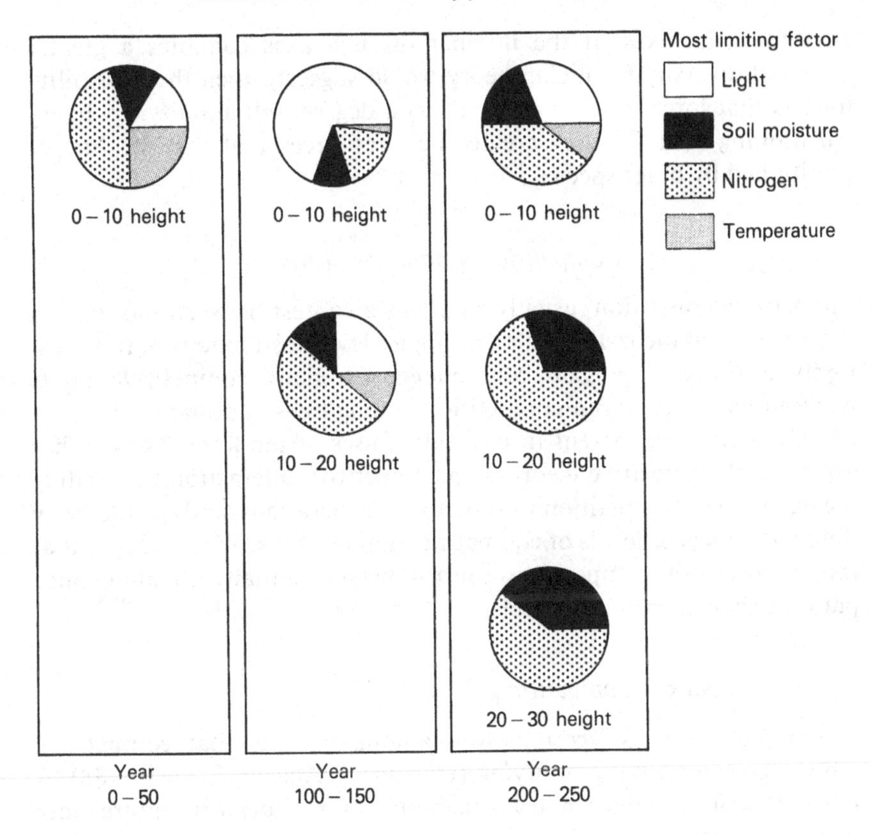

FIG 11.4 The factors most limiting to the growth of a tree as a function of height and time in a simulated succession (starting with bare soil at year 0), produced with an individual-based model of northern hardwood forest in the United States (based on Pastor & Post 1986). The sizes of the sections in the circles reflect the numbers of trees in the sample which are most limited by the factor indicated.

The overall pattern of the factors most limiting growth seen in Fig. 11.4 is a complex function of species/environment interactions, tree size and stand age. A dominant tree over the time periods shown would probably be nitrogen-limited through its entire life, while a subordinate tree (that might replace the dominant in the next generation) would have run a gauntlet of several factors limiting to a similar degree. It is expected that trees of the same species but of different heights would be limited by a variety of factors that changed in relative importance over successional time.

An interesting implication from these results is that trees can diversify their micro-environment in both time and space. In this way, the presence of trees adds to the dimensionality of the important environmental axes. If the number of such axes connotes a greater potential diversity (as niche theory would suggest), then the possibility follows that forest tree diversity is, to a degree, self-amplifying. Every partitioning of a resource creates new resources that may be further partitioned by other species.

Competition among forest trees

The term 'competition' usually connotes a contest between individuals that ends when the individuals die; higher-level abstractions of this term imply a simple aggregate consequence of many competitive bouts between individuals. But competition among trees has nuances that are not immediately apparent in traditional uses of the term. These other aspects include positive feedback in competitive interactions, extending the outcome of competition to subsequent generations, and qualitatively different modes or levels of competition among trees. We will argue that these aspects of competition contribute substantially to abundance patterns characteristic of forests.

Positive feedback in competition

A striking feature of competition among trees is that competitive advantage can be self-amplifying (Huston, DeAngelis & Post 1988). A tree that is slightly taller than a neighbour (competitor) will capture more light, and, other factors being equal, will grow better in that year. The next year, its height advantage will have increased, and it will be at even more of a competitive advantage. This effect is easy to visualize in the case of light, where competition is asymmetric and exploitative. But, if below-ground resource capture is proportional to tree size, we would expect the same sort of effect in competition for water or nutrients.

This positive feedback effect is not confined to competition among trees, but its consequences are especially obvious in forests. One consequence is the generation, in time, of size hierarchies in forests (Huston & DeAngelis 1987; Huston, DeAngelis & Post 1988). Spatial pattern in plant communities can also be influenced by self-amplifying competition: a frequent result is a uniform distribution of large trees, even though small trees may be randomly distributed. Size distributions as well as spatial pattern may be influenced by initial stand conditions (Huston, DeAngelis & Post 1988). In stands where there is little variation in height among plants, no clear competitive advantage can be developed, and a generally stunted stand of similarly sized trees develops.

With greater initial variation in height, a dominant tree can assert itself, and this leads to the uniform spatial pattern and size distribution typically associated with mature forests.

Extending competition to the next generation

The couplings between tree mortality and regeneration that defined functional roles of trees can extend competition to more than one generation. In this respect, the success of a tree in occupying a plot of land can set the stage for the success or failure of its progeny. As noted previously, both role-1 and role-4 species have patterns of mortality that favour their own regeneration. This sort of coupling can occur with resources other than light.

J. Pastor & W. M. Post (personal communication) have used their LINKAGES model in unpublished studies to illustrate the feedbacks that can control forest dynamics over several generations of trees. They have shown how nutrient sensitivity of tree species (i.e. the magnitude of a tree's response to relative nutrient availability) can interact with leaf litter quality (lignin content, C : N ratio) to generate self-amplifying species-composition patterns. For example, on infertile sandy soils in Wisconsin, their model grows pine species, which produce a low-quality litter that does little to improve the soil, which in turn prevents nutrient-demanding species from invading these sites, thus maintaining the pine forest. Conversely, on richer silt-loams more demanding species such as sugar maple (*Acer saccharum*) come to dominate. This species produces high-quality leaf litter, which maintains or increases soil fertility, and this in turn gives a competitive advantage to species that are very responsive to high levels of soil nitrogen availability. Again, the species present on the site alter the site in a way that favours self-replacement. This phenomenon could easily be conjectured from empirical evidence, but the model provides a helpful illustration of the mechanisms at work over several generations.

Levels of competition among trees

The couplings between mortality and regeneration which are used to define the functional roles of trees imply a set of rules that govern patterns of species replacement in gap dynamics. Like any set of rules, these may be followed strictly, loosely or not at all. In the case of forests where pattern is determined by gap dynamics, i.e. by competition for and temporary occupancy of space by individual trees, qualitatively different modes of competition may occur. Trees may compete within the rules implied by the foregoing discussion, in which case roles 1, 2 and 4 are favoured roles in undisturbed temperate deciduous forest.

Under these rules, role 3 is a loser at gap dynamics, but such species can persist by not 'playing' by this set of rules. This is the case of 'fugitive' species, which concede gap-scale competitive ability to persist regionally (Marks 1974; Urban, O'Neill & Shugart 1987). In effect, such species 'opt' to compete in a different and larger arena.

Finally, trees may compete 'to change the rules' that we have made to govern competitive outcomes; this represents a third mode of competition among trees. Certain fire-adapted trees (role 3, typically) may alter mortality schedules by interacting with fires in ways which take the advantage away from longer-lived, larger role-1 and role-2 trees (Heinselman 1981; Foster & King 1986). Environmental factors can also change the rules of competition. At high latitudes the low temperatures, short growing season and low sun angles interact to rule out role 1 as a possible strategy; boreal forests are dominated by role-3 species, i.e. birches, larches and pines, and role-2 species, i.e. spruces. Of course, frequent disturbances can shift the competitive advantage to role 3 in many forests.

Competition of the first sort (competition within the rules we have set) is the usual case for competitive interactions. In this case, one would expect the species that are the most similar to compete with each other most strongly. In the latter modes of competition (side-stepping the rules, or competing 'to change the rules'), one would expect the species that are the most dissimilar to compete most strongly. This suggests that the two levels of competition might result in cases where similar species could be competitors at one level (within their particular role) while being mutualists on a higher level (by both competing with other roles). This possibility adds another level of complexity to studies of forest composition based on inferences about tree-to-tree interactions.

SYSTEM-LEVEL IMPLICATIONS

The nature of the couplings among demographic processes has been a theme of much of this discussion. In this final section, we discuss some of the implications of these couplings, and of the tree–environment and tree-to-tree interactions discussed in the foregoing sections. We are especially concerned with dynamical patterns that might be construed as emergent properties of forests.

Species roles and forest dynamics

Simulation models provide a convenient means of examining the behaviour of different kinds of trees under various environmental regimes. One instructive example has been the simulation of forests comprised of species representing the four typal roles shown in Fig. 11.2, either singly or in combinations. Such experiments with models provide reference standards against which to evaluate patterns in real forests.

Mono-role forests

Trees with different roles produce essentially different dynamics of biomass and numbers when single-species stands are simulated with small model plots (about 0.1 ha). The long-term behaviour of numbers and biomass for species of the four types are shown in Fig. 11.5; each of the roles is reflected in a fundamentally different signature of numbers and biomass dynamics. The dynamics of a role-1 stand features explosive increases in biomass and numbers following the death of a large canopy tree and gives an even-eyed character to the stand, reflecting the tendency for episodic recruitment following a mortality of a large tree. Role-2 stands have mixed-age and mixed-size structures even at small scales, with episodes of enhanced recruitment associated with falls in biomass resulting from mortality of large trees. Role-3 stands feature episodes of recruitment that are preceded by the senescence of the previous cohort and an associated drop in biomass. The dynamics of role-4 stands result in mixed-age and mixed-size forests with little variance in either numbers or biomass dynamics, except when simultaneous deaths of several trees produce a transient response (as occurs about year 1200 in Fig. 11.5 (d)).

While these modes of stand dynamics are intended as typal cases, there are real forests that display similar kinetics. For example, Shugart (1987) found that the role-3 dynamics of numbers and mass was in agreement with the dynamics of *Pinus silvestris* stands at high latitude reported by Zyabchenko (1982). It is intriguing to speculate that forests under different environmental regimes or in different biogeographic realms might vary systematically in the relative abundance of functional roles of trees.

Mixed-role forests: succession

In this case the interactions among the species of different roles can produce a great variety of patterns in numbers and biomass. Huston &

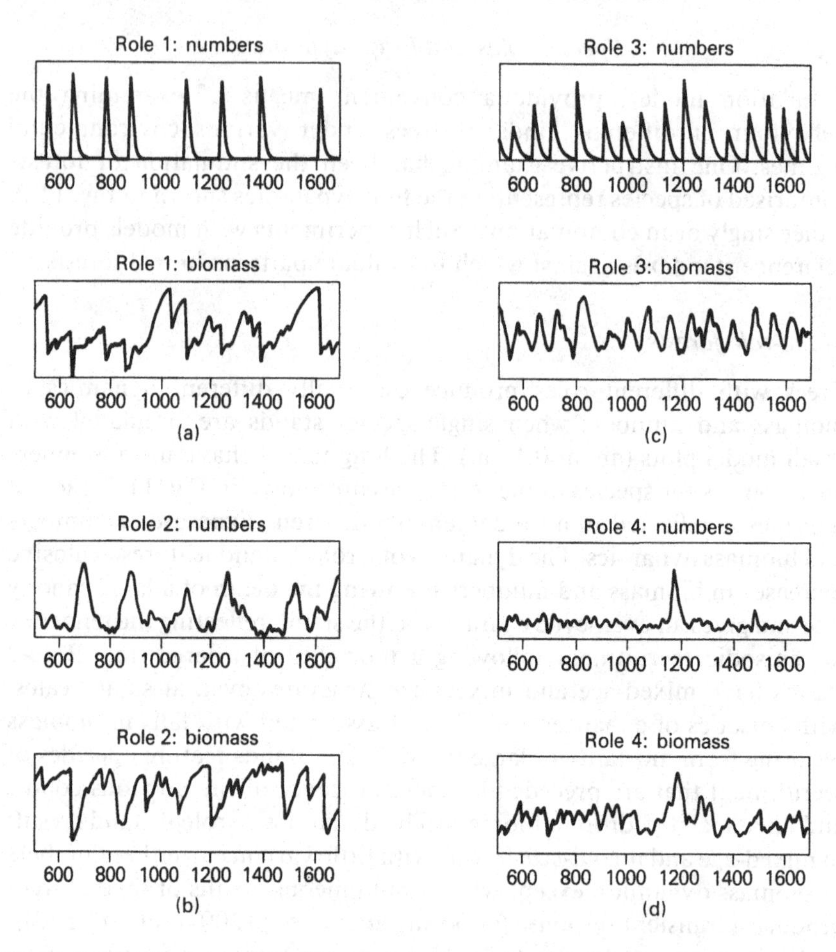

FIG 11.5 The dynamics of numbers and biomass in simulated forests of single species representing the four roles characterized in Fig. 11.2: (a) *Liriodendron tulipifera,* and (b) *Fagus grandifolia* (both simulated by the FORET model; Shugart & West 1977), (c) *Alphitonia excelsa,* and (d) *Baloghia lucida* (both simulated by the KIAMBRAM model; Shugart, Hopkins, Burgess & Mortlock 1981). Simulations were run for 2000 years, and the results for only part of that period are shown.

Smith (1987) used the ZELIG model in a series of experiments to allow species with a spectrum of life-history attributes to compete against one another. They found that the different combinations of species (and the life-history strategies these species represent) could produce a rich array of successional pathways. The relative success of species with different roles depended on their context. A species with a given set of life-history attributes might be judged successful (i.e. persist) under one set of environmental conditions and competitors, but lose under different conditions. A species could function as a seral species under some circumstances, yet act as a 'climax' species in other cases. The logical trade-offs in life-history and physiological features of trees prevented the emergence of a set of attributes that was always successful.

Similarly, Tilman (1982, 1988) has used biological trade-offs as a basis for understanding the diversity of plant species with regard to the division of resource axes. It is gratifying that his results (based on models that seem appropriate to populations that are well mixed and without significant size structure) are in qualitative agreement with those obtained with individual-based models that are not subject to these restrictions.

Frequency responses of forests

The temporal scales of demographic processes in forests, and the couplings among these processes, result in system-level patterns in forest dynamics that might not be obvious from a superficial consideration of the underlying mechanisms. These patterns include lagged or inertial dynamical responses to environmental variation, the possibility of hysteretic behaviour and multiple stable states, and resonance between the natural frequencies of tree life-history processes and environmental patterns.

Davis & Botkin (1985) used the JABOWA model to examine the sensitivity of forest systems to change in temperature. They drove the model with abrupt temperature changes of varying duration, and found that forests moderated these changes in two ways. Change in forest composition lagged behind the change in temperature by 100–200 years; the lag effect reflected the coupling of temperature-induced mortality and subsequent regeneration, and produced a gradual forest response to the abrupt environmental change. Forests were also shown to dampen the short-term environmental variation, smoothing year-to-year fluctuations by integrating their responses over longer time spans.

Model experiments involving these sorts of interactions can produce dynamics which feature hysteresis (Shugart, Emanuel & DeAngelis 1980). An example was found in a model forest dominated by one species role under one environmental regime (a particular temperature regime), but by another species role under a different environmental regime. Forest composition under either regime was self-reinforcing. Under a slowly changing climate, the system had two quasi-stable states, and changed abruptly from one state to the other. The point at which this transition occurred depended on the direction of change in the climate: the switch-point in a cooling climate was different from that in a warming climate. This sort of complex dynamical behaviour would correspond, in the spatial domain, to the occurrence of quasi-stable patches of forest that might change abruptly and radically in response to very slight environmental changes or perturbations.

Experiments indicate that the addition or deletion of a dominant species of a different role changes the frequency spectrum of a modelled forest. Role-1 species tend to amplify exogenous perturbations in a given frequency range, and role-2 species filter out frequencies in certain ranges (Emanuel, Shugart & West 1978). This is consistent with the concept of species of different roles competing by 'changing system-level rules'. One example of where this may occur in nature is in fire-prone forests. The presence of certain tree species in a forest increases the spread or intensity of wildfire, and these species may require fire-generated conditions for successful regeneration (Gill 1981). These positive feedbacks between the occurrence of fire and the biological features of a species can alter the fire frequency of the ecosystem. Such a change would correspond to a change in the frequency spectrum for the ecosystem.

It has been suggested that vegetation pattern may reflect a resonance between natural frequencies of plant life-cycles and the frequencies of environmental drivers such as meso-scale weather patterns (Neilson & Wullstein 1983; Neilson 1986). In this case the biogeographic distributions of taxa should correspond with large-scale (subcontinental) patterns in atmospheric flow structure. These spatial patterns have characteristic temporal scales, which would be expected to interact with the time-scales of demographic processes (time to reproduction, lifespan) to constrain the kinds of plants able to persist under a given environmental regime. Neilson suggests that these time-scales could 'resonate' to amplify vegetation responses, entraining demographics to the natural frequencies of the environmental drivers.

Such resonances would be a difficult notion to demonstrate empirically, but results of simulation models suggest that forests can dampen or amplify environmental drivers of certain frequencies. The ZELIG model, for example, dampens a sinusoidal driver with periods of less than about 50 years, but tracks very long-period drivers in a stable fashion, while periodic drivers at frequencies corresponding to the lifespans of the dominant trees are actually amplified (Fig. 11.6). This result suggests that forests may indeed have the potential to resonate with their constraining environmental regime.

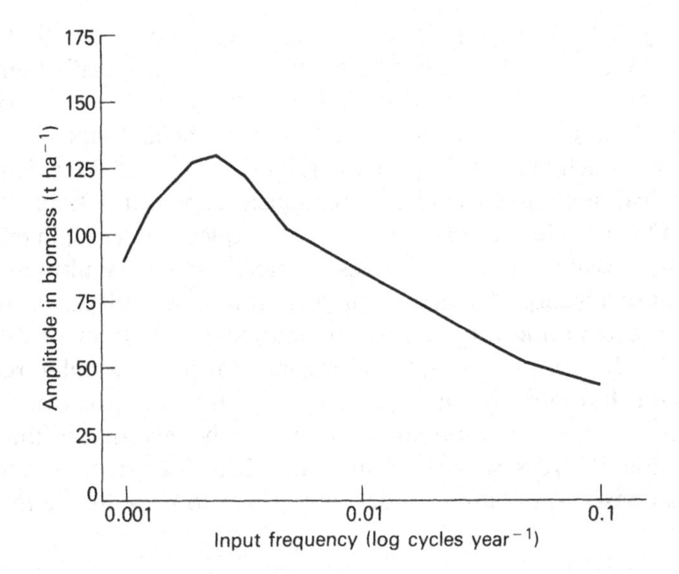

FIG 11.6 The relationship between the amplitude of variation in biomass and the frequency of return of an environmental 'driver', based on the effects of a sinusoidally varying driver in a simulation with the ZELIG model. The amplitude of variations under a regime of Gaussian 'noise' (with the same amplitude as the sine waves) was about 50 t ha^{-1}.

CONCLUSION

In many respects, the interactions among environmental conditions, the natural history of different species and tree–environment feedbacks make forests difficult ecosystems to study. Nevertheless, progress in understanding the compositional dynamics of forests in the past decade has been considerable. The relative abundances of forest trees are controlled by a multiplicity of factors operating at disparate time and space scales. The differences in regeneration success, growth rates of the individual trees in different environments, and mortality rates all influence the abundance of trees of different species. These sources of differential abundance are, in turn, influenced by a range of factors, some of which are controlled by the biology and ecology of the species involved.

Our theme in this paper has concerned the consequences of couplings and interactions between the demographic processes of tree growth, mortality and regeneration. We hope we have been successful in illustrating the utility of individual-based simulation models as tools for integrating these demographic processes at a common scale, that of the forest gap. We have argued that these couplings, in a specified environmental context, can explain several characteristic patterns observed in forests, including some very complex dynamical behaviours.

We have much to learn. In particular, there is much to do to define the physiological, anatomical and morphological constraints (and others?) that limit the possible strategies of a tree; this question seeks, in effect, to define the possible functional roles of trees, which is ultimately an evolutionary question. Given this understanding, we still have much to explain in accounting for the relative abundances of trees of different roles under different environmental regimes (or biogeographic realms). There is much to discover about tree–environment relations, and about the subtleties of competition among trees. We believe that a functional classification of trees will streamline this learning process, and that individual-based simulation models will prove to be valuable heuristic tools.

ACKNOWLEDGMENTS

Our research was supported in part by the US National Aeronautics and Space Administration (Grant NAG–5–1018), the US Environmental Protection Agency (Grant CR–814610–01–0), and the US National Science Foundation (Grant BSR–8702333).

REFERENCES

Anderson, M. C. (1964). Light relations of terrestrial plant communities and their measurement. *Biological Reviews,* **39**, 425–86.

Aubréville, A. (1933). La forêt de la Côte d'Ivoire. *Bulletin du Comité d'Études Historiques et Scientifiques de l'Afrique Occidentale Française, Paris,* **15**, 205–61.

Aubréville, A. (1938). La forêt coloniale: les forêts de l'Afrique occidentale française. *Annales, Académie des Sciences Coloniales, Paris,* **9**, 1–245.

Baker, F. S. (1949). A revised tolerance table. *Journal of Forestry,* **47**, 179–81.

Bormann, F. H. & Likens, G. E. (1979). *Pattern and Process in a Forested Ecosystem.* Springer, New York.

Botkin, D. B., Janak, J. F. & Wallis, J. R. (1972). Some ecological consequences of a computer model of forest growth. *Journal of Ecology,* **60**, 849–72.

Box, E. O. (1981). *Macroclimate and Plant Forms: an Introduction to Predictive Modelling in Phytogeography.* Junk, The Hague.

Brokaw, N. V. L. (1987). Gap-phase regeneration of three pioneer tree species in a tropical forest. *Journal of Ecology,* **75**, 9–19.

Challinor, D. (1968). Alteration of soil surface characteristics by four tree species. *Ecology,* **49**, 286–90.

Chesson, P. L. (1986). Environmental variation and the coexistence of species. *Community Ecology* (Ed. by J. Diamond & T. J. Case), pp. 240–56. Harper & Row, New York.

Chesson, P. L. & Warner, R. R. (1981). Environmental variability promotes coexistence in lottery competitive systems. *American Naturalist,* **117**, 923–43.

Cowan, I. R. (1968). The interception and absorption of radiation in plant stands. *Journal of Applied Ecology,* **5**, 367–79.

Cowan, I. R. (1971). Light in plant stands with horizontal foliage. *Journal of Applied Ecology,* **8**, 579–80.

Davis, M. B. & Botkin, D. B. (1985). Sensitivity of cool-temperate forests and their fossil pollen record to rapid temperature change. *Quaternary Research,* **23**, 327–40.

Denslow, J. S. (1980). Gap-partitioning among tropical rainforest trees. *Biotropica,* **12** (Supplement), 47–55.

Ellenberg, H. (1963). *Vegetation Mitteleuropas mit den Alpen.* Ulmer, Stuttgart.

Emanuel, W. R., Shugart, H. H. & West, D. C. (1978). Spectral analysis and forest dynamics: long-term effects of environmental perturbations. *Time Series and Ecological Processes* (Ed. by H. H. Shugart), pp. 194–210. Society of Industrial and Applied Mathematics, Philadelphia.

Farquhar, G. D. & von Caemmerer, S. (1982). Modelling of photosynethic response to environmental conditions. *Encyclopaedia of Plant Physiology,* **II**, 549–87.

Foster, D. R. & King, G. A. (1986). Vegetation pattern and diversity in SE Labrador, Canada: *Betula papyrifera* (birch) forest development in relation to fire history and physiography. *Journal of Ecology,* **74**, 465–83.

Franklyn, J. F. & Dyrness, C. T. (1969). Vegetation of Oregon and Washington. *United States Department of Agriculture Forest Service Research Paper PWN*-**80**, i–vi & 1–216.

Gill, A. M. (1981). Adaptive responses of Australian vascular plant species to fire. *Fire and the Australian Biota* (Ed. by A. M. Gill, R. H. Groves & I. R. Noble), pp. 243–72. Australian Academy of Science, Canberra.

Goff, F. G. & West, D. C. (1975). Canopy-understory interaction effects on forest population structure. *Forest Science,* **21**, 98–108.

Grime, J. P. (1977). Evidence for the existence of three primary strategies in plants and its relevance to ecological and evolutionary theory. *American Naturalist,* **111**, 1169–94.

Grime, J. P. (1979). *Plant Strategies and Vegetation Processes.* Wiley, Chichester.

Grubb, P. J. (1977). The maintenance of species-richness in plant communities: the importance of the regeneration niche. *Biological Reviews,* **52**, 107–45.

Harcombe, P. A. (1987). Tree life tables. *BioScience,* **37**, 557–68.

Harper, J. L. (1977). *Population Biology of Plants.* Academic Press, London.

Harper, J. L., Clatworthy, J. N., McNaughton, I. H. & Sagar, G. R. (1961). The evolution and ecology of closely related species living in the same area. *Evolution,* **15**, 209–27.

Heinselman, M. L. (1981). Fire and succession in the conifer forests of northern North America. *Forest Succession: Concepts and Application* (Ed. by D. C. West, H. H. Shugart & D. B. Botkin), pp. 374–405. Springer, New York.

Helvey, J. D. & Patric, J. H. (1965). Canopy and litter interception by hardwoods of eastern United States. *Water Resources Research,* **1**, 193–206.

Hett, J. M. & Loucks, O. L. (1968). Applications of life-table analyses to tree seedlings in Quetico Provincial Park, Ontario. *Forest Chronicle,* **34**, 1–4.

Holdridge, L. R. (1967). *Life Zone Ecology,* revised edn. Tropical Science Center, San Jose, Costa Rica.

Hubbell, S. H. & Foster, R. B. (1986). Biology, chance, and history and the structure of tropical rain forest communities. *Community Ecology* (Ed. by J. Diamond & T. J. Case), pp. 314–29. Harper & Row, New York.

Huston, M. H. & DeAngelis, D. L. (1987). Size bimodality in monospecific plant populations: a critical review of potential mechanisms. *American Naturalist,* **129**, 678–707.

Huston, M. H., DeAngelis, D. L. & Post, W. M. (1988). New computer models unify ecological theory. *BioScience,* **38**, 682–91.

Huston, M. H. & Smith, T. M. (1987). Life history, competition, and plant succession. *American Naturalist,* **130**, 168–98.

Jarvis, P. G. & Leverenz, J. W. (1983). Productivity of temperate, deciduous and evergreen forests. *Encyclopaedia of Plant Physiology,* **12D**, 233–80.

Kira, T., Shinozaki, K. & Hozumi, K. (1969). Structure of forest canopies as related to their primary productivity. *Plant and Cell Physiology,* **10**, 129–42.

Kohyama, T. (1987). Significance of architecture and allometry in saplings. *Functional Ecology,* **1**, 399–404.

Landsberg, J. J. (1986). *Physiological Ecology of Forest Production.* Academic Press, London.

Lorimer, C. G. (1980). Age structure and disturbance in history of a southern Appalachian virgin forest. *Ecology,* **61**, 1169–84.

Marks, P. L. (1974). The role of pin cherry (*Prunus pensylvanica*) in the maintenance of stability in northern hardwood ecosystems. *Ecological Monographs,* **44**, 73–88.

Mooney, H. A. & Godron, M. (Eds) **(1983).** Disturbance and Ecosystems: Components of Response. Springer, Berlin.

Munro, D. D. (1974). Forest growth models: a prognosis. *Growth Models for Tree and Stand Simulation* (Ed. by J. Fries), pp. 7–21. Research Note 30. Department of Forest Yield Research, Royal College of Forestry, Stockholm.

Neilson, R. P. (1986). High-resolution climatic analysis and southwest biogeography. *Science,* **232**, 27–34.

Neilson, R. P. & Wullstein, L. H. (1983). Biogeography of two southwest American oaks in relation to atmospheric dynamics. *Journal of Biogeography,* **10**, 275–97.

Noble, I. R. & Slatyer, R. O. (1980). The use of vital attributes to predict successional change in plant communities subject to recurrent disturbance. *Vegetatio.* **43**, 5–21.

Pastor, J. & Post, W. M. (1986). Influence of climate, soil moisture, and succession on forest carbon and nitrogen cycles. *Biogeochemistry,* **2**, 3–27.

Penridge, L. K. & Walker, J. (1986). Effect of neighbouring trees on eucalypt growth in a semi-arid woodland in Australia. *Journal of Ecology,* **74**, 925–36.

Pickett, S. T. A. & White, P. S. (Eds) (1985). *The Ecology of Natural Disturbances and Patch Dynamics.* Academic Press, New York.

Shear, G. M. & Stewart, W. D. (1934). Moisture and pH studies of the soil under forest trees. *Ecology,* **15**, 350–8.

Shugart, H. H. (1984). *A Theory of Forest Dynamics.* Springer, New York.

Shugart, H. H. (1987). Dynamic ecosystem consequences of tree birth and death patterns. *BioScience,* **37**, 596–602.

Shugart, H. H., Emanuel, W. R. & DeAngelis, D. L. (1980). Environmental gradients in a simulation model of a beech-yellow poplar stand. *Mathematical Bioscience,* **50**, 163–70.

Shugart, H. H., Hopkins, M. S., Burgess, I. P. & Mortlock, A. T. (1981). The development of a succession model for subtropical rain forest and its application to assess the effects of timber harvest at Wiangaree State Forest, New South Wales. *Journal of Environmental Management,* **11**, 243–65.

Shugart, H. H. & West, D. C. (1977). Development of an Appalachian deciduous forest succession model and its application to assessment of the impact of the chestnut blight. *Journal of Environmental Management,* **5**, 161–70.

Shugart, H. H. & West, D. C. (1980). Forest succession models. *BioScience,* **30**, 308–13.

Smith, T. M. & Urban, D. L. (1988). Scale and resolution of forest structural pattern. *Vegetatio,* **74**, 143–50.

Sterner, R. M., Ribis, C. A. & Schatz, G. E. (1986). Testing for life historical changes in spatial patterns of four tropical tree species. *Journal of Ecology,* **74**, 621–35.

Swift, M. J., Heal, O. W. & Anderson, J. M. (Eds) **(1979).** *Decomposition in Terrestrial Ecosystems.* British Ecological Society Symposium, 24. Blackwell Scientific Publications, Oxford.

Tilman, D. (1982). *Resource Competition and Community Structure.* Princeton University Press.

Tilman, D. (1988). *Plant Strategies and the Dynamics and Structure of Plant Communities.* Princeton University Press.

Urban, D. L., O'Neill, R. V. & Shugart, H. H. (1987). Landscape ecology. *BioScience,* **37**, 119–27.

Waring, R. H. (1987). Characteristics of trees predisposed to die. *BioScience,* **37**, 569–74.

Watt, A. S. (1925). On the ecology of British beechwoods with special reference to their regeneration. II. The development and structure of beech communities on the Sussex Downs. *Journal of Ecology,* **13**, 27–73.

Watt, A. S. (1947). Pattern and process in the plant community. *Journal of Ecology,* **35**, 1–22.

Wells, P. V. (1976). A climax index for broadleaf forest: an *n*-dimensional ecomorphological model of succession. *Central Hardwood Conference* (Ed. by J. S. Fralish, G. J. Weaver & R. C. Schlesinger), pp. 131–76. Department of Forestry, Southern Illinois University, Carbondale.

Wheelwright, N. T. (1986). A seven-year study of individual variation in fruit production in tropical bird-dispersed tree species in the family Lauraceae. *Frugivores and Seed Dispersal* (Ed. by A. Estrada & T. H. Fleming), pp. 21–35. Junk, Dordrecht.

White, P. S. (1979). Pattern, process and natural disturbance in vegetation. *Botanical Review,* **45**, 229–99.

Whitmore, T. C. (1982). On pattern and process in forests. *The Plant Community as a Working Mechanism* (Ed. by E. I. Newman), pp. 45–59. Special Publication of the British Ecological Society, 1. Blackwell Scientific Publications, Oxford.

Whittaker, R. H. (1953). A consideration of climax theory: the climax as a population and pattern. *Ecological Monographs.* **23**, 41–78.

Woodward, F. I. (1987). *Climate and Plant Distribution.* Cambridge University Press.

Zinke, P. J. (1962). The pattern of influence of individual forest trees on soil properties. *Ecology,* **43**, 130–3.

Zyabchenko, S. S. (1982). Age dynamics of scotch pine forests in the European north. *Lesovedenie,* **2**, 3–10.

12. INSECT HERBIVORY AND PLANT DEFENCE THEORY

P. J. EDWARDS
Department of Biology, University of Southampton, Southampton S09 5NH, UK

INTRODUCTION

In 1964 Ehrlich & Raven published a paper which has proved to be a corner-stone in our understanding of insect–plant relationships. In this paper, they presented a systematic evaluation of the families of plants fed upon by the larvae of various groups of butterflies, and showed that related groups of butterflies tend to feed upon related groups of plant species. For example, most larvae in the subfamily Pierinae feed on members of the Capparidaceae, Cruciferae and a few other apparently related families; members of the Danainae feed predominantly on the Apocynaceae and Asclepiadaceae. In seeking to explain these patterns of association between plants and insects, Ehrlich & Raven concluded that secondary plant chemistry played a leading role in determining plant utilization by insects. They suggested that plants have, through occasional mutations and recombinations, produced chemical compounds not directly related to basic metabolism. Some of these served to protect the plant against phytophagous insects, and plants thus defended entered a 'new adaptive zone'. However, insects can evolve to overcome such physiological obstacles. Having done so, an insect is free to diversify, largely in the absence of competition from other feeders. Thus they postulated a process of coevolution in which plants and their herbivores are continually adapting to changes in each other, in what came to be described as a kind of evolutionary arms race (Feeny 1976).

Since Ehrlich & Raven's seminal paper there has been an enormous growth of interest in insect–plant relationships, and recent research has led in distinct directions which have different emphases and which to some extent have reached different conclusions about the importance of insect herbivores in plant evolution. One important line of research concerns the structure of insect communities on plants, and the factors which influence insect populations (Strong, Lawton & Southwood 1984). The results of this approach cast doubt upon the universal importance of insects as a selective force in plant evolution. Populations of many plant-feeding insects are either small or very variable because of the harshness and vagaries of the environment and the impact of natural enemies. Invertebrate herbivores that are regulated at low densities by natural enemies appear to have virtually no impact on plant dynamics, and it seems improbable that they have a significant selective impact on plants (Crawley 1983). Strong, Lawton & Southwood (1984) concluded that although there are a few examples where the evidence of coevolution is compelling, most insect species probably do not interact sufficiently with their hosts to bring about evolutionary change.

A separate line of research, which provides the main theme of this paper, has focused on plant defences, and particularly upon the defensive role of secondary plant substances. There have been considerable efforts toward developing a general theory of plant defence which can explain the types and quantities of secondary compounds and other defences in plants in terms of their need of protection from insect herbivores.

This article examines some of the underlying assumptions of plant defence theory in an attempt to answer the central question: to what extent are plants defended against herbivores? It will argue that the uncritical use of the concept of strategies of defence may prove seriously misleading, and that more attention should be paid both to phylogenetic evidence and to understanding the influence of grazing upon plant fitness. Throughout this article the role of phytophagous insects will be emphasized because these are the most important invertebrate herbivores, although other groups — notably molluscs — may also be significant in some communities.

ESTABLISHED THEORIES ABOUT PLANT DEFENCES

The apparency hypothesis

The first general theory of plant defence was developed independently by Feeny (1976) and by Rhoades & Cates (1976). These authors emphasize the risk of discovery by herbivores as being the major factor influencing the extent of commitment by a plant to defence, and also the type of defence. It is argued that plants of some species are more vulnerable to insect attack than others simply because they are more abundant, or large, or conspicuous in some other way. Feeny refers to these plants which are highly susceptible to discovery by grazers as *apparent* plants. In contrast, individuals of other species are likely to escape discovery by insects (unapparent plants), because they are small or have a very patchy or unpredictable distribution.

The apparency hypothesis suggests that plants which have a high risk of discovery, such as large trees and late successional species, will have a large investment in broadly effective defences (quantitative defences) which reduce growth rate and fitness of all enemies. Such chemical defences are also generally associated with low nutritive value of plant tissues for herbivores and with relatively tough leaves. In contrast, unapparent plants, which are likely to escape grazing, require much lower investment in defences, and may employ relatively small concentrations of toxins (qualitative defences) which, although highly effective against non-adapted enemies, are in an evolutionary sense easy to overcome.

Growth rate hypothesis

An alternative explanation for the differences between plant species in their patterns of defence is that they are related to the intrinsic rate of growth of plants (i.e. not *r* but rates of growth of individual plants; Coley, Bryant & Chapin 1985). Much of the evidence for this hypothesis is provided by the important series of papers by Coley (1980; 1983; 1988) describing the patterns of insect herbivory in lowland rain forest on Barro Colorado Island in Panama. In this work, Coley chose forty-six species of trees which she divided into two life-history groupings: pioneers (twenty-two species), whose saplings occur only in gaps, and persistents (twenty-four species), whose saplings occur both in gaps and in shade. She measured the rates of herbivory on young and mature leaves of 1–2-m tall saplings of these species growing in gaps in the forest by recording both the total leaf area of marked leaves and the area of damage at the beginning of the study period and three weeks later. She also determined several physical and chemical properties of the leaves which might influence insect grazing, including concentrations of simple and condensed phenolics, measures of fibre content and leaf toughness, pubescence, and water and nitrogen contents.

Rates of herbivory of mature leaves varied considerably between species, but overall those of pioneer species were grazed about six times faster than those of persistents (0.24 and 0.04% leaf area grazed per day respectively). Over 70% of the variation in rates between plant species could be explained by the leaf variables measured, of which leaf toughness was the most important factor, followed by fibre content and nutrient value. There was also a strong positive correlation between leaf pubescence and herbivory. In general, the pioneers had less tough leaves which were often pubescent, lower concentrations of fibre and phenolics, and higher concentrations of nitrogen and water.

Coley found no evidence that pioneer species escape discovery by herbivores because of their spatial distribution, and concluded that apparency was not the factor determining their defensive characteristics. Instead, she proposed that habitat quality is the major influence behind the evolution of both the type and amount of anti-herbivore defences. In habitats with high levels of available resources, rapid growth is possible and can compensate for relatively high levels of herbivory. In contrast, in habitats where growth is restricted by low levels of resources, the consequences of herbivory are likely to be more severe and plants must invest more in defence. According to this hypothesis, the type of defence is also influenced by the growth characteristics of the species, and in particular by leaf longevity (Coley 1988). Long leaf lifetimes, characteristic of slow-growing species, favour selection for 'immobile' defences such as tannins and lignins which are metabolically inactive. Such materials have high initial construction costs, but do not have continued metabolic costs of turnover, so that their cost is independent of leaf lifetime. In contrast, shorter leaf lifetimes favour defences by 'mobile' compounds of low molecular weight such as alkaloids and cardiac glycosides which, since they have rapid turnover rates, must be continually synthesized. Since a forest is sufficiently complex to provide a range of micro-habitats which vary in levels of resources, this hypothesis can account for the difference observed between pioneer and persistent species.

The underlying assumptions of defence theory

Both the apparency hypothesis and the growth rate hypothesis point to two contrasting types or strategies of defence, although the conclusions about their ecological significance differ. Both hypotheses imply that natural selection has led to a high degree of optimization of defences so that the commitment by a plant to defence is a reflection of the selection pressure by herbivores in that environment; this idea is clearly reflected, for example, in the writings of Rhoades (1979, 1983). Thus both hypotheses rest upon two assumptions: (i) that insect herbivory has been a major selective factor in plant evolution, and (ii) that there has been sufficient genetic variation to produce a high degree of optimization in plant defences.

A serious problem with both hypotheses is that we do not know the general validity of these assumptions. Many of the data collected to test these hypotheses come from observational studies which correlate levels of grazing damage sustained by plants with chemical and anatomical characteristics of their foliage. However, such data are irrelevant for the purposes of demonstrating whether or not a plant has evolved defences. As Myers (1987) remarks, 'low levels of herbivore damage can indicate that herbivores are rare, that plants are well defended or that herbivores avoid leaves with low levels of nutrients'. To help establish whether plants are defended we must have independent evidence about the role of insect herbivores upon plant fitness. If we are to understand how lack of suitable genetic variation may have constrained the evolution of plant defences, then we must take an evolutionary approach and examine the phylogenetic evidence.

DEFENCE AND RESISTANCE

Despite the contradictory evidence that comes from many studies of insect populations (Strong, Lawton & Southwood 1984), there is now widespread acceptance of the central premise of plant defence theory, that insect herbivory has been a powerful selective force upon plants. Thus in the literature of insect–plant interactions, the word 'defence' has gradually come to replace the word 'resistance' that was used by Beck (1965) in describing the characteristics of plants which make them difficult for insects to exploit. However, there are many examples of plant species which appear very well defended (in the sense that they have tough leaves, low nutrient contents, high levels of condensed tannins, etc.), but which occur in habitats where there is no evidence to suggest that herbivory is a major selective factor. For example, Cottrell (1986) noted the apparent absence of coevolutionary associations between butterflies and plants in the fynbos of South Africa, and concluded that the exceptionally nutrient-poor quality of the diet available to insects prevented extensive phytophagy. The insect biomass in fynbos is extremely low and to a large extent the plants avoid being grazed through their very low nutritional value (Campbell 1986).

In recent literature on insect–plant relationships the word defence has been used much too freely. We must distinguish, at least in concept, between morphological or chemical attributes which protect a plant against herbivory and have evolved in response to herbivore pressure, and features which incidentally make a plant resistant to herbivores but which have evolved in response to other factors. In this paper, I shall therefore make a distinction between 'defence' (by which I mean adaptations which minimize grazing) and 'neutral resistance', which includes other plant properties which tend to reduce grazing. Some of the factors which may have led to the evolution of neutral resistance are reviewed here.

Factors selecting for characteristics that provide neutral resistance

Drought

Leaf toughness often proves to be the most important single factor in explaining the distribution of insect damage, e.g. in the thorough studies of Coley (1983). However, although tough leaves with a high proportion of lignin certainly are difficult for insects to feed upon, this is not evidence that they are defences. It is well known that plants of dry habitats exhibit so-called xeromorphic characters of hard, rigid leaves with a high proportion of skeletal material. The primary significance of these characters in xerophytes may be in providing support to the leaf so that there is not irreparable damage to mesophyll cells or cracking of the cuticle when leaf turgor is lost during desiccation (Grubb 1986).

Nutrient deficiency

In general, plants from resource-rich environments tend to have thin leaves with a low proportion of skeletal material and a high nitrogen concentration. In the absence of toxins such foliage is of high food value for insects, and rates of population increase are potentially high (Southwood 1973). In contrast, plants of nutrient-poor habitats (especially those deficient in nitrogen and phosphorus) commonly have tough leaves with many of the characteristics of hard-leaved xerophytes (Loveless 1961). The significance of small, hard leaves with thick cell walls in poor nutrient conditions may simply be that plants can maintain growth only by producing tissues with a high proportion of cell wall to cytoplasm, and therefore with a low nutrient content. For example, among the plants with the hardest leaves of all are those of tropical heath forest, which have exceptionally low nutritive value (Grubb 1986); only in the fynbos are there plants lower in nutritive value, with leaf nitrogen contents as low as 5 mg g^{-1} (Specht & Moll 1983).

It has recently been suggested that the high concentrations of carbon-based secondary substances (phenolics, terpenes, resins, etc.) commonly found in plants of nutrient-poor habitats may not be primarily defensive, but rather a metabolic consequence of nutrient deficiency (Tuomi *et al.* 1988). Chapin (1980) argued that every plant has an optimal carbon/mineral nutrient balance, at which growth rates are maximal. According to the 'nutrient stress' hypothesis (Tuomi *et al.* 1984, 1988), carbon is allocated to growth whenever there are sufficient mineral nutrients for new cells, and only surplus carbon is used for allelochemicals or for storage. Thus the total level of carbon-based metabolites is a function of the resources that cannot be used for primary metabolism. For example, in birch (*Betula pubescens*) nutrient deficiency leads to the accumulation of polyphenols (Tuomi *et al.* 1984), while application of nitrogen fertilizer promotes more rapid growth of tissues with low concentrations of phenolics. Similarly, the accumulation of high concentrations of tannins in deciduous trees towards the end of the growing season can be explained as the result of reduced availability of nitrogen for new growth.

This minimal interpretation of the significance of resins, terpenes, etc., is not easy to reconcile with the elaborate structures in which such compounds are accumulated in some species, or with the clear evolutionary trends in the disposition of such structures, e.g. resin canals of conifers first found only in leaves and in pith and cortex of stems (Rothwell 1982), then in the primary xylem as well, and finally in the secondary xylem too (Stewart 1983). However, even in these cases insect herbivory was not necessarily the primary selective factor.

Ultraviolet radiation

A problem faced by all terrestrial plants is the potentially damaging effects of u.v. radiation to macromolecules. It is probable that the universal occurrence of flavonoids in land plants is partly due to their high absorptivity of u.v. (Swain 1975). For example, *Psilotum nudum*, which has many primitive morphological characteristics, has biflavones in its cell walls which act as u.v. screens. The ubiquity of flavonoids may also be due to their anti-oxidant properties which reduce adventitious photo-oxidation of sensitive compounds. In general, tropical and high-altitude plants have relatively high concentrations of flavonoids in their immature growing tissues. Other secondary substances with apparent anti-herbivore properties may also confer resistance to the harmful effects of u.v. For example, Tallamy & Krischik (1989) showed that plants of *Cucurbita moschata* with low levels of cucurbitacins suffered reduced growth when grown under conditions of high u.v. Furthermore exposure to u.v. stimulated the synthesis of cucurbitacins.

Micro-organisms

Fungi and bacteria are ubiquitous and must have represented a threat to plants long before the first insects evolved. The widespread occurrence of extremely toxic phytoalexins, which are induced in plant cells by contact with fungal hyphae (Harborne 1988), is evidence of the importance of fungi as a selective pressure on plants. Many attributes of plants commonly regarded as anti-herbivore defences also confer resistance to micro-organisms; indeed there is a broad concordance between plant resistance factors for insects and for plant pathogens. Thus, secondary substances such as gallic acid, mustard oil glycosides, alkaloidal saponins and cyanogenic glycosides have been shown to confer resistance against fungi as well as against insects (Mansfield 1983). Similarly, characters associated with tough leaves, such as a thick cuticle, the presence of silica, and a high proportion of lignin, are also resistance factors against fungi, and many filamentous fungi elicit lignification in plant tissues, even without wounding (Ride 1983).

In practice, it is usually impossible to make a clear distinction between defence and resistance because many different selection pressures may act upon any plant attribute. Commenting upon the striking correlation of leaf hardness with leaf longevity in wet, moist and dry habitats, Grubb (1986) suggested that leaf hardness should be interpreted as a protection against an array of hazards, including penetration by micro-organisms and physical breakage by wind or raindrops, and not solely against damage by animals. Although carbon-based resistance may be primarily related to nutrient deficiency, selection by herbivores may have influenced the chemical form of stored allelochemicals or their distribution within the plant. However, in some environments, for example in conditions of shade and high humidity in tropical rain forest, micro-organisms may have been a more important selective factor. Given these problems of interpretation, it would be prudent to treat as resistance factors all properties which make plants difficult for insects, until there is convincing evidence that they are defences. Furthermore, even evolved defences are likely to be a product of diffuse coevolution (Janzen 1980; Fox 1981; Strong, Lawton & Southwood, 1984), implying that they have evolved in response to many pressures, including not only grazing by many insect species but also attack by micro-organisms.

INSECT HERBIVORY AND PLANT FITNESS

Fitness is a measure of the contribution that a genotype makes to the gene pool of the next generation, and depends on the fertility and the viability of the genotype in relation to other genotypes in the same population (Roughgarden 1979). We know remarkably little about the influence of insect herbivory upon either plant population dynamics or plant fitness in natural communities (Crawley 1983), largely because of the practical difficulties of devising suitable experiments. One difficulty when working with perennial species, and especially with long-lived plants, is the time-scale over which observations are needed. Inevitably ecologists rely upon measures such as the rates of defoliation, photosynthesis, increase in stem diameter and seed production, which are assumed to be correlates of fitness. For example, many studies have shown reduced plant growth and reduced rates of seed production as a result of insect defoliation, e.g. those of Waloff & Richards (1977), Crawley (1983) and Edwards & Gillman (1987). Less conspicuous forms of insect feeding, which have often been ignored, may also have substantial effects upon plant performance. For example, Dixon (1971) showed that there was a strong negative correlation between spring populations of the aphid *Drepanosiphum platanoides* and the production of new leaf area and new wood in the sycamore (*Acer pseudoplatanus*). However, we face serious difficulties in interpreting such effects in terms of plant fitness, particularly for trees and other long-lived plants. This is not simply because of the time span over which observations are needed, but also because of the possibility that reduced growth resulting from herbivory may be compensated to some degree by increased longevity. Thus, it is often far from clear how the immediately observable effects of herbivory upon plant growth affect the biological fitness of plants (Crawley 1983).

Another serious practical difficulty is that of manipulating insect populations for experimental purposes. Much of the available evidence comes from situations in which either all insects have been eliminated with the use of insecticides, or an insect species has been introduced into a new environment where its populations are not controlled by its usual predators and parasites. Such studies have shown convincingly that insects may have a major influence upon plant population dynamics, as in the case of *Hypericum perforatum* (Huffaker & Kennett 1959), *Melampyrum lineare* (Cantlon 1969), *Happlopappus squarrosus* (Louda 1982) and various early-successional herbs (Brown 1985). Similarly, the relative abundances of plant species in vegetation can be markedly affected (Brown 1982; McBrien, Harmsen & Crowder 1983). Few of these studies have considered specifically the influence of herbivory upon plant fitness.

One study which showed clearly how insect herbivory may act selectively within a population comes from an unplanned experiment: the introduction of the gypsy moth in the United States of America, where it devastated forests in New England. The United States Department of Agriculture Gypsy Moth Laboratory at Melrose Highland in Massachusetts monitored the effects of gypsy moth on forest vegetation for 20 years, and much of their huge body of data has recently been re-analysed (Campbell & Sloan 1977). Between 1911 and 1921 there were several outbreak years with high tree mortality, especially among the favoured species. Over the decade the relative abundances of tree species changed, and the most favoured species (white, red, scarlet and black oaks; grey and paper birch) declined from 47% to 36% of all individuals. However, there was evidence of strong selection within species; some trees were consistently grazed more than others, and subsequent mortality was highest amongst these individuals. Thus, over the period of the study there was progressive elimination of individuals most prone to defoliation. Presumably this was a major reason for the gradual decline in the virulence of outbreaks: in the period 1922–1931 the mean annual defoliation of the four oak species was 5.4%, compared with 37.9% during the previous decade.

This study provides dramatic evidence of the capacity for an insect species to act as a powerful selective agent, albeit in the rather special circumstances of an introduced species, but it is difficult to deduce from studies of this kind the effects upon plant fitness which might result from more subtle changes in levels of grazing. For most natural communities we have very little understanding of the role of insect herbivores in natural selection. It is improbable that insect herbivory is equally important as a selective influence in all environments, but there is no firm basis for drawing conclusions about its relative importance in particular communities.

One factor which may be particularly important in determining the importance of insect herbivory is the influence that grazing damage has upon the competitive balance between plants. Several lines of evidence suggest that grazing by invertebrates has a much greater effect upon plants when they are subject to competition (Harper 1977; Whittaker 1979). For example, it seems that success in the biological control of weeds using insects is often due not simply to grazing but to the influence that grazing damage has upon the competitive status of plants. The successful control of *Hypericum perforatum* in California was attributed to the intense competition from range plants that prevented the weed re-establishing itself after heavy grazing by the beetle *Chrysolina quadrigemina* (Holloway & Huffaker 1952). Similarly, the control of *Clidemia hirta* in Fiji by *Liothrips urichi* was supposed by Simmonds (1933) to have resulted from the insect checking the growth of the plant and allowing rival weeds to outgrow it.

An elegant illustration of the potential importance for plant fitness of grazing in competitive situations is an experiment in which slugs (*Agriolimax reticulatus*) were allowed to graze in trays of seedlings of

Lolium perenne (Harper 1977). After 5 days the slugs were removed and the plants were left to grow until they were harvested a few days later. Although most of the damaged plants survived, they were very much smaller than undamaged plants. For example, 22.8% of plants in one experiment were damaged and contributed 1.7% of the total dry weight at harvest. Thus competition accentuated the effects of grazing damage leading to a highly skewed distribution of individual plant weights, with the damaged plants among the smallest individuals.

A more elaborate experiment was performed by Bentley and Whittaker (1979), who investigated the effects of grazing by the chrysomelid beetle *Gastrophysa viridula* upon competition between *Rumex obtusifolius* and *R. crispus*. It was found that levels of grazing which had no significant effect upon either species when grown alone, led to substantially reduced performance when the two species were competing. A similar synergistic effect of grazing and competition was demonstrated in a field experiment to investigate effects of *G. viridula* grazing upon *Rumex obtusifolius* (Cottam, Whittaker & Malloch 1986). Again it was shown that grazing significantly reduced the growth of *R. obtusifolius* only when it was in competition with other plants (in this case *Agrostis capillaris* and *Festuca rubra*).

These experiments and observations support the hypothesis that grazing affects fitness of plants most strongly when they are competing for resources. Harper (1977) has discussed how such effects might arise. For example, under conditions in which light is limiting, two genotypes may be equal competitors. Growing in uncrowded conditions, individuals of each genotype might recover quickly from above-ground herbivory, but when growing together in dense mixtures, whichever genotype is subject to most herbivory will lose its position in the height hierarchy and be quickly overtopped and shaded. However, other work has not shown an interaction of this kind between the effects of grazing and of competition. Fowler & Rausher (1985) grew *Aristolochia reticulata* alone and in competition with either or both *Schizachyrium scoparium* var. *virile* and *Rubus trivialis*. Herbivory on *A. reticulata* by the pipevine swallowtail butterfly, *Battus philenor*, was simulated by clipping. Both competition and clipping individually had the expected negative effects on growth and reproduction of *A. reticulata*, but their joint effect was best described as additive.

In the absence of comparative field experiments to investigate the influence of grazing upon plant fitness we can only speculate about its importance in different environments. Coley, Bryant & Chapin (1985) and Coley (1988) suggested that the relative impact of herbivory was likely to be higher in intrinsically slow-growing plants characteristic of resource-poor environments, because the removal of a certain amount of tissue represents a larger fraction of net production than for fast-growing plants. However, this argument neglects the influence that grazing may have upon the balance of competition between plants. Although competition is important in determining the plant species composition of vegetation in most habitats, it is probably in resource-rich conditions that the synergism between insects and competition is most potent. One reason for this is that in conditions of ample water and nutrients, light is the factor which limits growth. The value of foliage to a plant is then not merely for photosynthesis but also as a weapon in competition, since leaves can shade those of neighbours (Edwards, Wratten & Gibberd 1989). A second reason is that in conditions where rapid growth is possible a plant can quickly benefit from the increased resources which become available when a neighbour is damaged, giving it an advantage which is never reversed. In contrast, in conditions where plants are slow-growing additional resources are not rapidly translated into increased growth. Clearly, a sustained difference between neighbours in the rate of grazing might give one individual an advantage. However, since in many environments insect populations fluctuate widely, occasional differences in grazing may have little effect upon the balance of plant competition. Thus, we might predict that in an environment rich in resources a certain rate of grazing damage (measured as a proportion of plant growth) will have a more severe effect upon a plant's fitness than the same rate of damage to a plant in a resource-poor environment.

TRENDS IN BIOCHEMICAL EVOLUTION

Ecologists have shown rather little interest in the evolutionary history of plants when trying to understand the present day characteristics of species (Hodgson 1986). To some extent, this neglect can be attributed to the absence of a clear consensus about the pathways of plant evolution, especially amongst flowering plants, which appear in great diversity in the Cretaceous with few clues in the fossil record as to their origins. However, the great value of an evolutionary perspective is that it provides evidence of the direction of ecological trends, and also of the constraints that have limited adaptation. Even our present, very rudimentary knowledge of ecological trends in plant evolution can profoundly influence our conclusions about adaptation and adaptive strategies (Stebbins 1974).

Secondary plant chemistry is of particular value in considering phylogeny, since the direction of evolution is often clear and corresponds with the sequence of steps in biosynthetic pathways. Recently, Gottlieb (1982) has presented a fascinating analysis of the evolution of secondary plant substances which reveals not only clear directions of evolutionary advance in the efficacy of toxins, but also strong phylogenetic constraints which limit the kinds of substance that plants can produce. It is clear that the evolution of new secondary compounds has been important for the success of particular taxa. There are many examples in which the acquisition of new biosynthetic pathways appears to have contributed to a phase of rapid evolution within a taxon. A good example is found within the Umbelliferae, which produce three types of coumarins — hydroxycoumarins, linear furanocoumarins, and angular furanocoumarins — which are biosynthetically related and represent a sequence which shows evolutionary advancement (Berenbaum 1983). They also represent a sequence of increasing toxicity to insects. The evolution of greater toxicity in the Umbelliferae has been associated with more rapid speciation; in consequence the genera with angular furanocoumarins are more diverse than those with linear furanocoumarins, which in turn are more diverse than those with only hydroxycoumarins. In some cases there is a strong geographic pattern in the modern distribution of certain compounds within a taxon (e.g. in the distribution of certain quinilozidine alkaloids in the Papilionoideae) suggesting that its spread into new regions was conditioned by the acquisition of a particular biosynthetic pathway (Gottlieb 1980). Evidence of this kind supports the hypothesis of Ehrlich & Raven (1964) that plants which acquire a novel defensive compound 'enter a new adaptive zone' where they are free to diversify.

An example of the constraints which limit the kinds of secondary substances in plants is provided by the types of alkaloids and their distribution among flowering plants. A close link exists between secondary metabolism and the primary metabolic pathways of a plant, so that the kinds and amounts of secondary compounds produced by plants are strongly affected by other aspects of metabolism. In herbaceous plants the shikimate pathway which leads to the production of lignin is much less important than in woody plants, and we find notably that in herbaceous species there is much less accumulation of condensed tannins, which are also products of the shikimate pathway (Gottlieb 1982). Instead we find secondary compounds, including alkaloids, produced from metabolites which are earlier steps in the shikimate pathway. For example, within the families of the Magnoliidae, there is a close negative relationship between the proportion of woody species in a family and the diversity of alkaloids (measured as the number of different skeletal types of benzylisoquinoline alkaloids). In more advanced herbaceous families, the importance of the shikimate pathway is further reduced and there are alkaloids which are derived, not from this pathway, but from acetate derivatives. These include the extremely toxic iridoid alkaloids produced by members of the Gentianiflorae, and the steroid alkaloids produced by the Solaniflorae. As Gottlieb writes, 'With the indolo-iridoid and steroid alkaloids diversification in search of alkaloidal potency seems to have reached a climax.'

The constraints to biochemical evolution mean that, although there are periods of rapid biochemical evolution within particular taxa, there are also long periods with little change. A useful marker in interpreting the rate and direction of change is the oxygenation level of secondary plant metabolites, which has tended to increase with evolutionary advancement. It seems that, throughout land plant evolution, enzyme systems have continued to evolve and thus permitted a gradually increasing efficiency in the utilization of oxygen. For example, Gottlieb (1982) showed that, within the families of the Magnoliidae, there is a strong positive correlation between the level of oxygenation of benzylisoquinoline alkaloids and the 'index of advancement' defined by Sporne (1980). Similarly, there is a close correlation between oxygenation state and skeletal advancement among the indole alkaloids of the Apocynaceae, among the sesquiterpene lactones of the Compositae (Emerenciano *et al.* 1986), and within various classes of secondary compounds in many other taxa.

The evidence from oxygenation states tells us that the secondary metabolites of many plant species have changed little over long periods of time, presumably because of the absence of suitable genetic variation. Indeed, it is the evidence of stasis which is perhaps the most important lesson for ecologists. Coevolution theory has suggested that plants are caught up in an evolutionary arms race with their insect herbivores, with continual adaptation and counter-adaptation. The biochemical evidence suggests that, although biosynthetic breakthroughs do occur which permit periods of rapid evolution, these events occur infrequently and the rate of chemical change within most taxa is generally slow.

DISCUSSION

Evolution of anti-herbivore defences

One of the problems we face in developing a general theory of plant defence is the conflicting nature of evidence about the importance of herbivory as an agent of natural selection. On the one hand, some kind of coevolutionary process between plants and insects appears to provide the best explanation for the diversity of secondary compounds and their distribution among terrestrial plants (Ehrlich & Raven 1964; Harborne 1988); such an explanation requires that insect herbivory has been a major selective force in plant evolution. On the other hand, it seems doubtful that insects are an important factor affecting plant fitness in all environments. The biochemical evidence also casts doubt upon the pace of the arms race for many plant taxa. Although there has been great diversification of secondary plant compounds and a general increase in the effectiveness of compounds as toxins, there exists among present-day plants a very great range in the evolutionary advancement of plant secondary chemistry; thus a failure to keep up in the 'arms race' does not inevitably lead to extinction. To reconcile these conflicting strands of evidence we must consider the development of insect–plant interactions against a broader background of ecological trends in evolution of the terrestrial plants.

One of the most important ecological trends has been progress in the abilities of plants to grow rapidly in conditions where resources of nutrients and water are abundant. The few surviving members of groups such as the Cycadales and most Gnetales exhibit characteristically relict distributions, and occupy deeply shaded, arid or nutrient-poor habitats where productivity is very low. Similarly, the most primitive of modern flowering plants have highly disjunct or restricted distributions and occur in relatively unproductive vegetation. Among flowering plants it seems clear that the pace of evolution has been generally faster in fertile environments, in which a capacity for rapid growth is important. Thus, despite the considerable adaptive radiation which has occurred within the largest families, it is generally true that slow-growing species prevail in the more ancient families (Stebbins 1974), while within recent families there are relatively more fast-growing species able to compete effectively in fertile habitats. For example, in a study of the ecological characteristics of plants in the Sheffield region, Hodgson (1986) found that the species which have been most successful in exploiting highly fertile habitats created by man, having the potential for rapid growth and a capacity for prolific regeneration by seed, belong mainly to families such as the Compositae and Labiatae, which are evolutionarily advanced according to the index of Sporne (1980).

The evolution of more rapid growth and greater competitive ability in resource-rich habitats presumably resulted from more efficient capture of resources of light, water and nutrients. To take extreme examples, it is easy to see that the organization of photosynthetic tissues in a tree such as sycamore (*Acer pseudoplatanus*) is much more effective in the capture of light than that of the primitive relict species *Araucaria araucana.* There have also been improvements in the efficiency of root systems. For example, the magnolioid root system is coarsely branched and highly dependent upon conversion to a mycorrhiza for effective uptake of phosphate (Baylis 1975). In contrast, many more advanced flowering plants have very fine, much-branched root systems with well-developed root hairs, and mycorrhizal infection is often unnecessary.

It is interesting to speculate that the improved ability of plants to capture plant nutrients must itself have affected the environment in which they lived. The return to the soil of more nutrient-rich litter would have led to more rapid mineralization, and thus to a greater availability of plant nutrients. However, perhaps the most significant event for soil fertility was the development of highly efficient symbioses between flowering plants and nitrogen-fixing prokaryotes which made possible rates of nitrogen fixation which greatly exceeded those achieved by free-living bacteria and blue-green algae, or by cycads which have blue-green algae as symbionts (Sprent 1979).

The evolutionary trend toward more rapidly growing plants must have had profound implications for insect–plant relationships. Whereas slow-growing species of infertile habitats tend to have a high level of neutral resistance, the less tough leaves and higher nutrient concentrations of plants with a high intrinsic rate of growth makes them more vulnerable to insects. Thus, during the course of terrestrial plant evolution, there was probably a gradual increase in the amount of herbivory and in the importance of insects as a factor affecting plant fitness. Whereas previously plants may have escaped grazing through a high degree of neutral resistance, there was probably an increasing need for effective defence as selection for rapid growth affected the structure and nutritional quality of their foliage. This selective pressure for improved defence occurred chiefly in conditions of abundant resources where the potential for rapid growth was greatest.

Defence and resistance among modern plants

Among present-day plants, we can expect to find a broad spectrum in the extent and effectiveness of defences against insect herbivores. Figure 12.1 illustrates how selection for defence is likely to vary in relation to the quality of the habitat for plant growth, and also in relation to factors which affect the size and stability of insect populations. At one extreme are those plants growing in very poor habitats where growth is slow

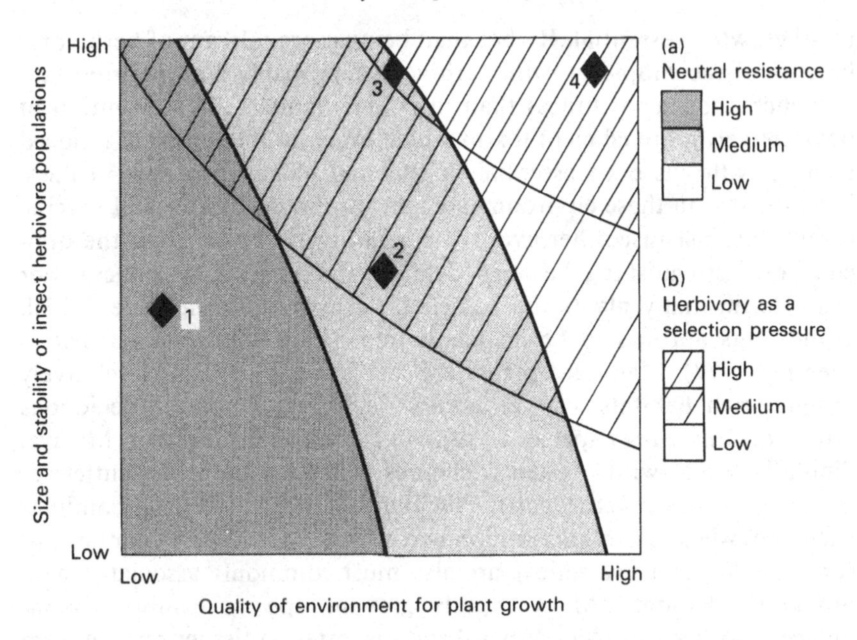

FIG 12.1 Diagram to show the hypothetical variation in (a) the neutral resistance of plants to insect herbivores, and (b) the importance of insect herbivory as a selective pressure, in relation to the quality of the environment for plant growth, and factors affecting the size and stability of insect populations. Plants characteristic of four types of habitat are located on this diagram as examples of the kind of variation which might be expected: (1) fynbos; (2) persistent trees of temperate deciduous forest; (3) persistent trees of tropical rain forest; (4) pioneer trees of tropical rain forest.

because of adverse conditions such as low temperatures, drought or low nutrient availability. Such plants typically have long-lived leaves which must be resistant to the range of hazards that they may encounter over an extended lifespan, such as desiccation, frost, wind and attack by micro-organisms. Although selection by herbivores may have accentuated those characteristics such as toughness or low nutrient content which make such plants resistant to insects, herbivory is unlikely to be the major selective factor. It is misleading to argue that such plants have a high investment in anti-herbivore defences. Indeed to survive in such conditions it is not essential to possess evolutionarily advanced defences, and many of the more primitive modern land plants persist in unproductive habitats.

At the other extreme are those plants which have exploited environments with high levels of resources where competition is intense and rapid growth is essential. Rather than being more tolerant of herbivores because of their rapid growth, herbivory is probably a more important influence upon plant fitness than in poorer habitats. The evolution of progressively more efficient toxins which exact little in terms of reduced plant growth has been essential in enabling plants to persist in these habitats. It is in these environments that we should look to find specific defence against insect herbivores (though always recognizing the difficulty of distinguishing between defence and neutral resistance). For example, in many plants the possession of extrafloral nectaries which attract ants appears to be an adaptation against herbivorous insects (Bentley 1977). Many ant plants are fast-growing species of relatively resource-rich habitats, e.g. *Viburnum opulus* in temperate deciduous forest, or *Macaranga* species in gaps in lowland rain forest in Malaya. Similarly leaves with hooked trichomes which puncture the cuticle of insects (e.g. *Phaseolus vulgaris*; Pillemer & Tingey 1976), or glandular trichomes whose secretions immobilize insects (e.g. the glandular hairs of certain wild potato varieties), are also most commonly associated with productive habitats. Many rapidly growing species exhibit wound-induced synthesis of secondary substances, often in tissues remote from the site of damage, e.g. proteinase inhibitors in tomato and other species (Green & Ryan 1972; Ryan 1979) and cucurbitacins in *Cucurbita moschata* (Tallamy 1985). There is growing evidence that this phenomenon also represents a defence against insect herbivores (Edwards & Wratten 1985).

We can expect the highest degree of optimization of plant defences in environments where selective pressures from insects are greatest. In conditions which both are resource-rich for plants and permit stable populations of insects, the influence of herbivory via natural selection will be most intense and we can expect to find the most developed defences. For example, in tropical rain forests of Central America populations of butterflies in the genus *Heliconius*, whose larvae feed on species of *Passiflora* are remarkably stable (Gilbert 1975), and it is not surprising that *Heliconius* and *Passiflora* provide one of the most convincing examples of coevolution between insects and plants. In contrast, insect populations on trees of temperate deciduous woodland fluctuate widely (Strong, Lawton & Southwood 1984) because of climatic variation, and particularly the severity of winters. In such conditions, where the selective pressures faced by plants may vary widely in successive years and herbivory is a major factor only occasionally (Edwards & Gillman 1987), we are less likely to find highly evolved defences.

CONCLUSION

In this article I have sought to challenge the assumptions which are implicit in the concept of plant defence strategies, and argue that ecologists must pay more attention to the directions of plant evolution. Some plant species evolved a long time ago and have changed little, not because they are ideally suited to their environment, but because of lack of suitable genetic variation (Bradshaw 1984). Other taxa are clearly advancing, often because of some evolutionary breakthrough in secondary plant chemistry which permits them to exploit new habitats or compete more effectively. Thus the plant world is a complex mixture of the advanced and the primitive, the efficient and the inefficient. In many environments selection pressures from herbivores are either not sufficiently intense or not consistent enough to produce a single, highly optimized solution or strategy. For these reasons a classification of plants according to defensive strategies may be seriously misleading since it attributes to plant characteristics an adaptive significance for which there is no supporting evidence. In assuming that patterns of resource allocation to defence reflect a high degree of optimization, we neglect the evidence for a broad spectrum in the effectiveness of plant defences among present-day plants. Thus I believe that, in focusing upon adaptive strategies, ecologists have been asking the wrong question: we should be interested, not in how plants in different environments are adapted to insect herbivory, but in how insect herbivory has affected (and is affecting) the course of evolution. The evidence outlined in this paper suggests that herbivory has indeed been an important factor in flowering plant evolution, but that its influence has been greatest in resource-rich habitats.

Finally, it is worth emphasizing that the most serious impediment to an understanding of the evolutionary significance of insects for plants is our lack of knowledge about the influence of grazing upon plant fitness. A major obstacle is the rather crude form of quantification of herbivory usually adopted, often in terms of total dry weight of leaf area consumed. We need more exact studies of the type and timing of damage, the tissues affected, and their spatial distribution within the plant. In particular we need to understand better how herbivory can affect the competitive balance between plants. As Janzen wrote in 1979, 'We badly need field experiments with wild plants designed to show how the fitness of these plants is affected by herbivores of different types, applied in a variety of competitive and edaphic circumstances.'

REFERENCES

Baylis, G. T. S. (1975). The magnolioid mycorrhiza and mycotrophy in root systems derived from it. *Endomycorrhizas* (Ed. by F. E. Sanders, B. Mosse & P. B. Tinker), pp. 373–90. Academic Press, London.

Beck, S. D. (1965). Resistance of plants to insects. *Annual Review of Entomology*, **10**, 207–32.

Bentley, B. L. (1977). Extrafloral nectaries and protection by pugnacious bodyguards. *Annual Review of Ecology and Systematics*, **8**, 407–27.

Bentley, S. & Whittaker, J. B. (1979). Effects of grazing by a chrysomelid beetle, *Gastrophysa viridula*, on competition between *Rumex obtusifolius* and *Rumex crispus*. *Journal of Ecology*, **67**, 79–90.

Berenbaum, M. (1983). Coumarins and caterpillars: a case for coevolution. *Evolution*, **37**, 163–79.

Bradshaw, A. D. (1984). The importance of evolutionary ideas in ecology — and vice versa. *Evolutionary Ecology* (Ed. by B. Shorrocks), pp. 1–25. Symposia of the British Ecological Society, 23. Blackwell Scientific Publications, Oxford.

Brown, V. K. (1982). The phytophagous insect community and its impact on early successional habitats. *Proceedings of the 5th International Symposium on Insect Plant Relationships, Wageningen* 1982 (Ed. by J. H. Visser & A. K. Minks), pp. 205–13. Pudoc, Wageningen.

Brown, V. K. (1985). Insect herbivores and plant succession. *Oikos*, **44**, 17–22.

Campbell, B. M. (1986). Plant spinescence and herbivory in a nutrient poor ecosystem. *Oikos*, **47**, 168–72.

Campbell, R. W. & Sloan, R. J. (1977). Forest stand responses to defoliation by the gypsy moth. *Forest Science*, **23**, (Supplement) 1–32.

Cantlon, J. E. (1969). The stability of natural populations and their sensitivity to technology. *Brookhaven Symposium in Biology*, **22**, 197–205.

Chapin F. S. (1980). The mineral nutrition of wild plants. *Annual Review of Ecology and Systematics*, **11**, 261–85.

Coley, P. D. (1980), Effects of leaf age and plant life history patterns on herbivory. *Nature*, **284**, 545–6.

Coley, P. D. (1983). Herbivory and defensive characteristics of tree species in a lowland tropical forest. *Ecological Monographs*, **53**, 209–33.

Coley, P. D. (1988). Effects of plant growth rate and leaf lifetime on the amount and type of anti-herbivore defense. *Oecologia*, **74**, 531–6.

Coley, P. D., Bryant, J. P. & Chapin, F. S. (1985). Resource availability and plant antiherbivore defense. *Science*, **230**, 895–9.

Cottam, D. A., Whittaker, J. B. & Malloch, A. J. C. (1986). The effects of chrysomelid beetle grazing and plant competition on the growth of *Rumex obtusifolius*. *Oecologia*, **70**, 452–6.

Cottrell, C. B. (1986). The absence of co-evolutionary associations with *capensis* floral element plants in the larval/plant relationships of South Western Cape butterflies. *Species and Speciation* (Ed. by E. S. Vrba). Transvaal Museum, Pretoria.

Crawley, M. J. (1983). *Herbivory: the Dynamics of Animal-Plant Interactions*. Blackwell Scientific Publications, Oxford.

Dixon, A. F. G. (1971). The role of aphids in wood formation. I. The effect of the sycamore aphid *Drepanosiphum platanoides* (Schr.) (Aphididae) on the growth of sycamore *Acer pseudoplatanus*. *Journal of Applied Ecology*, **8**, 165–79.

Edwards, P. J. & Gillman, M. P. (1987). Herbivores and plant succession. *Colonization,*

Succession and Stability (Ed. by A. J. Gray, M. J. Crawley & P. J. Edwards), pp. 295–314. Symposia of the British Ecological Society, 26. Blackwell Scientific Publications, Oxford.

Edwards, P. J. & Wratten, S. D. (1985). Induced plant defences against insect grazing: fact or artefact. *Oikos,* **44**,70–4.

Edwards, P. J., Wratten, S. D. & Gibberd, R. M. (in press). The impact of inducible phytochemicals on food selection by insect herbivores and its consequences for the distribution of grazing damage. *Phytochemical Induction by Herbivores* (Ed. by M. J. Raupp & D. W. Tallamy). Wiley, New York.

Ehrlich, P. R. & Raven, P. H. (1964). Butterflies and plants: a study in plant coevolution. *Evolution,* **18**, 586–608.

Emerenciano, V. D. E. P., Kaplan, M. A. C., Gottlieb, O. R., de Bonfanti, M. R. M., Ferreira, Z. S. & Comegno, L. M. A. (1986). Evolution of sesquiterpene lactones in Asteraceae. *Biochemical Systematics and Ecology,* **14**, 585–9.

Feeny, P. (1976). Plant apparency and chemical defenses. *Recent Advances in Phytochemistry,* **10**, 1–41.

Fowler, N. L. & Rausher, M. D. (1985). Joint effects of competitors and herbivores on growth and reproduction in *Aristolochia reticulata. Ecology,* **66**, 1580–7.

Fox, L. R. (1981). Defense and dynamics in plant-herbivore systems. *American Zoologist,* **21**, 853–64.

Gilbert, L. E. (1975). Ecological consequences of a coevolved mutualism between butterflies and plants. *Coevolution of Animals and Plants* (Ed. by L. E. Gilbert & P. R. Raven), pp. 210–40. University of Texas Press, Austin.

Gottlieb, O. R. (1980). Micromolecular systematics: principles and practice. *Chemosystematics: Principles and Practice* (Ed. by F. A. Bisby, J. G. Vaughan & C. A. Wright). pp. 329–52. Academic Press, London.

Gottlieb, O. R. (1982). *Micromolecular Evolution, Systematics and Ecology.* Springer, Berlin.

Green, T. R. & Ryan, C. A. (1972). Wound-induced proteinase inhibitor in plant leaves: a possible defense mechanism against insects. *Science,* **175**, 776–7.

Grubb, P. J. (1986). Sclerophylls, pachyphylls and pycnophylls: nature and significance of hard leaf surfaces. *Insects and the Plant Surface* (Ed. by B. Juniper & Sir Richard Southwood), pp. 137–50. Arnold, London.

Harborne, J. B. (1988). *Introduction to Ecological Biochemistry.* Academic Press, London.

Harper, J. L. (1977). *The Population Biology of Plants.* Academic Press, London.

Hodgson, J. G. (1986). Commonness and rarity in plants with special reference to the Sheffield flora. Part III: Taxonomic and evolutionary aspects. *Biological Conservation,* **36**, 275–96.

Holloway, J. K. & Huffaker, C. B. (1952). Insects to control a weed. *Yearbook of Agriculture,* 135–40.

Huffaker, C. B. & Kennett, C. E. (1959). A ten year study of vegetational changes associated with biological control of Klamath weed. *Journal of Range Management,* **12**, 69–82.

Janzen, D. H (1979). New horizons in the biology of plant defenses. *Herbivores: their Interactions with Secondary Plant Metabolites* (Ed. G. A. Rosenthal & D. H. Janzen), pp. 331–50. Academic Press, New York.

Janzen, D. H. (1980). When is it coevolution? *Evolution,* **34**, 611–12.

Louda, S. M. (1982). Limitations of the recruitment of the shrub *Haplopappus squarrosus* (Asteraceae) by flower- and seed-feeding insects. *Journal of Ecology,* **70**, 43–53.

Loveless, A. R. (1961). A nutritional interpretation of sclerophylly based on differences in the chemical composition of sclerophyllous and mesophytic leaves. *Annals of Botany,* **25**, 168–84.

McBrien, H., Harmsen, R. & Crowder, A. (1983). A case of insect grazing affecting plant succession. *Ecology*, **64**, 1035–9.

Mansfield, J. W, (1983). Antimicrobial compounds. *Biochemical Plant Pathology* (Ed. by J. A. Callow), pp. 237–66. Wiley, Chichester.

Myers, J. H. (1987). Nutrient availability and the deployment of mechanical defenses in grazed plants: a new experimental approach to the optimal defense theory. *Oikos*, **49**, 350–1.

Pillemer, E. A. & Tingey, W. M. (1976). Hooked trichomes: a physical plant barrier to a major agricultural pest. *Science*, **193**, 482–4.

Rhoades, D. F. (1979). Evolution of chemical defence against herbivores. *Herbivores: their Interrelationships with Plant Secondary Metabolites* (Ed. by G. A. Rosenthal & D. H. Janzen), pp. 3–54. Academic Press, New York.

Rhoades, D. F. (1983). Herbivore population dynamics and plant chemistry. *Variable Plants and Herbivores in Natural and Managed Systems* (Ed. by R. F. Denno & S. McClure), pp. 155–220. Academic Press, New York.

Rhoades, D. F. & Cates, R. G. (1976). Toward a general theory of plant antiherbivore chemistry. *Recent Advances in Phytochemistry*, **10**, 168–213.

Ride, J. P. (1983). Cell walls and other structural barriers in defence. *Biochemical Plant Pathology* (Ed. by J. A. Callow), pp. 215–36. Wiley, Chichester.

Rothwell, G. W. (1982). New interpretations of the earliest conifers. *Review of Palaeobotany and Palynology*, **37**, 7–28.

Roughgarden, J. (1979). *Theory of Population Genetics and Evolutionary Ecology: an Introduction.* Macmillan, New York.

Ryan, C. A. (1979). Proteinase inhibitors. *Herbivores: their Interaction with Secondary Plant Metabolities* (Ed. by G. A. Rosenthal & D. H. Janzen), pp. 599–618. Academic Press, New York.

Simmonds, H. W. (1933). The biological control of the weed *Clidemia hirta* D. Don in Fiji. *Bulletin of Entomological Research*, **24**, 345–8.

Southwood, T. R. E. (1973). The insect-plant relationship — an evolutionary perspective. *Symposia of the Royal Entomological Society, London*, **6**, 3–30.

Specht, R. L. & Moll, E. J. (1983). Mediterranean type heathlands and sclerophyllous shrublands of the world: an overview. *Mediterranean-type Ecosystems: the Role of Nutrients* (Ed. by F. J. Kruger, D. T. Mitchell & J. U. M. Jarvis), pp. 41–65. Springer, Berlin.

Sporne, K. R. (1980). A re-investigation of character correlations among dicotyledons. *New Phytologist*, **85**, 419–49.

Sprent, J. I. (1979). *The Biology of Nitrogen-fixing Organisms.* McGraw-Hill, London.

Stebbins, G. L. (1974). *Flowering Plants — Evolution above the Species Level.* Harvard University Press, Cambridge, Massachusetts.

Stewart, W. N. (1983). *Palaeobotany and the Evolution of Plants.* Cambridge University Press.

Strong, D. R., Lawton, J. H. & Southwood, Sir Richard (1984). *Insects on Plants: Community Patterns and Mechanisms.* Blackwell Scientific Publications, Oxford.

Swain, T. (1975). Evolution of flavonoid compounds. *The Flavonoids* (Ed. by J. B. Harborne, T. J. Mabry & H. Mabry), pp. 1096–129. Chapman & Hall, London.

Tallamy, D. (1985). Squash beetle trenching behaviour: an adaptation against induced cucurbit defenses. *Ecology*, **66**, 1574–9.

Tallamy, D. W. & Krischik, V. A. (in press). Variation and function of cucurbitacins in *Cucurbita:* an examination of current hypotheses. *Phytochemical induction by Herbivores* (Ed. by M. J. Raupp & D. W. Tallamy). Wiley, New York.

Tuomi, J., Niemela, P., Chapin, F. S., Bryant, J. P. & Siren, S. (1988). Defensive responses of trees in relation to their carbon/nutrient balance. *Mechanisms of Woody Plant Defenses against Insects: Search for Pattern* (Ed. by W. J. Mattson, J. Levieux & C. Bernard-Dagan), pp. 57–72. Springer, New York.

Tuomi, J., Niemela, P., Haukioja, E., Siren, S. & Neuvonen, S. (1984). Nutrient stress: an explanation for plant anti-herbivore responses to defoliation. *Oecologia,* **61**, 208–10.

Waloff, N. & Richards, O. W. (1977). The effect of insect fauna on growth, mortality and natality of broom, *Sarothamnus scoparius. Journal of Applied Ecology,* **14**, 787–98.

Whittaker, J. B. (1979). Invertebrate grazing, competition and plant dynamics. *Population Dynamics* (Ed. by R. M. Anderson, B. D. Turner & L. R. Taylor), pp. 207–22. Symposia of the British Ecological Society, 20. Blackwell Scientific Publications, Oxford.

13. BUTTERFLY-ANT MUTUALISMS

N. E. PIERCE
Department of Biology, Princeton University, Princeton, New Jersey 08544–1003, USA

INTRODUCTION

At some stage in every elementary biology course the idea of interaction between species is introduced. This is sometimes accompanied by a chart consisting of a 3 × 3 matrix labelled with two mythical entities, 'species A' and 'species B'. The matrix is composed of pluses, minuses and zeros: 00 is the trivial case in which there is no interaction; +− (or −+) denotes parasitism or predation; − − signifies competition, and −0 (or 0−), interference; +0 (or 0+) indicates commensalism; and ++ is mutualism. The category 00 is invoked rather more often than is probably justified. If one is not interested in interaction *per se*, the temptation is strong to treat one's organism in isolation. This, almost without exception, is misleading. A hypothetical 100-year-old ecologist, who joined this Society in the year of its founding and who would be duly honoured at an anniversary meeting such as this one, would immediately understand why. He would have witnessed the development of ecology from glorified natural history to a major scientific discipline and would be fully aware that a feature of the evolution of ecology has been an increasing recognition that the natural world is complex, and that food webs are intricate networks. Research into +− interactions has been a popular study area ever since the fundamentals of ecological theory were formulated in the early days of the discipline. Competition (− −) and interference (−0) have not been lagging far behind: again, many of the most far-reaching results in early ecology, both theoretical and empirical, directly addressed these issues, especially competition.

So what of ++ and +0? Our old man (his sex is assumed on the basis of the make-up of the founding membership of this Society) will remember little emphasis on these topics throughout his career as an ecologist. Mutualism (++) is here taken simply to denote an interaction in which the fitness of each party is increased by the action of its partner. There has, however, been a flurry of theoretical interest in mutualism over the past decade or so (see Boucher 1985), and this stimulated development of models of the population dynamics of mutualists (May 1981; Addicott 1984; Wolin 1985; Pierce & Young 1986; reviewed in

Boucher, James & Keeler 1982) and of their evolution (e.g. Trivers 1971; Roughgarden 1975; Wilson 1980; Axelrod & Hamilton 1981; Keeler 1981, 1985; Maynard Smith 1982; Axelrod 1984; Vandermeer 1984; Law 1985; Templeton & Gilbert 1985). Commensalism (+0) has yet to generate an extensive literature of its own, perhaps because it is rather uninteresting: after all, effectively only one party is doing the interacting.

The reasons why mutualism has attracted so much theoretical interest lie, I believe, primarily with the emergence of social biology. Altruistic social interactions that appear to contradict the central survival-of-the-fittest dogma of a Darwinian theory of evolution with its emphasis on the fitness of the *individual* have always been an enigma to evolutionary biologists. It was only with the work set in motion by Hamilton (1964) that we have been able to feel comfortable in an established evolutionary framework with such social 'anomalies'. However, kin selection arguments necessarily only apply within species; alternative models to explain the evolution of co-operation between species have been necessary (Trivers 1971; Axelrod & Hamilton 1981). In fact, interspecific mutualism provides us with the ideal opportunity to explore such proposed mechanisms of social evolution as reciprocal altruism, in the absence of the complicating factor of kin selection. Only recently (May 1981) have refinements to standard Lotka–Volterra models made possible the mathematical description of mutualistic systems (see Addicott 1984 for a discussion).

From an empirical point of view, mutualism is amenable to study for a number of reasons. First, it is common. With the exception of pollination syndromes, the phenomenon may not be as obvious as predation, but mutualism, it is becoming apparent, is a major evolutionary theme: from mycorrhizal fungi associated with plant roots to micro-organisms in the guts of termites. The extent of mutualism in nature will not, I think, be fully realized until we have a more complete understanding of microbial ecology. Second, the strength of these associations is highly variable: many occasionally ant-tended aphid species survive well in the absence of ants, while neither the termite nor its gut micro-organisms can survive independently when forcibly parted. This range of relationship, from loosely facultative to strictly obligate, gives us a corresponding range of systems suitable for different kinds of experimental approach. If a given group displays a trend in the strength of its mutualism, then it may be possible through the use of comparative studies to identify those ecological correlates that have been critical in driving its evolution. Third, because mutualists are often highly dependent upon each other, it is likely that the selective forces shaping the association are strong and therefore *identifiable.* It is possible to recognize key components of an organism's biology simply by elucidating what costs and benefits it experiences from associating with its partner. Thus, in the case of the aphid–ant relationship, we can surmise that defence from predators and parasites is a significant evolutionary 'problem' for the aphid. Fourth, although a mutualism is, by definition, two-sided, there exist, in many cases, asymmetries between the two parties involved. For instance, for some lycaenid butterfly species the presence of tending ants is imperative if the larvae are to survive while the ants, although they benefit nutritionally from the lycaenid larvae, can survive in their absence. This facilitates experimental manipulations because ant exclusion results only in the extinction of the butterflies, permitting quantification of ant performance in the absence of the mutualist. Assuming a hypothetical situation in which the asymmetry was reversed so that the ants could not survive without the butterflies, it would be possible to reconstruct the 'other side of the story', thereby completing an overall picture. Finally, it is worth adding that the output of evolutionary and ecological theoreticians in the form of testable predictions and models is in itself major incentive for empiricists to study mutualism.

A conceptual framework for studies of mutualism has only recently been available. Such a framework is necessary because mutualisms are necessarily complex: co-operation invariably entails exchange and communication, both of which are likely to confound simple analysis. This complexity does not stop at the straightforward level of the relationship between the two parties involved, but other, extrinsic, factors also intrude. A study of the termite–micro-organism mutualism should go beyond a mere analysis of the actual relationship and include, for example, the impact of the termite's diet on the 'ecology' of the protozoa. Thus a study of mutualism must usually entail the kind of multidimensional exercise envisioned by Price *et al.* (1980) in their model of interaction among three trophic levels, or Janzen (1985) in his discussion of diffuse effects of mutualism. It is important to recognize that trade between the two participants generally does *not* involve the transfer of a common currency. Thus, while ants derive nutrition from their aphid mutualists, the reward to the aphids is protection from natural enemies — the life/dinner principle for mutualists. It is extremely difficult to compare, on a simple cost–benefit basis, the relative contribution of each species to the other's fitness. Although a number of studies of mutualism have discussed this problem (for example Schemske

1983; Addicott 1986), I know of none that has satisfactorily quantified the costs and benefits to each party in terms of each one's reproductive success.

Lycaenid butterflies and their attendant ants exemplify all these advantages and disadvantages, as will become clear in what follows. They have, however, some special features which, to my mind, make them especially suitable for studies of mutualism. Even within a single butterfly genus, the types of association can be quite varied. Species may be completely untended; they may have the ability to deter ants from attacking them without actually being tended; they may be loosely facultative in being tended only occasionally by a number of different ant species; or they may be obligately tended by members of a single species of ant. Various permutations of these categories also exist. Lycaenids are particularly amendable to large-scale comparative analysis because, as butterflies, they have traditionally attracted the attention of naturalists, which means that extensive records of their natural histories exist. For the same reasons, they have, unlike many groups of insects (including, unfortunately, the ants), been reasonably well classified, although debate still exists concerning the placement of the riodinids, which I here include in my discussion with the cautionary note that they might yet prove to be a separate family (Eliot 1973). It is also possible to culture many of the relevant species of both ants and butterflies in the laboratory, permitting the kind of carefully controlled experimental manipulation that I consider to be an essential complement to field studies.

What follows is an account of my work on such a lycaenid–ant mutualism, that of *Jalmenus evagoras*, a lycaenid ranging from the temperate south of Australia to subtropical regions just north of Brisbane, and its main attendant ant species, *Iridomyrmex anceps*. Although I am not entirely sure that the methods that I will describe together constitute the correct prescription for 'an exact ecology', my collaborators and I have, by means of a combination of experimental, comparative and biochemical techniques, been able to build up a reasonable picture of the dynamics of this mutualism. We have attempted to import the essential elements of the field into the laboratory, and we have employed a number of laboratory-style manipulations of the system in the field. My intention was first to identify costs and benefits of the association, and second to provide a qualitative sense of the direction and magnitude of those costs and benefits through the use of quantitative techniques.

NATURAL HISTORY

The Australian genus *Jalmenus* contains at least nine species whose larvae associate with dolechoderine ants (Common & Waterhouse 1981). One of the these species, *Jalmenus evagoras*, has been the focus of investigation for the past 5 years (Kitching 1983; Pierce 1983, 1984, 1985; Pierce & Elgar 1985; Pierce & Young 1986; Pierce *et al.* 1987; Elgar & Pierce 1988; Smiley, Atsatt & Pierce 1988). Species in the genus *Jalmenus* have interesting and variable life-histories. For example, at least five of the species appear to have obligate, species-specific associations with ants, in the sense that larvae are never found without ants in the field and are only ever found with one species of ant, whereas others, including *J. evagoras*, are known to associate with several species of congeneric ants. Two 'species groups' of ants in the genus *Iridomyrmex*, *I. anceps* and *I. rufoniger* are particularly important associates of *J. evagoras* in our study sites. The taxonomy of these groups has yet to be resolved. For convenience, I will refer to them here as *I. anceps* and *I. rufoniger* (although this nomenclature is misleading in that we know that the *I. rufoniger* tending *J. evagoras* is distinct morphologically from the *I. rufoniger* tending *J. daemeli.* In all cases, we have deposited reference specimens with the Australian National Insect Collection). Like their lycaenid partners, attendant ant species also vary in many aspects of their biology, such as body size, tending behaviours and nest construction. Larvae and pupae often aggregate on host plants, which include approximately twenty species in the genus *Acacia*, and are also extremely localized where they occur, most likely as a result of their dependence upon ants. Thus individuals in natural populations are easily marked and observed in the field for their entire lifetimes. We have had no difficulty in rearing species of *Jalmenus* with their attendant ants on potted food plants in the laboratory.

COSTS AND BENEFITS FOR LYCAENIDS

The general approach that we have used in assessing the costs and benefits of the association for each party involved has been through the use of exclusion experiments, conducted both in the field and the laboratory. These have used sticky barricades to confine the distributions of ants, potted host plants to culture lycaenid larvae and manipulate their distributions in the field, and artificial nest boxes to house queenright (i.e. containing a queen) attendant ant colonies and control their distributions in the field.

Our first ant exclusion experiments were designed to assess the benefit that lycaenid butterflies receive from associating with ants, and in particular we wanted to see whether attendant ants protected larvae from parasites and predators. Ant defence of larvae has been demonstrated for a number of other species (Ross 1966; Pierce & Mead 1981; Pierce & Easteal 1986; DeVries 1987). To do this, we excluded ants from tending larvae in the field by applying a sticky barricade of Tanglefoot (The Tanglefoot Company, Grand Rapids, Michigan) around the bases of the larval host plants. Controls were treated in the same manner, except that Tanglefoot was smeared around only half the stem, allowing ants continued access to the larvae. In addition, we secured clear, plastic drop cloths coated with Tanglefoot beneath each tree to catch any larvae or pupae that might drop off. In this way, we could be sure that disappearances from the trees were due to differential predation by aerial predators and not simply due to larvae dropping off plants without ants (Pierce & Easteal 1986).

We have now repeated the ant exclusion experiments five times over three seasons at three different sites at Mount Nebo, Queensland (152° 47′E/ 27° 23′S), in order to assess how patchiness in space and time of predators and parasitoids might influence the outcome of the experiment. What we have discovered so far is that although the natural enemies of *J. evagoras* do indeed vary from site to site and year to year, the net effect of ant removal is always the same (Fig. 13.1): larvae and pupae deprived of attendant ants cannot survive (Pierce *et al.* 1987, and unpublished).

The benefit that juveniles of *J. evagoras* receive from associating with ants is survival, and we reasoned that, unless the lycaenids were actually fooling their attendant ants and parasitizing them in some way, the cost that the lycaenids would be able to pay for their association might be considerable. In particular, since larvae and pupae produce food secretions for attendant ants, we suspected that maintaining attendant ants might affect larval development. To look at this question, we compared the development of larvae raised with and without attendant ants in the laboratory.

Our experiment revealed both a benefit and a cost of ant attendance for lycaenid larvae. The benefit is that, in addition to guarding juveniles, the presence of attendant ants shortens larval duration, thereby reducing the time that larvae are exposed to the threat of predators and parasitoids (but see Henning 1984). Thus larvae with ants took approximately 23 days to pupate, whereas those without ants took about 29 days. The cost, however, is expressed as a reduction in adult size. For example, females

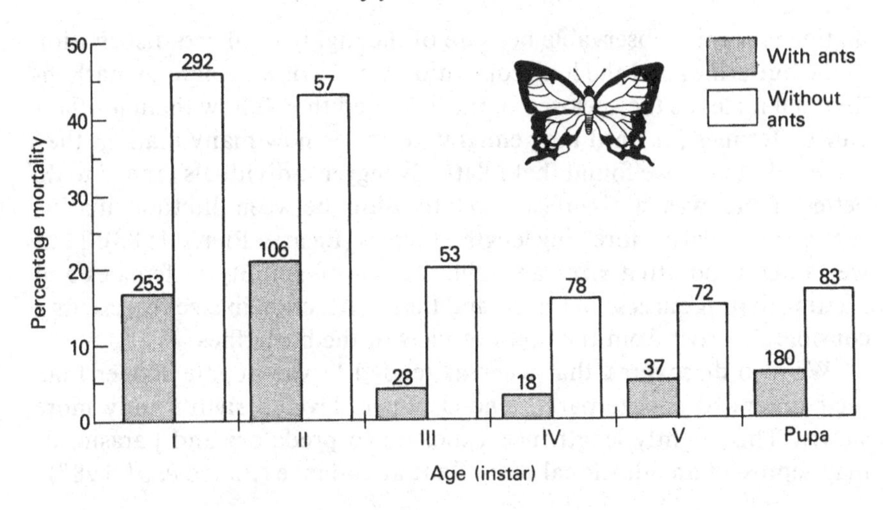

FIG 13.1 Benefit to butterflies. Diagram depicts the age-specific mortality due to predation of juveniles of *Jalmenus evagoras* in an ant-exclusion experiment at Mount Nebo, Queensland; sample sizes are given above each bar (redrawn from Pierce *et al.* 1987).

that are tended by ants pupate at a weight that is approximately 20% lighter than their untended counterparts. Since ant attendance also shortens development time, we found no significant difference in growth rates between larvae raised with and without ants. In the presence of ants, larvae simply pupate earlier at a much smaller size (Pierce *et al.* 1987).

It remained to determine whether smaller size indeed represented a cost to the butterflies. We examined females and males separately in this analysis. First, we raised freshly mated females in the laboratory, collected all of the eggs laid by each individual, and found a highly significant correlation between size and fecundity in this species. Second, we measured the mating success of individual males in the field, and compared this with their forewing length relative to all other males involved in competition for female mates at the same time. When a pupa is about to eclose, as many as twenty males may gather around it, forming a 'mating ball'. The males engage in a frenzied scramble as the pupa ecloses and copulation takes place before a teneral female has even had time to expand her wings. Pairs remain mating on a tree for several hours. Dissections of field-caught females indicate that females mate only once, although in the laboratory we have been able to induce them to mate more than once on rare occasions. Mating in *J. evagoras* does not always involve the formation of a visually dramatic mating ball, and eclosing females are frequently found by single males. However, virtually every mating is readily observable because of the highly localized distribution of the butterflies and the long copulation time. We were able to mark individual males as they eclosed in the field, and then follow them for their entire lifetimes (or until they emigrated) to see how many matings they achieved. Again, we found that relatively bigger individuals tended to do better: there was a significant relationship between lifetime mating success and relative forewing length in males (Elgar & Pierce 1988). Thus we can conclude that size can influence both fecundity in females and lifetime mating success in males, and that a reduction in size represents a considerable cost from the point of view of the butterflies.

We also discovered that whereas tended larvae pupate sooner than their untended counterparts, tended pupae develop significantly more slowly. This slightly lengthened exposure to predators and parasitoids may represent an additional cost of ant attendance (Pierce *et al.* 1987).

COSTS AND BENEFITS FOR ATTENDANT ANTS

By comparison with the lycaenids, which are readily identifiable and easy to work with, we know very little about the costs and benefits of associating with *J. evagoras* for its attendant ants. Few quantitative data exist describing the benefits that ants receive in any of their apparently mutualistic interactions with other insects (but see Degen *et al.* 1986; Fiedler & Maschwitz 1988; and Buckley 1987 for review). It is easy to see why this is the case: ants that nest underground are difficult to work with, and, perhaps more importantly, they present a serious difficulty in deciding what unit to measure. Should we be assessing benefits to individual foraging ants, or to individual colonies? And, when we are working with a large, amoeboid polygynous and polydomous 'colony' that can extend as a single, self-compatible unit for distances of more than a kilometre, then what is the right unit to measure? Ecologists have approached this difficulty in different ways. Some have settled on individual foraging ants as the units of measurement (e.g. Lanza & Krauss 1984), whereas others have concentrated on colony and group level dynamics (e.g. Brian 1983; Sudd & Sudd 1985; Gordon 1986). We have attempted to study benefits at the levels of both the individual forager and the colony.

Several lines of evidence suggest that colonies of *I. anceps* receive substantial rewards for their efforts. First, simple inspection of the association indicates that this is the case: numerous attendant ants continuously groom and lick the larvae, and solicit secretions from a specialized organ on the seventh abdominal segment. Ants invariably establish nest extensions or 'bivouacs' containing brood at the base of trees containing larvae of *J. evagoras*, and it seems highly unlikely that they would move their 'central place' in this way unless the foraging rewards are high.

Second, we reasoned from our earlier work that tended pupae might develop slightly more slowly than their untended counterparts because of their need to feed attendant ants. Unlike larvae that alter their feeding behaviour in response to ants, pupae are essentially restricted in their expendable resources. To examine this possibility, we compared the weight loss of tended and untended pupae during a 5-day period. Pupae matched for age and size were divided into two groups, and half were placed on poles from which they could be tended by workers from a queenright laboratory colony of *I. anceps*. The remaining half were placed on adjacent poles where ants had been excluded. Pupae that were tended by ants for only 5 days lost 25% more wet weight than their untended counterparts. Thus pupae may supply rewards for ants by diverting metabolic resources from metamorphosis (Pierce *et al.* 1987).

Third, we attempted to measure the weight of the food harvested by attendant ants. A representative tree infested with sixty-two juveniles of *J. evagoras* was selected for observation, and the rate of ants travelling up and down the tree was measured at 2-hour intervals over a 24-hour period. We then took the wet and dry weights of each individual ant (Fig. 13.2). By comparing the dry weights of individual ants travelling to and from larvae of *J. evagoras*, we estimated that the daily biomass removal from a tree containing sixty-two juveniles of *J. evagoras* was about 400 mg. The mean dry weight of a worker of *I. anceps* is about 0.4 mg, so, if we use 10% as an estimate of biomass conversion from one trophic level to the next, then the net food removed from larvae on this single, representative tree was equivalent to the production of about 100 new workers of *I. anceps* in one day (Pierce *et al.* 1987).

Finally, we have measured how the secretions of lycaenid larvae contribute to colony growth and investment into reproductives. My student, David Nash, and I have collected nests of an attendant ant species, *I. rufoniger*, and are rearing them in the laboratory on water and an artificial diet (Bhatkar & Whitcomb 1970) supplemented by the secretions of differing numbers of larvae of *J. evagoras* feeding on potted host plants. The starting nests for this experiment are comprised of sister queens taken from polygynous colonies in the field, each provisioned by an equal number of workers and a similar weight of brood. The nests are housed in glass test tubes to allow observation of both growth rates and the investment into different castes. Each colony is fed equal and small

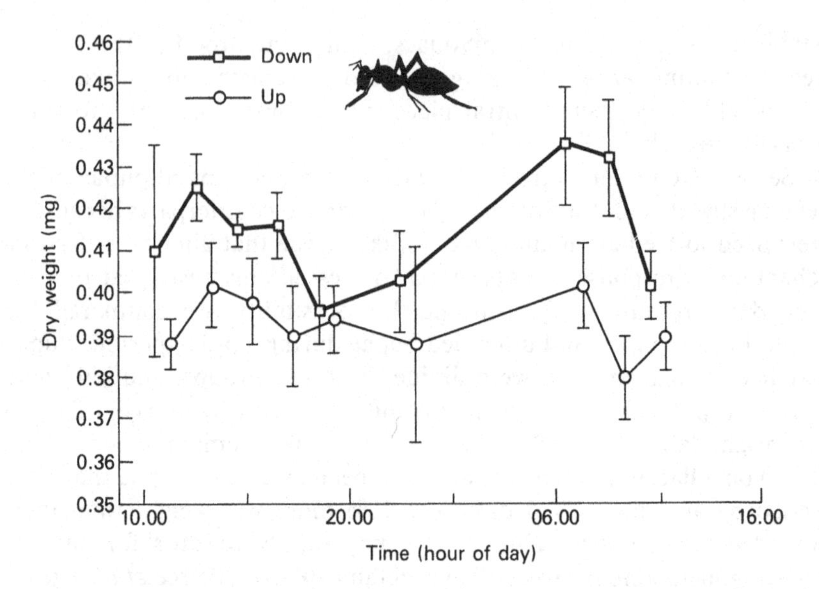

FIG 13.2 Benefit to ants. Graph illustrates differences in dry weights of ants foraging on a tree containing sixty-two juveniles of *Jalmenus evagoras* over a 24 h period. 'Up' weights are means for ants travelling up the tree and 'down' weights are means for ants travelling back down again. Bars represent standard errors; sample sizes are over twenty in all time zones, except 22:30, when they are both 7 (redrawn from Pierce *et al.* 1987).

amounts of ant diet to supplement the larval secretions. Although the nests are still growing, our preliminary results (after 40 days) indicate that those nests whose workers have been allowed access to host plants containing lycaenid larvae have significantly higher growth rates (as reflected by numbers of eggs laid) than nests whose workers are foraging on host plants with no larvae.

Although this experiment supports the idea that the interaction between *Jalmenus evagoras* and its attendant ants is indeed mutualistic, it raises many new questions. How would factors such as colony size, age and diet influence the results of the study? These characteristics are doubtless crucial in determining the effect that lycaenids have on the population dynamics of the ants, and they are characteristics which naturally vary in the field. However, they continue to present a challenge to our understanding of the interaction because they are difficult to study even in the laboratory, let alone under field conditions.

Finally, we are still in the process of measuring the costs of the association for the attendant ants. One of the methods that we are using is simply to measure the metabolic cost to the ants of foraging on larvae and pupae (Nielsen, Torben & Holm-Jensen 1982; Peters 1983; Dreisig 1988; Fiedler & Maschwitz 1988). We also hope to gain information about the nature of both the costs and the benefits involved through a consideration of the foraging decisions made by ants when presented with larvae placed at different distances from the nest.

THE BIOCHEMISTRY OF LARVAL SECRETIONS OF SPECIES OF *JALMENUS* AND THEIR ATTRACTIVENESS TO ANTS

Unlike the 'honeydew' of aphids, the secretions of the Lycaenidae come from specialized exocrine glands. In addition to having an unusually thick cuticle (Malicky 1970), as well as larval and pupal stridulatory organs (Kitching 1983; Pierce & Elgar 1985), the larvae of species of *Jalmenus* possess at least three sets of glands that appear to be adaptations for associating with ants, two of which are probably important in ant appeasement and reward (Kitching 1983): epidermal glands called 'pore cupolas' scattered all over the surface of the larva (Malicky 1969, 1970), and, in the middle of the seventh abdominal tergite, the 'dorsal organ' (Newcomer 1912; Maschwitz, Wust & Schurian 1975). In general, this secretion is produced only when the ants have signalled the caterpillar in the right way, although a few species are known to produce secretions in the absence of ants (Hinton 1951).

We sampled these secretions to determine whether they constituted a genuinely nutritional reward for the attendant ants or whether they were merely a form of chemical trickery whereby the lycaenids are able to fool the ants into tending them. The latter is clearly the case in parasitic lycaenids such as the Large Blue, *Maculinea arion*. We found that both the pore cupolas and the dorsal organ of *J. evagoras* produce free amino acids, especially serine (at a concentration of about 30 mM from the dorsal organ). By conducting choice experiments in the field, we were also able to show that serine was one of the preferred amino acids of *I. anceps*: workers could definitely distinguish between different amino acids, and alanine, histidine, leucine and serine were highly significantly preferred over water alone. Different colonies varied in the total amount that they drank and in their preferences, although the rankings of preference between colonies were similar.

As serine is the main amino acid secreted by *J. evagoras* and is also one of the preferred amino acids of *I. anceps*, it seemed possible that serine was an important currency in the interaction. By scoring the rate of ant attendance on pupae which differed in their attractiveness to ants, and by assaying the total amount of serine secreted by the pupae, we established that there was a strong correlation between the amount of serine secreted by a pupa and its attractiveness to ants. While this correlation does not establish cause and effect, it clearly suggests that serine has a critical role to play in this association (Pierce 1983; N. E. Pierce, unpublished).

Given the apparently stringent requirements of tended *J. evagoras* larvae in terms of amino acid output, it seemed likely that nitrogen limitation could influence lycaenid host plant choices. By releasing free-flying females of *J. evagoras* in a bush house, and providing them with a choice between fertilized (i.e. nitrogen-rich) and unfertilized potted host plants, my student Matthew Baylis and I found that females preferred to lay eggs on fertilized trees rather than on the unfertilized controls. Moreover, larvae that were raised on fertilized trees attracted more attendant ants and survived better than their counterparts on unfertilized controls. These results provided strong evidence that ovipositing females can respond to different levels of nitrogren in their host plants, though it is not clear what proximate cues they use in recognizing these plants, or whether other compounds that may covary with nitrogen are involved in the interaction (see Mattson 1980; Scriber & Slansky 1981; Myers 1985).

Further, if nitrogen is of particular importance to lycaenids that associate with ants, I reasoned that these lycaenids might have a predilection for feeding on protein-rich plants. Fortunately, excellent records of host plant use and ant association are available for Australia, South Africa and North America, and, for the species of these regions, there is indeed a strong correlation across all the lycaenids between ant association and feeding on relatively nitrogen-rich food plants, such as legumes (Pierce 1987).

SPECIES SPECIFICITY IN LYCAENID–ANT INTERACTIONS

Species specificity is of particular interest in the study of lycaenid–ant associations because in many ways it lies at the heart of the problem: an understanding of how and why interactions are species-specific will help resolve questions about the ecological mechanisms promoting both facultative and obligate mutualism, as well as the chemical communication between the lycaenids and their ant partners. We have approached this problem from two perspectives: the point of view of the ovipositing butterflies, and the point of view of their attendant ants. Although much of this work is still in preparation for publication, I present it here both as a direction for future research, and as an indication of some of the methodological problems inherent in studying a complex system.

Because attendant ants are essential for the survival of *J. evagoras,* we suspected that ants might play an important role in the host-finding behaviour of the female butterflies. The oviposition behaviour of females of *J. evagoras* has proved to be exceptionally tractable for field studies, in part because of the dense and localized occurrence of populations of *J. evagoras*, and in part because the larvae and pupae aggregate, and females are attracted to conspecific juveniles during oviposition (Pierce & Elgar 1985). In order to test whether females respond to ants during oviposition, we arranged potted host plants in a circular arena in the field. These were provisioned with equal numbers of late instar larvae, and ants were allowed to tend larvae on half of the plants for several days, whereupon treatments were switched to control for possible host plant or position effects. This experiment demonstrated that females did indeed use ants as cues during oviposition, and that they were far more likely to lay egg masses on plants with larvae and ants than on plants with larvae but without ants. We then repeated the same experiment, but used juveniles of the homopteran, *Sextius virescens,* to attract ants (rather than conspecific juveniles). Again, females preferred plants with homopterans and ants (Pierce & Elgar 1985).

This satisfied our curiosity about whether females responded to workers of *I. anceps* during oviposition, but it did not tell us whether females could distinguish between different ant species. In an 'ant smorgasbord' experiment, we offered females of *J. evagoras* a choice between host plants inhabited by different species of ants. The ants were introduced into the field in artificial nest boxes, and allowed to forage on honeydew produced by juveniles of *S. virescens,* which were placed on potted host plants adjacent to the ant nests and attached to them by stick bridges. Two potted plants were positioned beside each treatment as controls: one contained equal numbers of homopteran juveniles that were not tended by ants, and the other contained homopteran juveniles that were tended by workers of *I. anceps,* the species that commonly attends larvae of *J. evagoras* in the field. Twelve different ant species were simultaneously presented to ovipositing females of *J. evagoras* in this manner. Females actively avoided several of the ant species, treated a number as if there were no ants present (laying as many egg masses on plants with ants as on plants without ants), and preferred to oviposit only on those trees where the homopterans were tended by *I. anceps,* the common ant associate. However, females were indifferent to another attendant ant, *I. rufoniger,* that we have observed occasionally associating with larvae of *J. evagoras* in the field.

In a sequel to the ant smorgasbord experiment, we ran an experiment to determine whether workers of *I. anceps* were better tenders than workers of *I. rufoniger.* Three colonies of each species were arranged in the field in an area not inhabited by either ant species. Each colony was allowed to tend larvae of *J. evagoras* feeding on potted host plants. Twenty-five first instars were placed on each tree, and their survival was monitored daily for two weeks. At the end of two weeks, only about 15% of larvae tended by *I. rufoniger* remained on each plant, whereas about 60% of those tended by *I. anceps* were still surviving. Thus, under these experimental conditions, *I. anceps* was clearly a better tender than *I. rufoniger.* We concluded that selection has favoured species-specific identification of ants by ovipositing butterflies only for those species that provide adequate protection to insure high survival of their lycaenid associates (but see Law & Koptur 1985). The actual mechanism by which females discriminate different ant species in the field is still not known.

Our second approach to the question of species specificity took the perspective of the ants. Can ants distinguish between different species of Lycaenidae, and do they prefer larvae of the species they normally associate with? This work was done with four lycaenid/ant pairs: *J. evagoras/I. anceps, J. daemeli/I. rufoniger, J. pseudictinus/Froggatella kirbyi,* and *J. ictinus/I. purpureus.* In each case, ants were housed in artificial nest boxes in the laboratory, and allowed access to larvae feeding on potted plants of a single species, *Acacia irrorata.* In a mix-and-match experiment, five first instars of each lycaenid were offered to each ant species, and the survival of these larvae was monitored every day for 2 weeks. This demonstrates that, in addition to their normal lycaenid associate, there was little latitude in acceptance between ants and lycaenids. Workers of *I. rufoniger* accepted *J. daemeli* (their normal associate) and *J. evagoras,* and workers of *F. kirbyi* accepted *J. pseudictinus* (their normal associate) and *J. daemeli.* In all other cases, larvae of 'foreign' lycaenid species were attacked and consumed.

However, we also found that in another laboratory situation, a separate colony of *F. kirbyi* actively tended larvae of both *J. evagoras* and *J. daemeli.* The host plant in this instance was *A. decurrens* rather than *A. irrorata,* and the colony in the first experiment was reared on dilute honey and chopped cockroaches, whereas the colony in the second was reared on an artificial diet (Bhatkar & Whitcomb 1970). In other respects, our experimental protocol was very similar. These apparent experimental inconsistencies point toward an important consideration: variation in colony responses to larval secretions. Thus our future experiments in this area will include not only replicates of larvae, but also replicates of attendant ant colonies, and consideration of possible host plant effects. Such preference tests should also probably be run under field conditions, since colony diet may well influence ant behaviour. Likewise, the ant smorgasbord experiment should probably be run again using multiple ant colonies, rather than simply one colony per species.

Once we had managed to culture the four different species of *Jalmenus* in the laboratory, we were interested in comparing the secretions produced by each species for its attendant ants. We were able to raise all four species successfully on potted plants of *A. irrorata,* hoping to control for possible differences in the secretions caused by host plant species effects. The amino acid profile secreted by each species was unique. The secretions of *J. evagoras* were the simplest, containing primarily serine and small amounts of leucine, whereas those of *J. pseudictinus* contained a complex blend of histidine, arginine, serine, leucine, alanine and others. The secretions of *J. daemeli* did not contain amino acids, but showed a consistent, broad peak in the profile that probably corresponds to a small peptide.

If different species of *Jalmenus* secrete unique amino acid profiles, do the respective ant associates of these species prefer different combinations of amino acids? We had to return to the field to answer this question, and we set up 'drinking-straw' experiments in the field for each of these species. The different ant associates clearly varied in their amino acid preferences, both quantitatively and qualitatively. One species, *I. rufoniger,* was less attracted to amino acids and much preferred to forage on sucrose solutions instead. Not too surprisingly, this was the species whose usual lycaenid partner, *J. daemeli,* did not secrete amino acids. However the remaining two species, *I. purpureus* and *F. kirbyi,* both foraged actively on amino acids, and on different ones from those preferred by *I. anceps.* Moreover, although each species differed significantly in its amino acid preferences, the colonies within each species also varied significantly in the total amount consumed, as well as in preferences for certain amino acids. This was strong evidence again that colony level variation *must* be taken into account in studies of lycaenid–ant interactions. How did the preferences of the ants match up with the secretions of their lycaenid associates? For the remaining two species pairs, *J. ictinus/I. purpureus* and *J. pseudictinus/F. kirbyi,* the Spearman rank correlation between the relative concentrations of amino acids produced by larvae and the amino acids preferred by their attendant ants was significant, indicating that the lycaenids may indeed secrete those amino acids that their ant associates prefer. Now the compelling question remains: why do the different ant species differ in their preferences for different amino acids? Doubtless their preferences are related in some way to their metabolic needs, and, since ants rely so heavily on chemical communication, my guess is that the preferred amino acids may be important building blocks for commonly used pheromones. For example, serine is a direct precursor for formic acid, and thus might be of particular importance for formicine ants that secret formic acid in relatively large quantities.

A POSSIBLE PRE-ADAPTATION ON THE PART OF ANTS FOR TENDING LYCAENID LARVAE

In addition to being a highly desirable food source, the amino acids secreted by different species of lycaenids might also be involved in recognition: that is, the food itself might act as a communication signal for the ants. It is clear from the results of the bioassays described above that ants can distinguish between different amino acids, and hence it seems possible that amino acids could also act as discriminating substances in chemical communication. The unique amino acid profiles secreted by each species of *Jalmenus* could therefore be not only the product of the food preferences of their associated ants, but also the cue that the ants use in favourably recognizing the larvae so that they choose to tend them rather than attack them.

As a signal, epidermal secretions of lycaenid butterflies operate on several levels of recognition (Hölldobler & Michener 1980). Larvae of all parasitic and many mutualistic species are somehow able to ensure species-specific recognition by their ant associates, and larvae of parasitic species that are carried into the nest by ants are clearly capable of mimicking ant brood signals. There is no evidence, however, that the initial appeasement and adoption of lycaenid larvae by host ants is ever colony-specific, although it is conceivable that, once a colony has adopted a larva, it can somehow impart to it a colony-specific odour (Vander Meer & Wojcik 1982).

As mentioned before, larvae of many species of lycaenids are tended by only one species of ant or several closely related species of ants. This is particularly true of parasitic larvae that are carried into the ant nest (Cottrell 1984). In these unusual species, such as the Large Blue, *Maculinea arion*, the larvae become carnivorous on the ant brood, completing their development and pupating in the ant nest. Larvae of these species are invariably carried into the brood chambers of their host's nest where they are treated as if they were brood; clearly they are able to mimic not only species-specific signals, but also brood-specific signals of their host ants. We shall discuss these two types of signals as distinct from one another, although it is possible that any discriminating substances that ensure brood recognition are the same as those responsible for species recognition, but vary, for example, in their relative concentrations. In this context, we use the term 'discriminating substances' suggested by Hölldobler & Michener (1980) to describe exocrine secretions that identify individuals between species, or brood within species. It is still unknown whether identification is based on odours that are genetically controlled, environmentally conditioned, or a combination of both.

As yet we have little evidence that amino acids in the epidermal secretions of lycaenids could act as communication signals for attendant ants (beyond their nutritive function), but two observations indicate that this is possible. First, when we examined the combined amino acid profiles of approximately fifty pupae from each of five colonies of *I. anceps,* we found that their profiles matched those obtained from larvae and pupae of *J. evagoras* in the sense that they all contained significant amounts of serine (as well as a variety of unidentified peptides.) Second, we noticed that workers of *Pheidole megacephala,* a species of ant whose workers are hostile to larvae of *J. evagoras* and attack them upon encounter, were completely indifferent to a solution of the amino acid serine, whereas workers of *I. anceps* literally stood on top of each other to drink this solution. Other studies (Inouye & Waller 1984; Lanza & Krauss 1984) have also shown that honey-bees and ants (species of *Leptothorax* and *Monomorium*) can distinguish between different amino acids. These behaviours demonstrate two things: (1) there are differences in the dietary preferences of particular ant species, and (2) the striking contrast in response suggests that amino acids at least have the potential to function as discriminating substances.

It is possible that the recognition and nutritive functions of amino acid secretions are interrelated, the former being a ritualized adaptation of the latter. Regurgitation of amino acids by larvae has been considered important for dietary reasons in the evolution of the Hymenoptera (Maschwitz 1966; Wust 1973; Hunt 1982), and perhaps epidermal secretions of amino acids by the larvae and pupae of ants have also become important in species or brood recognition. Interestingly, Walsh & Tschinkel (1974) found that the absence of protein in the diet of *Solenopsis invicta* resulted in an inconsistent response by workers to their own brood.

Several researchers (Fielde 1903; Jaisson 1975; Le Moli 1978; see

Carlin 1988 for review) have shown that, in certain situations, brood of one species of ant will be tolerated by workers of a wide variety of alien ants, and Hölldobler (1973) has postulated that brood-tending pheromones may occupy a high position in a hierarchical order of recognition pheromones. This may also be true of the 'brood' substances produced by lycaenids. *Maculinea teleius,* a lycaenid that in Japan is always associated with ants in the genus *Myrmica,* has also been discovered in the brood chambers of a nest of *Lasius niger* (Fukuda *et al.* 1978). In this instance brood-recognition signals must have overridden any other signals produced by the larvae and induced workers of *L. niger* not only to carry larvae of *M. teleius* into the brood chamber of their nest, but also to treat them there as brood, even though *Lasius* and *Myrmica* are members of different ant subfamilies.

For parasitic species of Lycaenidae that are actually carried into the ant nest, it is not difficult to imagine that the larvae are mimicking an ant brood signal. My suggestion here, however, is that even those species that are not carried into the nest may be mimicking some critical *portion* of the brood stimulus that causes ants to care for, groom and protect them exterior to the nest.

This analogy between the behaviour shown by ants towards lycaenid larvae and their brood relies in part on the idea that there exists a generalized component of the brood signal that is identifiable to many species of ants. If lycaenid larvae do in fact mimic such a generalized signal, this could also account for why many lycaenids are accepted by a wide variety of tending ants. For example, as many as eleven different species of ants have been recorded as tending the larvae of *Plebejus icarioides* (Downey 1962). These putative gustatory and/or olfactory signals may act in conjunction with tactile stimuli in insuring brood recognition. For example, Brian (1975) has shown that features such as size, shape, turgidity and hairiness of brood are important variables in brood recognition by various species of *Myrmica.*

A number of similarities exist between ant–brood and ant–lycaenid associations even among those lycaenids whose larvae are not carried into the ant nest. The chemical attractants in both cases appear to be non-volatile 'surface' attractants (Wilson 1971; Walsh & Tschinkel 1974; Brian 1975); adult ants must make contact with or come very close to making contact with larvae or pupae before recognizing them. In addition, these attractants are stable in nature. Walsh & Tschinkel (1974) observed that workers of *S. invicta* would respond normally to dead brood for at least 21 hours, and Robinson & Cherrett (1974) also found that workers of *Atta cephalotes* responded to brood killed by freezing.

Similarly, I have observed that attendant ants persist in tending carcasses of the larvae of *Glaucopsyche lygdamus* and *J. evagoras* for up to a week following death.

Moreover, the behaviour shown by tending ants toward lycaenid larvae is also remarkably similar to the behaviour shown toward brood. In particular, ants spend considerable amounts of time licking and grooming larvae, just as they do their brood. They sometimes appear to show great fidelity to larvae; in one case I observed several marked individual workers continue to tend the same larva of *Glaucopsyche lygdamus* for 10 days. This kind of fidelity has also been documented for ants tending homopterans (Ebbers & Barrows 1980). Finally, ants tending lycaenid larvae also spend considerable time grooming themselves, as do ants in nest brood chambers, perhaps thereby spreading substances gleaned from the larvae over their own bodies.

As more is learned about the nature of ant pheromones, the mechanism by which lycaenid larvae have broken the communication codes of their host ants will surely be elucidated. One of the greatest difficulties in determining the nature of ant brood pheromones has been distinguishing whether workers are responding to a food stimulus or to an actual communication signal. Bioassays that can discriminate between these responses may be extremely difficult to design. It is possible, for example, that substances such as amino acids that originally acted as phagostimulants for ants have evolved to function as communication signals. Glancey *et al.* (1970) observed that workers of *Solenopsis invicta* placed corn grits treated with homogenized extracts of their own juveniles with the brood in their nests. Henning (1983) also used the technique of Glancey *et al.* to examine the response of attendant ants to the tissue of lycaenid larvae (*Aloeides dentatus* and *Lepidochrysops ignota*) extracted with dichloromethane, with similar results. However Walsh & Tschinkel (1974) were unable to repeat the assay of Glancey *et al.* when they had modified the nest design to create a separate, discrete brood chamber, and suggested that the brood response may have been confounded with a food response.

There are many gaps in our understanding of the proximate mechanisms that maintain the association between lycaenids and ants, particularly in our understanding of the biochemical nature of the secretions of lycaenid larvae. The notion that free amino acids can be secreted in particular combinations and concentrations to create unique, recognizable profiles provides an attractively parsimonious mechanism for a chemical communication code in an animal that can distinguish the difference between different amino acids. Nevertheless, it is hard to imagine that this information alone could signal ants to tend larvae rather than attack them, and it seems likely that larvae produce compounds other than amino acids that are also involved in the recognition process. Recent work has focused on the possible significance of cuticular hydrocarbons produced by social insect hosts and their guests. The termitophilous beetle *Trichopsenius frosti* synthesizes a hydrocarbon pattern identical to that of its host, *Reticulitermes flavipes* (Howard, McDaniel & Blomquist 1980), and the myrmecophilous beetle *Myrmecaphodius excavaticollis* appears to acquire species-specific hydrocarbons from each of at least four different species of *Solenopsis* hosts (Vander Meer & Wojcik 1982).

TOWARD A MORE EXACT ECOLOGY

What can an analysis of lycaenid butterflies and ants tell us about mutualism in general? In essence, this study has at least three long-term goals: (1) to measure the costs and benefits for both partners in a mutualism; (2) to consider some of the pre-adaptations of both parties that may have promoted the evolution of the mutualism; and (3) to assess possible evolutionary consequences of the interaction. Clearly, the association that lycaenid butterflies have with ants has profoundly shaped their evolution and subsequent diversification, and I have discussed these evolutionary considerations elsewhere (Pierce 1987). However, more empirical studies are needed before ecologists will be able to reach any general appreciation for the importance of mutualism in generating or maintaining diversity, or otherwise structuring natural communities.

What we require is a comparative framework: we need to have more studies of particular mutualistic systems from which to generalize. Such a framework will be necessary in order to assess whether there are particular features of organisms that predispose them to associate symbiotically, and whether such associations are then likely to have characteristic evolutionary outcomes (c.f. Law 1985). Consider two possible examples. First, mutualisms often appear to give rise to parasitism, especially in situations where the pay-offs are highly asymmetrical, as exhibited by a number of lycaenid–ant interactions. To predict why and when this might occur, we need to know more about selective forces that determine the *degree* of association found between different mutualists: under what circumstances and how commonly can we expect obligate interactions to arise? How can an obligate dependence on the part of one or both partners affect population structure and subsequent evolution of both? Second, the results shown here suggest that ant mutualists are 'keystone' participants in the interaction between lycaenids, their host plants, parasites and predators. If we were to remove ant mutualists from our Australian *Acacia* communities, diversity would decrease significantly. While the ecological literature abounds with discussions of whether competition and/or predation are important in structuring communities, relatively few attempts have been made to assess the relative importance of mutualism as a mechanism in either generating or maintaining community diversity. In part this is because the necessary data do not yet exist.

We could speculate about the importance of learning more about the role of mutualism in natural communities by taking a hypothetical example. Consider an effort aimed at reforestation following the destruction of tropical rain forests. Even if we were able successfully to replant the appropriate species in their native habitats, such an effort at regeneration might nevertheless completely fail if the appropriate, mutualistic pollinators and seed dispersers for these plants had gone extinct in their absence. For specialized, obligate mutualists, this is clearly a possibility.

The results of this research show that working toward a more exact ecology can yield benefits in terms of a better understanding of complex species interactions. By 'exactness', I refer to several main approaches, one being the implementation of technological innovations such as HPLC in analysing and quantifying (in this case) lycaenid secretions, the second being a quantitative and experimental approach in weighing out costs and benefits for each partner and determining the mechanisms underlying species interactions, and the third being the use of comparative studies in looking for ecological correlates of particular life-history traits. For example, had we not discovered that lycaenids secrete amino acids as rewards for attendant ants, we might never have scrutinized the role of nitrogen in lycaenid–host plant interactions. And, had we not found that different amino acids were being secreted by different lycaenid species, we would never have enquired further into how the biochemistry of ant nutrition might be related to chemical communication. However, I thank Evelyn Hutchinson for providing what is certainly the most fitting summary to the work discussed here: when I told him about the theme of this symposium, he remarked, 'Well, it's all very well to aim toward a more *exact* ecology, as long as no one makes the mistake of thinking that, by doing so, they will discover a more *simple* ecology.'

ACKNOWLEDGMENTS

This work would have been impossible without the participation of many students and colleagues to whom I am very grateful, including T. Anderson, P. R. Atsatt, M. Baylis, K. Benbow, A. J. Berry, J. Braverman, R. C. Buckley, E. R. Carper, S. Easteal, M. A. Elgar, C. Hill, R. Keen, R. L. Kitching, P. Mead, D. R. Nash, B. Normark, P. J. Rogers, J. T. Smiley, T. Stanley, M. F. J. Taylor, and P. Vowles. Ideas presented in different parts of this manuscript were aided by helpful discussions with A. J. Berry, N. Carlin, P. J. DeVries, B. Hölldobler, H. S. Horn, R. L. Kitching, R. M. May and D. I. Rubenstein. This work is supported by a grant from the National Science Foundation.

REFERENCES

Addicott, J. F. (1984). Mutualistic interactions in population and community processes. *A New Ecology: Novel Approaches to Interactive Systems* (Ed. by P. W. Price, C. N. Slobodchikoff & B. S. Gaud), pp. 437–55. Wiley, New York.

Addicott, J. F. (1986). Variation in the costs and benefits of mutualism: the interaction between yuccas and yucca moths. *Oecologia*, **70**, 486–94.

Axelrod, R. (1984). *The Evolution of Cooperation.* Basic Books, New York.

Axelrod, R. & Hamilton, W. D. (1981). The evolution of cooperation. *Science*, **211**, 1390–6.

Bhatkar, A. & Whitcomb, W. H. (1970). Artificial diet for rearing various species of ants. *Florida Entomologist*, **53**, 230–2.

Boucher, D. H. (1985). The idea of mutualism, past and future. *The Biology of Mutualisms* (Ed. by D. H. Boucher), pp. 1–28. Croom Helm, Kent.

Boucher, D. H., James, S. & Keeler, K. H. (1982). The ecology of mutualism. *Annual Review of Ecology and Systematics*, **13**, 315–47.

Brian, M. V. (1975). Larval recognition by workers of the ant *Myrmica. Animal Behaviour*, **23**, 745–56.

Brian, M. V. (1983). *Social Insects: Ecology and Behavioural Biology.* Chapman & Hall, London.

Buckley, R. (1987). Interactions involving plants, Homoptera, and ants. *Annual Review of Ecology and Systematics*, **18**, 111–35.

Carlin, N. F. (1988). Species, kin and other forms of recognition in the brood discrimination behavior of ants. *Advances in Myrmecology* (Ed. by R. H. Arnett), pp. 267–95. Brill, Leiden.

Common, I. F. B. & Waterhouse, D. F. (1981). *Butterflies of Australia*, 2nd edn. Angus & Robertson, Sydney.

Cottrell, C. B. (1984). Aphytophagy in butterflies: its relationship to myrmecophily. *Zoological Journal of the Linnaean Society*, **79**, 1–57.

Degen, A. A., Gersanim M., Avivi, Y. & Weisbrot, N. (1986). Honeydew intake of the weaver ant *Polyrachis simplex* (Hymenoptera: Formicidae) attending the aphid *Chaitophorus populialbae* (Homoptera: Aphididae). *Insectes Sociaux*, **33**, 211–15.

DeVries, P. J. (1987). *Ecological aspects of ant association and hostplant use in a riodinid butterfly.* Ph.D. thesis, University of Texas, Austin.

Downey, J. C. (1962) Myrmecophily in *Plebejus (Icaricia) icarioides* (Lepidoptera:

Lycaenidae). *Entomological News,* **73**, 57–66.

Dreisig, H. (1988). Foraging rate of ants collecting honeydew or extrafloral nectar, and some possible constraints. *Ecological Entomology,* **13**, 143–54.

Ebbers, B. C. & Barrows, N. (1980). Individual ants specialize on particular aphid herds. *Proceedings of the Entomological Society of Washington,* **82**, 405–7.

Elgar, M. A. & Pierce, N. E. (1988). Mating success and fecundity in an ant-tended lycaenid butterfly. *Reproductive Success: Studies of Selection and Adaptation in Contrasting Breeding Systems* (Ed. by T. H. Clutton-Brock), pp. 57–95. Chicago University Press.

Eliot, J. N. (1973). The higher classification of the Lycaenidae (Lepidoptera): a tentative arrangement. *Bulletin of the British Museum (Natural History), Entomology,* **28**, 375–505.

Fiedler, K. & Maschwitz, U. (1988). Functional analysis of the myrmecophilous relationships between ants (Hymenoptera: Formicidae) and lycaenids (Lepidoptera: Lycaenidae). *Oecologia,* **75**, 204–6.

Fielde, A. M. (1903). Artificial mixed nests of ants. *Biological Bulletin, Marine Biological Laboratory, Woods Hole,* **5** (6), 320–5.

Fukuda, H., Kubo, K., Takeshi, K., Takahashi, A., Takahashi. M., Tanaka, B., Wakabayashi, M. & Shirozu, T. (1978). *Insects' Life in Japan.* III. *Butterflies.* Hoikusha, Tokyo.

Glancey, B. M., Stringerm C. E., Craig, C. H., Bishop, P. M. & Martin, B. B. (1970). Pheromone may induce brood tending in the fire ant, *Solenopsis saevissima. Nature,* **226**, 863–4.

Gordon, D. M. (1986). The dynamics of group behaviour. *Perspectives in Ethology,* **7** , 217–31.

Hamilton, W. D. (1964). The evolution of social behaviour. *Journal of Theoretical Biology,* **7**, 1–52.

Henning, S. F. (1983). Chemical communication between lycaenid larvae (Lepidoptera: Lycaenidae) and ants (Hymenoptera: Formicidae). *Journal of the Entomological Society of South Africa,* **46**, 341–66.

Henning, S. F. (1984). The effect of ant association on lycaenid larval duration (Lepidoptera: Lycaenidae). *Entomologist's Record and Journal of Variation,* **96**, 99–102.

Hinton, H. E. (1951). Myrmecophilous Lycaenidae and other Lepidoptera—a summary. *Proceedings of the London Entomology and Natural History Society,* **1951**, 111–75.

Hölldobler, B. (1973). Zur Ethologie der chemischen Verständigung bei Ameisen. *Nova Acta Leopoldina,* **9**, 259–92.

Hölldobler, B. & Michener, C. D. (1980). Mechanisms of identification and discrimination in social Hymenoptera. *Evolution of Social Behaviour: Hypotheses and Empirical Tests* (Ed. by H. Markl), pp. 35–58. Dahlem Konferensen, Weinheim.

Howard, R. W., McDaniel, C. A. & Blomquist, G. J. (1980). Chemical mimicry as an integrating mechanism: cuticular hydrocarbons of a termitophile and its host. *Science,* **210**, 431–3.

Hunt, J. H. (1982). Trophallaxis and the evolution of eusocial Hymenoptera. *The Biology of Social Insects* (Ed. by M. D. Breed, C. D. Michener & H. E. Evans), pp. 201–5. Westview Press, Boulder, Colorado.

Inouye, D. W. & Waller, G. D. (1984). Responses of honey bees (*Apis mellifera*) to amino acid solutions mimicking floral nectars. *Ecology,* **64**, 618–25.

Jaisson, P. (1975). L'impregnation dans l'ontogenèse des comportements de soins aux cocons chez la jeune fourmi rousse (*Formica polyctena* Foerst). *Behavior,* **52**, 1–37.

Janzen, D. H. (1985). The natural history of mutualisms. *The Biology of Mutualisms* (Ed. by D. H. Boucher), pp. 40–99. Croom Helm, London.

Keeler, K. H. (1981). A model for a facultative, non-symbiotic mutualism. *American Naturalist,* **118**, 488–98.

Keeler, K. H. (1985). Cost : benefit models of mutualism. *The Biology of Mutualisms* (Ed. by D. H. Boucher), pp. 100–27. Croom Helm, London.

Kitching, R. L. (1983). Myrmecophilous organs of the larvae and pupae of the lycaenid butterfly *Jalmenus evagoras* (Donovan). *Journal of Natural History,* **17**, 471–81.

Lanza, J. & Krauss, B. R. (1984). Detection of amino acids in artificial nectars by two tropical ants, *Leptothorax* and *Monomorium. Oecologia,* **63**, 423–5.

Law, R. (1985). Evolution in a mutualistic environment. *The Biology of Mutualisms* (Ed. by D. H. Boucher), pp. 145–70. Croom Helm, London.

Law, R. & Koptur, S. (1985). On the evolution of non-specific mutualism. *Biological Journal of the Linnean Society,* **26**, 23–67.

Le Moli, F. (1978). Social influence on the acquisition of behavioural patterns in the ant *Formica fusca. Bollettino di Zoologia, Pubblicato dall'Unione Zoologica Italiana,* **45**, 399–404.

Malicky, H. (1969). Versuch einer Analyse der ökologischen Beziehungen zwischen Lycaeniden (Lepidoptera) und Formiciden (Hymenoptera). *Tijdschrift Entomologische,* **112**, 213–98.

Malicky, H. (1970). New aspects on the association between lycaenid larvae (Lycaenidae) and ants (Formicidae, Hymenoptera). *Journal of the Lepidopterists' Society,* **24**, 190–202.

Maschwitz, U. (1966). Das Speichelsekret der Wespenlarven und seine biologische Bedeutung. *Zeitschrift für vergeichende Physiologie,* **53**, 228–52.

Maschwitz, U., Wust, M. & Schurian, K. (1975). Blaulingsraupen als Zuckerlieferanten für Ameisen. *Oecologia,* **18**, 17–21.

Mattson, W. J. (1980). Herbivory in relation to plant nitrogen content. *Annual Review of Ecology and Systematics,* **11**, 119–61.

May, R. M. (1981). Models for two interacting populations. *Theoretical Ecology: Principles and Applications* (Ed. by R. M. May), pp. 78–104. Blackwell Scientific Publications, Oxford.

Maynard Smith, J. (1982). *Evolution and the Theory of Games.* Cambridge University Press.

Myers, J. H. (1985). Effect of physiological condition of the host plant on the ovipositional choice of the cabbage white butterfly, *Pieris rapae. Journal of Animal Ecology,* **54**, 193–205.

Newcomer, E. J. (1912). Some observations on the relation of ants and lycaenid caterpillars, and a description of the relational organs of the latter. *Journal of the New York Entomological Society,* **20**, 31–6.

Nielsen, M. G., Torben, F. J. & Holm-Jensen, I. (1982). Effect of load carriage on the respiratory metabolism of worker ants of *Camponotus herculeanus* (Formicidae). *Oikos,* **39**, 137–42.

Peters, R. H. (1983). *The Ecological Implications of Body Size.* Cambridge University Press.

Pierce, N. E. (1983). *The ecology and evolution of symbioses between lycaenid butterflies and ants.* Ph.D. thesis, Harvard University, Cambridge, Massachusetts.

Pierce, N. E. (1984). Amplified species diversity: a case study of an Australian lycaenid butterfly and its attendant ants. *Biology of Butterflies* (Ed. by R. I. Vane Wright & P. R. Ackery), pp. 197–200. Symposium of the Royal Entomological Society of London, 11. Academic Press, London.

Pierce, N. E. (1985). Lycaenid butterflies and ants: selection for nitrogen-fixing and other protein-rich food plants. *American Naturalist,* **125**, 888–95.

Pierce, N. E. (1987). The evolution and biogeography of associations between lycaenid butterflies and ants. *Oxford Surveys in Evolutionary Biology* (Ed. by P. H. Harvey & L. Partridge), pp. 89–116. Oxford University Press.

Pierce, N. E. & Easteal, S. (1986). The selective advantage of attendant ants for the larvae of a lycaenid butterfly, *Glaucopsyche lygdamus. Journal of Animal Ecology,* **55**, 451–62.

Pierce, N. E. & Elgar, M. A. (1985). The influence of ants on host plant selection by *Jalmenus evagoras,* a myrmecophilous lycaenid butterfly. *Behavioral Ecology and Sociobiology,* **16**, 209–22.

Pierce, N. E., Kitching, R. L., Buckley, R. C., Taylor M. F. J. & Benbow, K. (1987). The costs and benefits of cooperation for the Australian lycaenid butterfly, *Jalmenus evagoras* and its attendant ants. *Behavioural Ecology and Sociobiology,* **21**, 237–48.

Pierce, N. E. & Mead, P. S. (1981). Parasitoids as selective agents in the symbiosis between lycaenid butterfly caterpillars and ants. *Science,* **211**, 1185–7.

Pierce, N. E. & Young, W. R. (1986). Lycaenid butterflies and ants: two-species stable equilibria in mutualistic, commensal, and parasitic interactions. *American Naturalist,* **128**, 216–27.

Price, P. W., Bouton, C. E., Gross, P., McPherson, B. A., Thompson, J. N. & Weis, A. E. (1980). Interactions among three trophic levels: influence of plants on interactions between insect herbivores and natural enemies. *Annual Review of Ecology and Systematics,* **11**, 41–65.

Robinson, S. W. & Cherrett, J. M. (1974). Laboratory investigations to evaluate the possible use of brood pheromones of the leaf-cutting ant *Atta cephalotes* (L.) (Formicidae, Attini) as a component in an attractive bait. *Bulletin of Entomological Research,* **63**, 519–29.

Ross, G. N. (1966). Life history studies on Mexican butterflies. IV. The ecology and ethology of *Anatole rossi,* A myrmecophilous metalmark (Lepidoptera: Riodinidae). *Annals of the Entomological Society of America,* **59**, 985–1004.

Roughgarden, J. (1975). Evolution of marine symbiosis—a simple cost-benefit model. *Ecology,* **56**, 1201–8.

Schemske, D. W. (1983). Limits to specialization and coevolution in plant–animal mutualisms. *Coevolution* (Ed. by M. H. Nitecki), pp. 67–109. University of Chicago Press.

Scriber, J. M. & Slansky, F. (1981). The nutritional ecology of immature insects. *Annual Review of Entomology,* **26**, 183–211.

Smiley, J. T., Atsatt, P. R. & Pierce, N. E. (1988). Local distribution of the lycaenid butterfly, *Jalmenus evagoras,* in response to host ants and plants. *Oecologia,* **77**, 416–22.

Sudd, J. H. & Sudd, M. E. (1985). Seasonal changes in the response of wood-ants to sucrose baits. *Ecological Entomology,* **10**, 89–97.

Templeton, A. R. & Gilbert, L. E. (1985). Population genetics and the coevolution of mutualism. *The Biology of Mutualisms* (Ed. by D. H. Boucher), pp. 128–44. Croom Helm, London.

Trivers, R. L. (1971). The evolution of reciprocal altruism. *Quarterly Review of Biology,* **46**, 35–57.

Vandermeer, J. (1984). The evolution of mutualism. *Evolutionary Ecology* (Ed. by B. A. Shorrocks), pp. 221–32. Symposia of the British Ecological Society, 23. Blackwell Scientific Publications, Oxford.

Vander Meer, R. K. & Wocjik, D. P. (1982). Chemical mimicry in the myrmecophilous beetle *Myrmecaphodius excavaticollis. Science,* **218**, 806–8.

Walsh, J. P. & Tschinkel, W. R. (1974). Brood recognition by contact pheromone in the red imported fire ant, *Solenopsis invicta. Animal Behaviour,* **22**, 695–704.

Wilson, D. S. (1980). *The Natural Selection of Populations and Communities.* Benjamin-Cummings, Menlo Park California.

Wilson, E. O. (1971). *The Insect Societies.* Harvard University Press, Cambridge,

Massachusetts.

Wolin, C. L. (1985). The population dynamics of mutualistic systems. *The Biology of Mutualisms* (Ed. by D. H. Boucher), pp. 248–69. Croom Helm, London.

Wust, M. (1973). Stomodeale und proctodeale Sekrete von Ameisenlarven und ihre biologische Bedeutung. *Proceedings of the VII Congress of the International Union for the Study of Social Insects*, pp. 412–18.

VI. ECOSYSTEM ECOLOGY

Whereas it is relatively easy to see how exactitude may be approached in the reductionist disciplines dealt with in Section II, it may readily be felt that the higher the level of organization the less applicable is the term exact. But systems analysis applied to whole ecosystems affords a tool which on the one hand offers a much greater precision in analysis of the whole, and on the other pinpoints those parts of the system requiring further study and more exact data.

Ulanowicz and Paul discuss marine and terrestrial ecosystems respectively. The challenge in both is to relate local, measurable phenomena to processes going on at a larger, even global scale. Interdisciplinary teams are required, and both authors range widely over the work of scientists contributing data and ideas from a multitude of viewpoints.

Ulanowicz filled an unavoidable gap in our programme at very short notice, for which we are most grateful. In his discussion of ocean ecosystems, he reminds us of a point which surfaces in several guises in this symposium: that the object of his study becomes something of an abstraction, and that part of our problem in studying complex systems at the ecosystem level is in formulating an adequate abstraction so as to permit the holist and reductionist to join forces in the pursuit of progress.

On the terrestrial front, Paul shows us how the soil must be considered not just as a major ecosystem component but as a controller of ecosystem processes. Not only do both the long- and the short-term changes taking place in soils determine the amount and nature of the soil organic matter, but these in turn act as a continuing influence on processes taking place in the soil. Describing soil as 'the only black box left in ecology', Paul goes a significant way toward unravelling its mysteries through the technique of concentrating on carbon and nitrogen interactions in the organic matter fraction and the associated microbial biomass.

14. ENERGY FLOW AND PRODUCTIVITY IN THE OCEANS

R. E. ULANOWICZ
Chesapeake Biological Laboratory, University of Maryland, Solomons, MD 20688, USA

INTRODUCTION

The term 'ecosystem' carries epistemological implications that are often not well appreciated. While most ecologists are aware that the concept embraces both the living and the abiotic entities of ecological ensembles (Tansley 1935; Waring 1989), fewer realize that some systems scientists, for example Klir (1981), draw a sharp distinction between the physical object of study and the associated mental image that the observer creates of that object — this latter abstraction being considered that which is properly termed 'the system'.

If 'systems' are actually abstractions of reality, it naturally follows that there will exist wide ranges of opinion concerning both (i) the utility of such abstraction, and (ii) the type of abstraction that should be drawn. These two issues pervade the literature of science in general and, *a fortiori,* ecology in particular. But disagreement and debate are preconditions to scientific understanding, so it is often helpful to chronicle recent advances in a particular discipline against the background created by these broader questions.

Concerning the utility of systems abstraction, the debate on this point translates into the familiar dialogue of reductionism vs. holism. At one extreme stand the nominalists, who are inclined to regard only individual, real objects as worthy of consideration. To the arch-nominalist any collection of real objects, say organisms, into categories, such as populations, guilds or communities, is futile, if not downright puerile. Such categories are believed to exist in name alone. Perhaps a larger fraction of ecologists is willing to entertain simple aggregations, such as populations, but holds that the direction of causality behind biological phenomena is almost exclusively from objects and events at smaller scales toward perceived larger entities and processes. Properties of these larger ensembles are assumed to be either accidental or epiphenomenal.

Of an opposing opinion is the holist, who is willing to reify and to attribute 'active agency' to relational entities, such as demes or ecosystems.

What happens at larger scales is believed to be independent to some degree of events transpiring at finer resolutions. Furthermore, the larger entities are thought to exert some influence upon what can occur among their parts. As I hope to demonstrate, investigators positioned all along the reductionist–holist axis have contributed toward a more quantitative marine ecology.

Given that some degree of system abstraction is both unavoidable and beneficial, there still remains an infinity of ways of forming a mental picture of the object of study, and the results are bound to differ markedly in nature and complexity among investigators. For example, Ashby (1953) pointed out that one's perception of the complexity of a sheep's brain depends upon whether one is a neurophysiologist or a butcher! The issue of how to abstract is understandably less general than the degree to which one should abstract, and the particular methods vary among the disciplines. In my opinion, two major schools of thought exist on how to form an image of an ecosystem in aquatic ecology and biological oceanography, partitioned loosely between British and American ecologists. The two schools are probably best associated with the names of Charles Elton and Raymond Lindeman respectively.

Charles Elton, emeritus of Oxford University, cast his descriptions of ecosystems primarily in terms of the numbers and sizes of organisms, as in his book of 1927. In particular, he noted that the animals doing the eating were usually larger than their prey. The prey were never so small that it took a long time for the predator to collect, nor were they so large that the prey were difficult to catch and overpower (Cousins 1985a). Although Elton was concerned mostly with terrestrial systems, the importance of body size to predator–prey relationships appears to be even stronger in the water, where, as the saying goes, 'Big fish eat little fish, eat smaller fish, etc.'

Raymond Lindeman's (1942) concept of ecosystem had an impact on American ecology that is out of all proportion to his tragically brief career. A graduate student of G. E. Hutchinson at Yale University, Lindeman advanced the notion of the ecosystem as a thermodynamical hierarchy, where plants occupy the lowest level and are fed upon by herbivores, who in turn are preyed upon by carnivores, and so on. The contents of the trophic levels are estimated in terms of material or energy; and, if in the latter, then the second law of thermodynamics dictates that progressively less energy is transferred to successively higher trophic levels in pyramidal fashion. Less well-known by those who have not read his original paper is the fact that Lindeman also connected the upper trophic levels with their base via a set of detrital feedbacks.

Number and size versus functional group and transformation—the two perspectives echo throughout contemporary ecology, and appear especially to structure dialogues among biological oceanographers. Whereas one might expect most Eltonian descriptions to be reductionistic and Lindemanian representations to be steeped in holism, one need not search long to discover significant exceptions. The interplay between factors in how one goes about abstracting from ecological collections into ecosystems is subtle and hinders the separation of projects into crisp epistemological categories. None the less, most advances in quantitative marine ecology do seem to be motivated (often unconsciously) by how the investigators stand in regard to these two issues. For this reason I have attempted to structure this review into four parts according to whether a particular project reflects more the approach of Elton or Lindeman, reductionism or holism.

ELTONIAN REDUCTIONISM

It is difficult to reconcile nominalism *ad extremum* with the larger thrust of science, replete as scientific thought is with laws and generalizations. But there do exist articulate prophets who journey the lecture circuit warning against unnecessary reification of higher-level abstractions. Simberloff (1983), for example, argues that species interactions are insignificant in comparison with the effects the physical environment has upon the fate of an organism, and foresees the time when community ecology should cease to exist. Lehman (1986) argues that classifying lakes according to their macroscopic properties (clarity, trophic status, mean temperature, etc.) is unlikely to yield robust prediction or understanding; the latter he feels is better sought by a clearer knowledge of the genetical properties of organisms. He allows, however, that integrating biological properties with food web dynamics does hold much promise. Others (Hughes 1985; Price 1986) do entertain the notion of an ecological community, but argue that the community can exhibit only mechanistic behaviour. They believe that the progressive elucidation of constituent mechanisms will eventually obviate any need to consider many-species phenomena, such as indirect effects (cf. also Schoener 1986).

Whilst many reductionist projects yield few quantitative results useful to others, there is one notable exception. If one focuses upon individual organisms, there is little left to quantify, save for an individual's size, behaviour and physiological rates. The Eltonian emphasis upon body size as the leading descriptor of an organism helped to initiate a search for allometric formulae that relate the rates of processes such as respiration, generation time and productivity (in plants) to the characteristic dimensions of the organism in question (Platt 1985). Haldane (1928) had commented earlier on how body size reveals much information about an organism's physiology, but it was Kleiber (1947) and especially Hemmingsen (1960) who found great regularity among data on how respiration rates vary with body size. In particular, they found that respiration rates vary as a power function of the organism's size or weight, i.e.

$$R = \alpha W^{\gamma}$$

where R is the respiration rate, W is the organism's weight and α and γ are constants.

The marine ecologist Fenchel (1974) was able to generalize this allometric formula for respiration to apply to Elton's speculation that generation time is also a function of body size. Briefly, Fenchel discovered that generation time also can be described by the same form of allometric equation as applies to respiration; that the multiplicative factor varies across groups of species, their ecologies and the types of physiological rates being described; but that the power γ was relatively constant over all the data. This virtual constancy bespeaks an underlying principle, and Platt & Silvert (1981) argued that there exists a universal, size-related time-scale in organisms to which all processes that can be expressed in units of time are related in a simple way.

ELTONIAN COMMUNITIES

The results from allometry did not derive in any way from the notions of community or ecosystems. However, the robustness and generality of allometric relations were most attractive to those seeking to describe the marine pelagic community in quantitative fashion. If Elton and Haldane were correct in citing body size as the paramount characteristic of an organism, then an organism's size should overshadow even its taxonomic designation. Ecological communities are conventionally described in terms of their taxonomic parts. Given the advantages of dealing with body size, might not a more effective quantitative theory ensue from classifying organisms according to size? Such a shift in emphasis would also afford significant practical advantages to limnologists and biological oceanographers. By employing inductive counting devices (such as the Coulter counter) it is relatively easy to acquire data on particle size distributions (psds) of planktonic communities in the range from 2 to 250 μm.

Evidence accumulated that the psds of planktonic communities in the ocean were relatively flat and lacking in definite structure, as shown in Fig. 14.1, based on the work of Sheldon, Prakash & Sutcliffe (1972). Platt (1985) noted how this relative flatness is the consequence of a simple transformation of the Eltonian pyramid, and had already been anticipated by H. T. Odum (1971). Platt & Denman (1978), employing the results of Fenchel, calculated that the density of living particles should decrease slowly as the negative one-fifth power of their body sizes.

If an organism's physiological rates are tied to its body size, and the distribution of size classes in an ecosystem is known, then by integrating the allometric formulae over the range of all sizes one should be able to estimate community parameters, such as gross respiration or total primary and heterotrophic production — rates which are central to a thermodynamic analysis of the community (Platt, Lewis & Geider 1984; Paloheimo 1988). Of course, system indices are rarely sufficient to describe interactions at the compartment level, and Silvert & Platt (1978, 1980) have modelled the transfers among the various size categories within the psd. Their first effort accounted only for transfers from smaller to larger size classes. It predicted that any transient in the distribution would propagate up the length scale with weak attenuation and no change in shape. Later refinements were to include 'diffusion' (i.e. propagation backward as well as forward) and feedback to smaller-size classes. Cousins (1983, 1985b), an ardent admirer of Elton, feels that some distinction based on function should be made among the types of particles comprising a community. He advocates the estimation of separate psds for autotrophs, heterotrophs and detritus. Presumably, differences in the dynamics of particles within each of these three categories are small in comparison with the differences in behaviour between the categories. Peters (1983) is optimistic that allometric and particle-size models offer the best starting-point for a predictive ecology.

The interest in particle size distributions of marine organisms has catalysed a number of new discoveries and practical applications. Most advances stem from efforts to obtain better information on plankton below 2 μm in effective diameter. Pomeroy (1974) suggested the existence of a widespread and diverse community of picoplankton, and this was verified by Waterbury *et al.* (1979) and Johnson & Sieburth (1979). Platt, Subba Rao & Irwin (1983), citing allometric and thermodynamic relationships, estimated that about half of the productivity of the world's oceans was generated by these smallest of micro-organisms. Just how much this incremental autotrophy affects heterotrophic production at macroscropic scales remains at issue. Everyone seems to agree

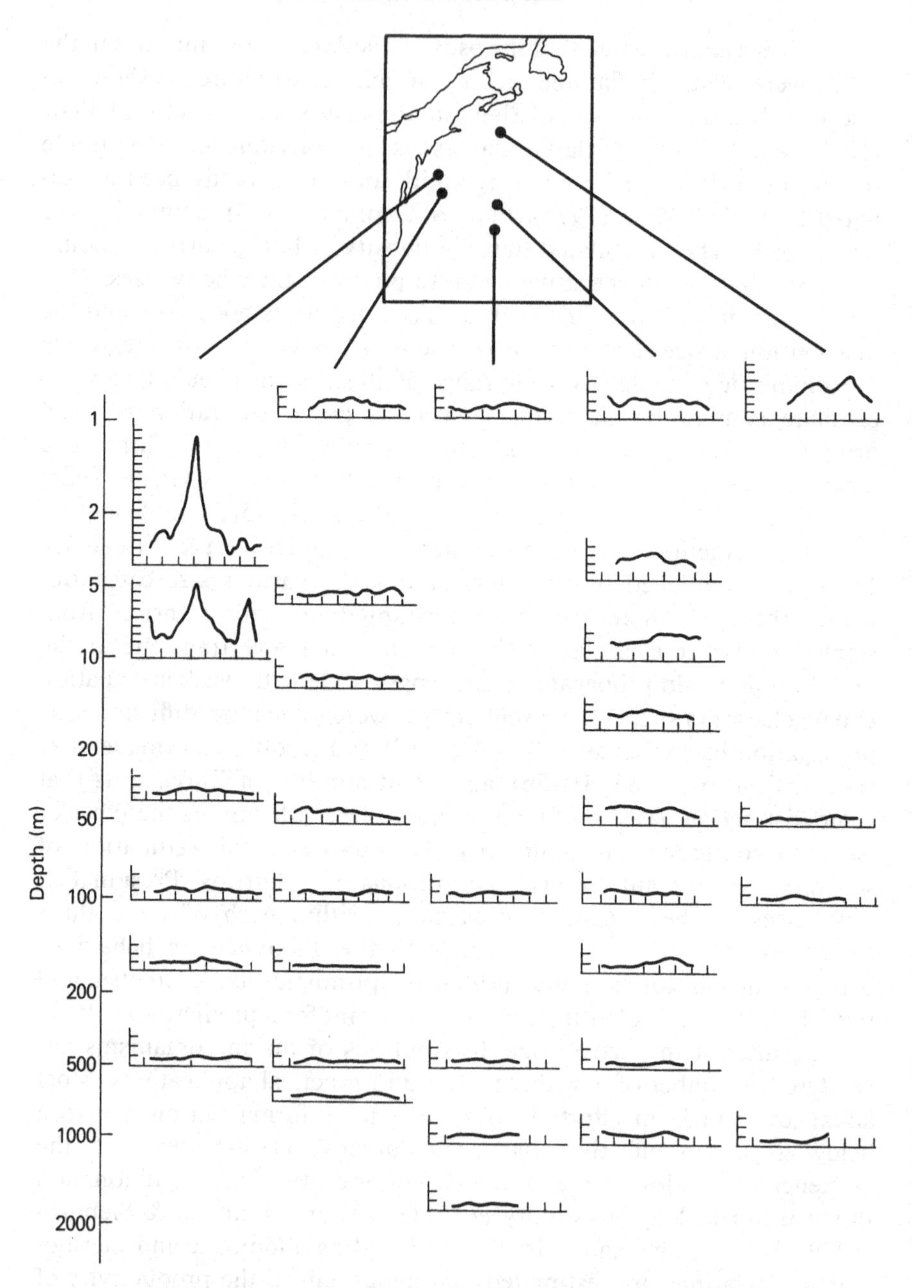

FIG 14.1 Size distributions of suspended particulate matter at various depths in the western North Atlantic (after Sheldon, Prakash & Sutcliffe 1972); in each graph the horizontal axis indicates particle diameter (with markers at 1, 2, 4, 8, 16, 32, 64 and 128 μm) and the vertical axis particle concentration by volume (ppm, with markers at intervals of 0.01).

there is a high degree of cycling and dissipation among the members of what has come to be known as the 'microbial loop' (Azam *et al.* 1983; Goldman 1984). Some, for example, Ducklow *et al.* (1986), argue that virtually all the picoplankton production is dissipated within the microbial loop, whereas others like Sherr, Sherr & Albright (1987) maintain that significant amounts are passed up the trophic web. In either event, it is highly probable that microbial metabolism is generating enormous quantities of dissolved organic carbon compounds (DOCs) within the upper layers of the world's oceans. New analytical methods point to DOC levels that are about four times those previously reported; the high levels result at least in part from the activity of the picoplankton (Sugimura & Suzuki 1988). The discovery of this new stock of organic carbon also ameliorates a previously disturbing imbalance in the global carbon cycle.

Knowledge about the nature of living particles below 2 μm in diameter is fast accruing now that the biomedical techniques of flow cytometry (Yentsch 1983) are being adapted for shipboard analysis of ocean plankton, pushing the limits of taxonomic detection down to 0.25 μm (Robertson & Button 1987) and possibly as low as 45 nm (Chisholm, Olson & Yentsch 1988). At the other end of the scale, side-scanning sonar appears to extend the acquisition of data on particle sizes well into the macroscopic ranges (Ehrenberg *et al.* 1981).

Paralleling the importance of organism size for ecosystem behaviour stands the possibility that physical phenomena of various characteristic scales also influence the make-up of the biotic community. That is, the measurements of spatial heterogeneities in physical variables, e.g. water movement, light and temperature, could be compared with the concomitant spatial structure of biotic distributions to investigate the potential coupling of physical and biological phenomena. To effect such a comparison marine ecologists have borrowed from their colleagues in physical oceanography, who are wont to represent their data on ocean currents in terms of 'power spectra'. Without going into the details of Fourier analysis, a power spectrum represents the aggregate amounts of kinetic power in fluid motions of successive characteristic lengths, e.g. the diameters of turbulent eddies. Typically, more kinetic energy is possessed by larger-scale currents, and this power is dissipated into heat as these motions degrade into smaller eddies. The same formal calculations can be applied to a temporal or spatial series of measurements on any other variable, such as chlorophyll concentration, to yield the 'power spectrum' of a biotic variable (Platt & Denman 1975).

It has been hypothesized by Okubo (1974) that when non-linear species interactions are affected by turbulent diffusion 'diffusive instabilities' could result. These spatial structures are to some degree independent of the underlying fluid motions. Whenever the slopes of the power spectra for water flow and biotic concentration are parallel, one is safe in assuming that the biota are being passively dispersed by the currents in true planktonic fashion. A departure of the two spectra could be evidence of a biological dissipative structure. The occurrence of dissipative structures seems tied to the boundaries of the water body, because Powell *et al.* (1975) observed dissonance between the physical and phytoplankton spectra at lengths greater than 100 m, whereas Gower, Denman & Holyer (1980) concluded that phytoplankton are being passively dispersed by ocean currents on scales as large as 10–100 km.

NON-HOLISTIC LINDEMAN SYSTEMS

Whereas Elton tried to make sense out of complex ecosystems by concentrating on the numbers of organisms and their sizes, Lindeman paid scant attention to these attributes and focused instead upon the functional and relational attributes of the ecological ensemble. It should be noted that Lindeman's motivations for describing his senescent lake systems the way he did derived largely from the physical science of thermodynamics. He was inclined to describe the parts of the system in physical terms, such as the amounts of material or energy embodied in them, and the relationships among his system parts likewise were gauged by the magnitudes of the transfers of material between components. Classical thermodynamics can be regarded as a self-consistent theory of relationships among macroscopic properties. Reference to microscopic entities such as molecules is not necessary and is even scorned in certain thermodynamic circles: whence derives Lindeman's casual attitude toward attributes that Elton regarded as paramount. Lindeman's emphasis was more on describing the configuration of the entire ecosystem and upon the functions of its parts within the context of the whole.

Of course, Lindeman was not the first to treat ecosystems almost as physical entities. Among his predecessors Lotka (1922), an actuary from Baltimore, captured the attention of many with, among other things, his quantitative description of predator–prey relationships. His name is most often connected with that of Volterra to identify their model of the interaction of sharks and fishes in the Aegean. They represented this predator–prey relationship using two coupled, ordinary, non-linear differential equations. It is interesting to note that Lindeman did not attempt to portray the trophic dynamics of his lake system in terms of differential equations. Perhaps he was unfamiliar with this type of mathematics; but, even if he were not, it is unlikely that such a model would have helped his narrative to any significant extent. Known analytical solutions to non-linear differential equations are sparse.

It was advances in automated computation that revived interest in simulating ecological dynamics with differential equations. Analogue electrical circuitry could be quickly fashioned to integrate systems of coupled differential equations without recourse to analytical methods. H. T. Odum (1971), another student of G. E. Hutchinson, employed analogue computers to simulate ecosystem behaviour using data on aquatic communities collected in the style of Lindeman. Meanwhile, digital computers evolved during the late 1960s that were capable of integrating the numerical difference counterparts to differential equations with much greater ease and accuracy than had been available with analogue computers. It was then possible to 'solve' systems comprised of hundreds of coupled differential equations.

Because of the potential that multivariate simulation modelling held for quantitative ecology, it became a key task in the North American contribution to the International Biological Programme. Unfortunately, the full promise of large-scale simulation models in biological oceanography has not yet been fulfilled, and there are those who say it probably never will be. Nevertheless, praiseworthy attempts at simulating marine ecosystems do exist, for example those of Steele (1974), Kremer & Nixon (1978), Longhurst (1978) and Cushing (1981), that afford penetrating insights into ecosystem mechanics and dynamics. As will become apparent presently, the data amassed in the course of these exercises are beginning to pay dividends in another analytical context.

In 1977 the governing body of the Scientific Committee on Oceanic Research (SCOR) commissioned a working group to assess the value of mathematical models for biological oceanography (Platt, Mann & Ulanowicz 1981). The committee, under the chairmanship of K. H. Mann, cautioned against over-confidence in the results of simulations of many coupled biotic processes. They cited uncertainties in the data, the amplification of these errors during the simulation process, and the possibility of deterministic chaos as difficulties plaguing many-variable simulations. But their chief criticism was that, despite the avowed intention of modellers to treat the entire system, the simulation paradigm does not address the system *as a whole.* That is, systems of coupled difference equations are reductionistic in the mechanistic sense.

In the process of simulation modelling one normally begins by defining the parts of the system, then identifies the connections between the parts and proceeds to describe each bilateral interaction in terms of some (usually fixed) mathematical function. Finally, these constitutive relations are incorporated into some balance scheme and numerically integrated. The dynamics of the ensemble of bilateral interactions is assumed to mimic the behaviour of the ecological community as a whole. Recent results from reconstructability analysis (Klir & Folger 1988) pinpoint why such representation often fails: it does not take account of possible higher-order interactions or allow for the evolution of the system topology (the network of interactions) and constitutive relations.

These criticisms apply only to systems of coupled processes, and the SCOR committee commended the success of some models that treated only a single species or process, such as those of Paloheimo & Dickie (1966), Parsons, Lebrasseur & Fulton (1967) and Jassby & Platt (1976). Simple models of nutrient fluxes (Moloney *et al.* 1986), trophic interactions (Frost 1987) and primary productivity have been highly useful tools for understanding and measuring these processes. Platt & Sathyendranath (1988) have combined a single-process model that predicts phytoplankton production from ambient light quality with several physical models of optics to infer the magnitude of primary production over large areas of the world's ocean from satellite data on ocean colour.

As for understanding the dynamics of the world's fisheries, single-process models have helped to structure dialogue for many years (Beverton & Holt 1957; Ricker 1975; Cushing & Horwood 1977), although most of these quantitative models belong more to the Eltonian school, cast as they are in terms of fish numbers and body lengths. Rothschild (1986) declared that 'A suitable formulation for linking whole ecosystem models with recruitment variability . . . is not yet available', due, in his opinion, to the exclusion of density dependence and amplifier effects in both ecosystem and recruitment models. He urged the incorporation of somatic/reproductive energetic ratios (a Lindeman–Elton hybrid) into recruitment models, and held some hope for an ecosystem approach to the problem of predicting fish population levels, because he felt that stock 'variability may arise as much from environmental variables as from biotic variables'.

This possibility that physical forces may dominate interbiotic effects in certain spatial domains makes feasible the prediction of biotic levels and distributions in these regions, providing of course that the driving physical forces are themselves known or predictable. Harris (1988), for example, documents the strong connection between phytoplankton ecology and physical driving forces. Nihoul (1986) presents examples of how the strong gradients in physical energies that exist near to physical interfaces (e.g. air–water, water–sediment, water–ice, etc.) can radically alter the rates of biological processes, such as production.

WHOLE-SYSTEM ECOLOGY

If the simulation modelling of ecological communities is not a wholly satisfactory quantification of ecosystems, then whither should we direct our efforts? One possibility is to enclose subsystems of a biotic community to be used as microcosm (Giesy 1980) or mesocosm (Grice & Reeve 1982) analogue models of the prototype. There have been some ambitious efforts to construct mesocosms of marine ecosystems, notably the CEPEX enclosures of oceanic planktonic communities (Grice *et al.* 1980) and the MERL tanks (Nixon *et al.* 1984) containing both pelagic and benthic elements of an estuarine ecosystem, as well as more modest endeavours to contain natural communities and replicate their environments, e.g. those of Lane & Collins (1985) and Frost (1987).

If fidelity to the dynamics of the prototype community is the hallmark of an exact ecology, then mesocosms offer perhaps the best available method for predicting ecological behaviour. They are capable of yielding a cornucopia of data on coupled processes that are more reliable than anything issuing from approaches of a more a priori nature. The use of mesocosms as both a management and a research tool definitely should be encouraged. Such promise notwithstanding, mesocosms are not the panacea for biological oceanography. If the objective of our research is to gain some understanding about the workings of ecosystems, observing mesocosms still leaves us with the problem of how to abstract from the events transpiring in the container. Secondly, any mesocosm is by definition a subsystem of the prototype, and one still faces the problem associated with reconstructability.

One approach to interpreting mesocosm data is purely phenomenological and inductive in nature — akin to what Peters (1983, 1986) calls ‘empirical limnology’. Synoptic time-series data on the principal biotic and physical elements in the mesocosm are accumulated and are then used as input to an algorithm that automatically parses the relationships among the variables (cf. Ulanowicz *et al.* 1978; Ivakhnenko, Krotov & Visotsky 1979; Klir 1981; Shaffer 1988; Vezina & Platt 1988). It is hoped that after a sufficient number of such exercises certain laws or principles of ecological behaviour will emerge, much in the way that the laws of thermodynamics evolved from quantitative observations on gases and steam engines.

Another direction toward understanding whole-systems behaviour has been promulgated by the SCOR Committee on Biological Modelling in the Oceans (Platt, Mann & Ulanowicz 1981). They urged that greater emphasis in biological oceanography be placed upon the elucidation and quantification of the flows of material and energy among the components of marine ecosystems. Three international workshops to foster interest in the topic of flows in marine ecosystems have been organized by members of the Committee (Fasham 1984a,b; Ulanowicz & Platt 1985; Wulff, Field & Mann 1989). Regardless of how it was prompted, there has recently been a distinct renewal of interest in quantifying fluxes of material and energy in marine systems (De Vries & Hopstaken 1984; Longhurst 1984; Fasham 1984a,b; Platt & Harrison 1985; Beddington 1986; Gordon *et al.* 1986; Sissenwine 1986; Peterson & Fry 1987; Cushing 1988; Baird & Ulanowicz 1989; Waring 1989). This renewal of interest builds upon an already interesting base of marine studies, for example those of Steele (1974), Jansson & Wulff (1976), Warwick, Joint & Radford (1979), Baird & Milne (1981) and Dame & Patten (1981). The monitoring of fluxes of carbon in the ocean has taken on a new urgency in the wake of ambiguities surrounding the global balance of this element. An international effort called the Joint Global Ocean Flux Study (JGOFS) has been mounted recently under the auspices of SCOR to 'determine and understand on a global scale the processes controlling the time-varying fluxes of carbon and associated biogenic elements in the ocean' (Brewer *et al.* 1986; JGOFS 1987).

A problem with the acquisition of flow data is that the resultant web of biotic and abiotic exchanges often becomes quite complicated. As an example, the flows of carbon among thirty-six major compartments of the ecosystem representing the mesohaline region of the Chesapeake estuary is shown in Fig. 14.2. The biomass values and flow magnitudes in this figure are averaged over the full extent of the mesohaline region. The annual flows are based on four separate seasonal networks estimated by Baird & Ulanowicz (in press). Spatial averaging was achieved by multiplying the rates for each exchange by that fraction of the mesohaline region over which the given transaction occurs. Temporal dynamics would be better resolved if data were available to represent the flow systems as twenty-six biweekly networks. Furthermore, it is likely that the ecosystems represented in the various tributaries and littoral zones and the channel region all differ in their network topologies, and comparisons among these should provide useful insights into the functioning of the larger mesohaline ensemble.

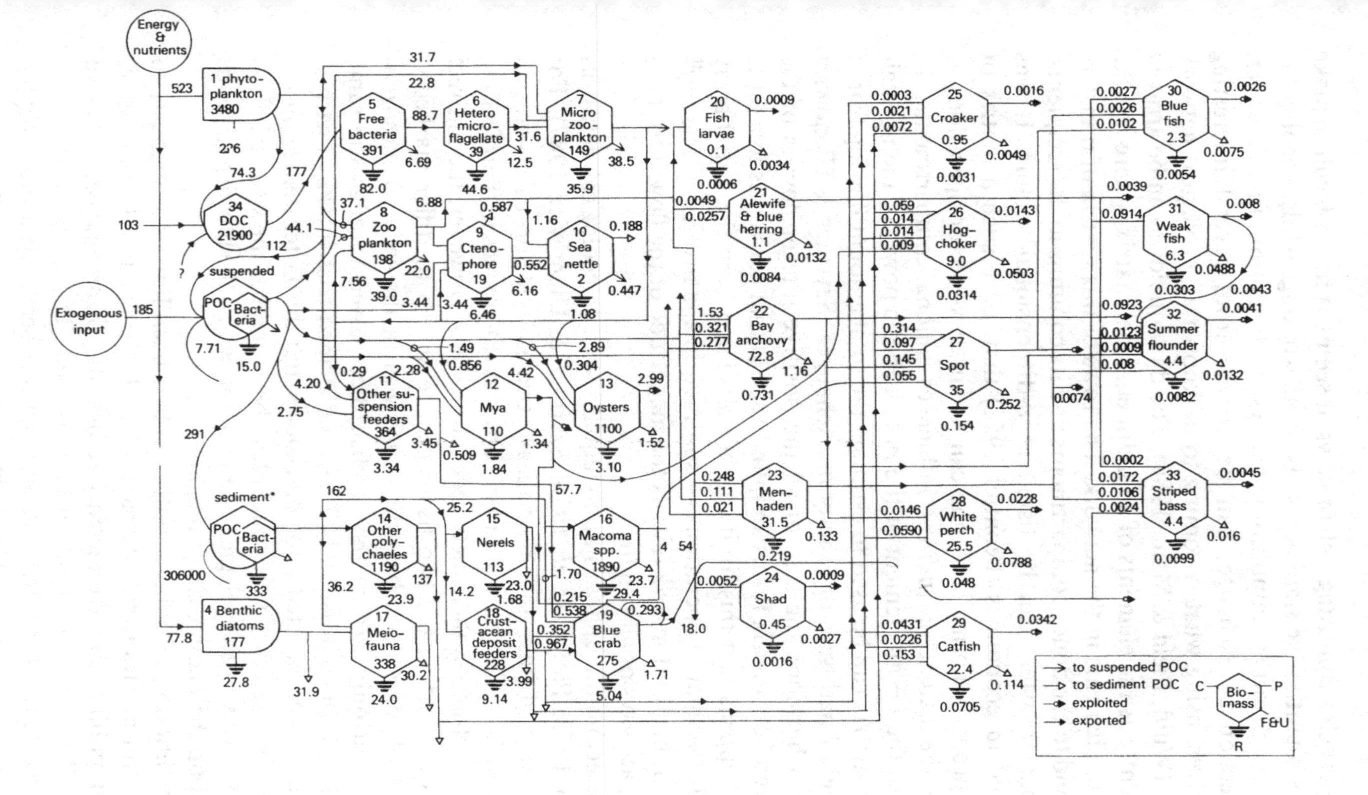

FIG 14.2 A schematic representation of the annual carbon flows among the principal components of the Chesapeake mesohaline ecosystem (based on Baird & Ulanowicz 1989). Standing crops of carbon are indicated within the compartments in mg m^{-2} and the indicated flows of carbon are in g m^{-2} $year^{-1}$.

Given such a confusing web of flows, it seems at first difficult to make definitive quantitative statements about its workings whilst keeping the number of a priori assumptions on the system to a minimum. However, the collection of exchanges can be viewed from several interesting perspectives, using what is coming to be known as 'ecological network analysis' (Wulff, Field & Mann 1989). The data on flows can be arrayed conveniently as the elements of an n-dimensional matrix, where n is the number of elements in the network. The rows of this matrix of transfers correspond to the donor compartments and its columns to the recipients. Exogenous transfers can be listed as n-dimensional vectors. It was discovered in economics (Leontief 1951) that, once such a matrix of flows is properly normalized, one can use matrix–vector operations to estimate the extent of *indirect* influences among compartments. In particular, the $i-j$th element of the mth algebraic power of the normalized flow matrix can be used to estimate the total amount flowing from i to j along *all* pathways of exactly m exchanges (Szyrmer & Ulanowicz 1987). Furthermore, the sum of all the positive integer powers of the normalized flow matrix forms an infinite series that converges to a finite limit. It happens that this limit can be calculated from the initial array using only two matrix operations — subtraction and inversion. The $i-j^{th}$ element of this limit matrix reveals the magnitude of the flow from i to j over all pathways of all lengths.

These economic analyses apply equally well to ecological networks (Hannon 1973; Patten *et al.* 1976; Szyrmer & Ulanowicz 1987). For example, striped bass and bluefish are two pelagic top carnivores in Chesapeake Bay. There are not many obvious differences in their feeding behaviours, although their diets differ somewhat. However, the indirect resources supporting the two fishes can be shown to differ markedly. Over 65% of the striped bass diet once was in the form of mesozooplankton, but only 29% of the bluefish diet passed through the same compartment. In contrast, only 2% of the striped bass intake of carbon was formerly incorporated into polychaetes, but this latter item supports 48% of the bluefish diet. One concludes, therefore that the striped bass is sustained mostly by the pelagic grazing chain, whilst the bluefish depends mostly upon the benthic detrital chain.

The estimation of indirect influence had been limited previously to 'vertical' trophic transfers of material and energy, but Ulanowicz & Puccia (unpublished) have recently reformulated the analysis to reveal 'horizontal' competitive relationships as well. The methodology now exists to gauge all *non-proximate* interactions occurring in the trophic network. Most exciting is the observation that indirect influences are sometimes greater in magnitude and even opposite in character to direct interactions (Higashi & Patten, in press). Hence, observations made of direct effects should be interpreted only in the full context in which they occur.

Lindeman's idea of assigning each taxon to a distinct trophic compartment certainly would fail to apply to the network in Fig. 14.2, where numerous heterotrophs feed along multiple pathways of differing trophic lengths (Cousins 1985b). However, if one is willing to consider a distributed mapping of species to trophic levels, difficulties quickly fade. For example, if a given taxon were to obtain 20% of its intake along trophic pathways one step removed from the autotrophs, 70% along pathways two steps removed and 10% along sequences three steps distant, then the activities and biomass of that taxon would be assigned in those given proportions to the herbivore, carnivore and top carnivore levels, respectively. The mechanics for mapping a complex web of ecosystem feeding relations into a straight chain of trophic transfers were developed by Ulanowicz & Kemp (1979). Recently it has become possible to add the abiotic detrital recycling pathways to the analysis (Ulanowicz, in press). Thus, one can now reconfigure the most complicated webs of trophic interactions into Lindeman-style trophic chains with detrital feedbacks.

The trophic level reformulation of the Chesapeake network is depicted in Fig. 14.3 and reveals several interesting features. Trophic pathways at least eight steps long are apparent among the Chesapeake biota; however, the amounts passing through the top three levels are vanishingly small. In fact, when the average levels at which each compartment feeds are calculated, they all fall below five — a limit discussed elsewhere (Slobodkin 1961; Pimm & Lawton 1977). Secondly, it is clear that production at higher levels is significantly dependent upon the recycling of carbon (and some of its chemically bound energy) through the non-living elements of the system. Detritivory exceeds herbivory by a ratio of more than 10:1, and 70% of all the inputs to the detrital pool derive from recycling. Finally, there is a trend (not without exception) for successively higher trophic levels to *decrease* in efficiency, contradicting what Lindeman (1942) had expected.

As cycling is such an important element of most systems, it should be helpful to analyse the pathways for cycling in greater detail. Elsewhere (Ulanowicz 1983) I have outlined methods both for enumerating all biogeochemical cycles and for extracting those feedbacks from their supporting web of dissipative transfers. Cycling of material usually indicates cybernetic control at work in the system (Odum 1971;

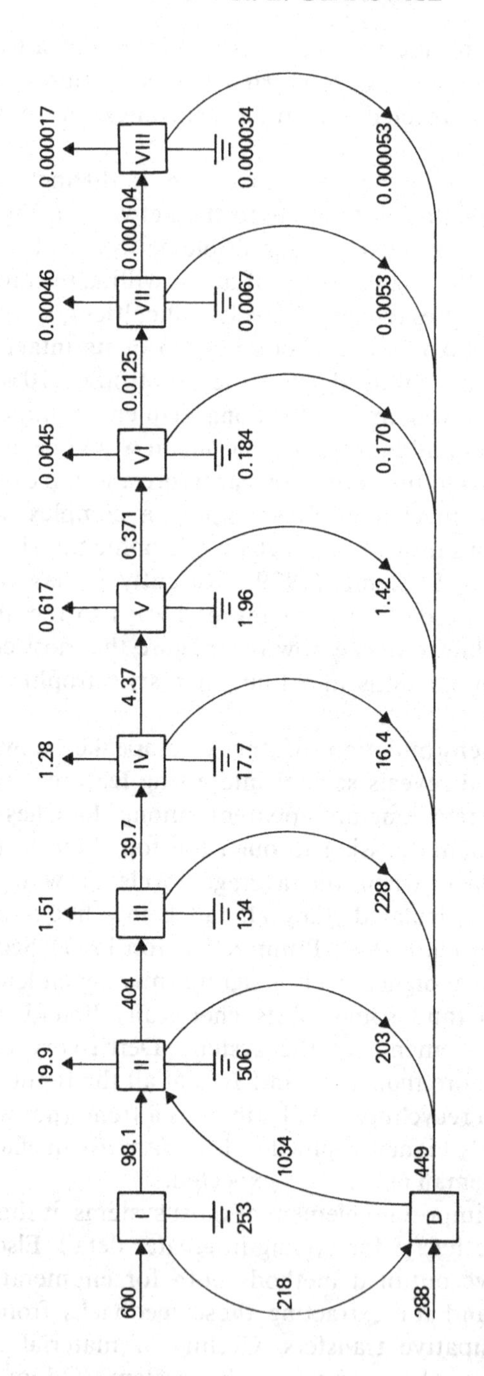

FIG 14.3 The aggregation of the flow web in Fig. 14.2 into a concatonated chain of transfers through eight trophic levels. Recycling of non-living material is through compartment *D*. Flows of carbon are in g m^{-2} $year^{-1}$.

Ulanowicz 1986), and the pattern of such recycling can help to define the function of a compartment in the context of the ecosystem.

For example, the cycles present in the Chesapeake network number only sixty-one. When properly weighted and aggregated they form the pattern shown in Fig. 14.4. The bipartite configuration of the cycles is surprising and suggests that control among the planktonic community is largely decoupled from that among the benthos and nekton. It is also interesting to note that fourteen of the compartments do not engage in any cycling of carbon. Eight of these non-participants are filter-feeding benthos or finfish. They may be regarded as 'boundary' populations, whose function is to shunt material and energy from the planktonic domain into the benthic-nektonic realm of control. Curiously, three elements of the microbial 'loop' also do not engage in any cycling. As regards carbon, the term 'loop' appears to be a misnomer. The microbiota function more to shunt excess primary production out of the ecosystem via their own respiration (Ducklow *et al.* 1986).

Indirect flow, trophic and cycle analyses address how a subsystem of a given ecological network can impose itself as a unit upon its constituent parts. But, if one's goal is to quantify the process of ecological succession, it becomes necessary to characterize the status of the ecosystem as a whole. E. P. Odum (1969) cited twenty-four community properties that he felt reflected the succession toward a more mature community. Most of Odum's system-level indices can be calculated from the same data used to create flow networks, e.g. overall $P:B$, $P:R$ or $R:B$ ratios. Others, such as nutrient retention time, require analytical interpretation (Hannon 1979), whilst those remaining, such as body size and life-cycle, are better included in Eltonian projects.

I have argued (Ulanowicz 1986) that all of the Lindemanian measures that appear on Odum's list are particular facets of a more general index called the network ascendancy. Ascendacy is a product of two factors, one quantifying the total activity of the system and the other specifying the level of organization inherent in the network structure. By organization is meant here the average degree of stenotrophic behaviour, or the extent to which those pathways of higher throughput efficiency dominate their more numerous, but less effective counterparts. The quantity 'average mutual information' as defined in information theory can be adapted to serve as the factor representing the organizational level of a given network (Rutledge, Basorre & Mulholland 1976). Any increase in the network ascendancy is presumed to mirror the growth and/or development of the whole system.

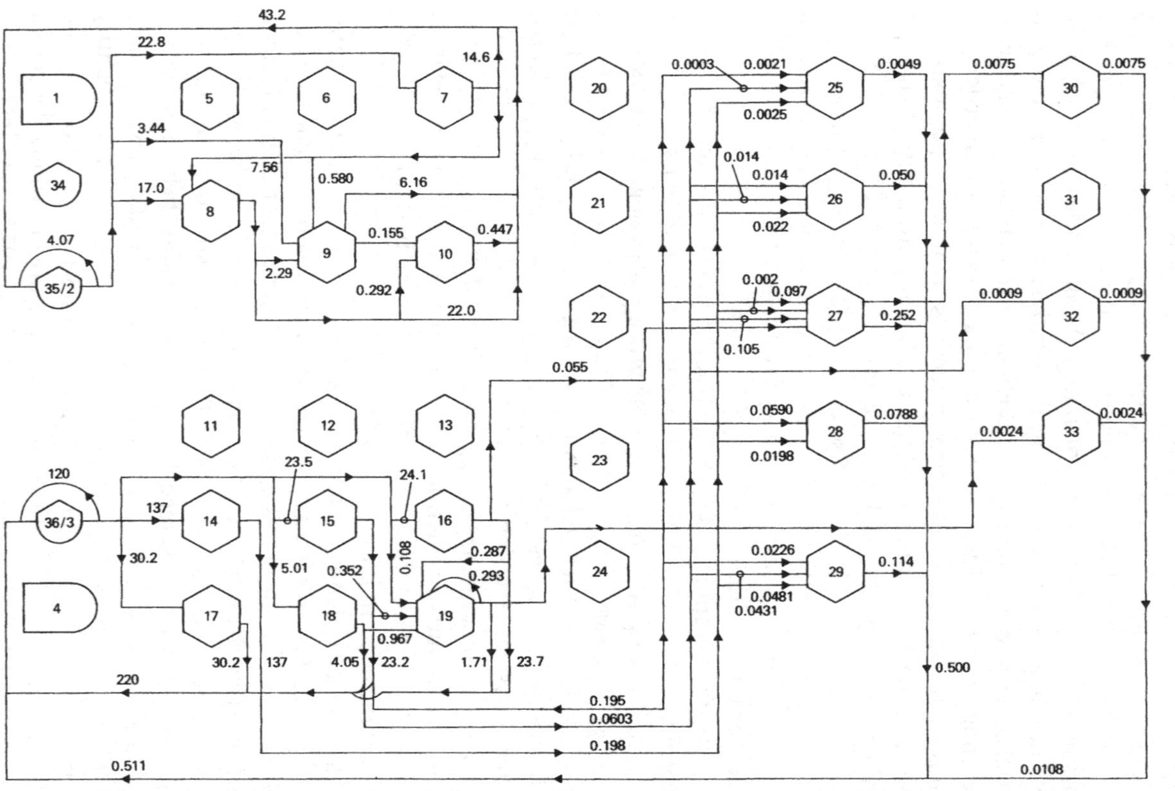

FIG 14.4 The composite cycling of carbon that occurs in the flow diagram in Fig. 14.2. Units of flow and numbering of the compartments are the same as in Fig. 14.2.

A remaining issue is whether systems-level organization, as quantified by ascendancy or any similar whole-system index, is purely epiphenomenal or results from an autonomous higher-level process. I have argued (Ulanowicz 1989) that autopoietic cycles as may exist in ecosystems characteristically endure longer than their components, and, more importantly, that the make-up of the autocatalytic cycle exerts a degree of selection upon its replacement parts. Hence, I regard cybernetic feedback in ecosystems as one of the origins of formal causes, *sensu* Artistotle, that effect the organization of ecosystems.

FUTURE DIRECTIONS

Left to the realm of academic speculation, opinions on how to abstract ecosystems are likely to continue to diverge. There are countervailing pressures favouring a more accommodating dialogue, however, as the world and its oceans are beset by a host of environmental problems. Humanity looks increasingly to ecologists to forecast future conditions, and offer suggestions on how to ameliorate the consequences of impacts on the biosphere. Those able to influence societal behaviour need to know what will be the effects of a slight rise in oceanic temperature, how low concentrations of pollutants might affect oceanic resources, or which biota will suffer as a consequence of a modest rise in sea-level.

It is unlikely that any of the avenues of research discussed in this paper would alone be sufficient to provide reliable answers to such questions. Simulation modelling *per se* is too mechanical an analogue ever to yield reliable predictions. Likewise, network analysis remains largely descriptive, providing only suggestions about ecosystem behaviour.

Until the advent of some radically new methodology, our present best hope for understanding ecodynamics and forecasting impacts would seem to lie in research on possible hybrids of existing tools. For example, Field, Moloney & Attwood (1989) have created a particle-size-based model simulating the evolution of a plankton community following an upwelling event astride the Banguella current off south-western Africa. The authors use their package to compute a temporal series of flow networks, each of which becomes the subject of network analysis. Thereby the authors are able to trace the course of ecosystem-level properties over time.

Although the efforts of Field, Moloney & Attwood (1989) combine most of the tools discussed in this review, even further amalgamation of methods is both feasible and desirable. Their system attributes result strictly from the characteristics of the driving model, whereas a realistic portrayal of events might arise through allowing feedback to occur between the ecosystems being simulated and elements at adjacent hierarchical levels. For example, the integration of the model could be interrupted at convenient intervals, and changed by stochastic influences emanating from the system's infrastructure (Allen & McGlade 1987). At somewhat longer intervals the status of the ecosystem flow network could be evaluated, and these results used to alter the dynamical structure of the model so as to follow more closely those system-level trends most commonly observed in nature (Cheung 1985; T. F. Fontaine, personal communication).

The oceans that gave birth to life continue to sustain it today. They also provide the background to one of the most challenging and promising of contemporary scientific endeavours.

ACKNOWLEDGMENTS

I was contacted close to the date of the jubilee to fill a vacancy on the list of speakers and found it necessary to emphasize the opinions of friends and colleagues over personal library research in order to meet the deadlines for this review. I am most grateful to Steven Cousins, Christopher D'Elia, Michael Fasham, John Field, Peter Grubb, Bengt-Owe Jansson, James Love, Scott Nixon, Trevor Platt, Brian Rothschild, William Silvert and Frederik Wulff for responding to my pleas for help. My special thanks go to James Love, Jacqueline McGlade and Frederick Wulff for reading the draft manuscript and suggesting most useful changes. Mrs Jeri Pharis and Mrs Frances Younger both laboured diligently under pressure to help me prepare the text and diagrams on time.

REFERENCES

Allen, P. M. & McGlade, J. M. (1987). Evolutionary drive: the effect of microscopic diversity, error making and noise. *Foundations of Physics*, Special Edition, July.

Ashby, W. R. (1953). Some peculiarities of complex systems. *Cybernetic Medicine*, **9**, 1–7.

Azam, F., Fenchel, T., Field, J. G., Gray, J. S., Reid-Meyer, L. A. & Thingstad, F. (1983). The ecological role of water column microbes in the sea. *Marine Ecology Progress Series*, **10**, 257–63.

Baird, D. & Milne, H. (1981). Energy flow in the Ythan estuary, Aberdeenshire, Scotland. *Estuarine Coastal Shelf Science*, **13**, 455–72.

Baird, D. & Ulanowicz, R. E. (in press). The seasonal dynamics of the Chesapeake Bay ecosystem. *Ecological Monographs*.

Beddington, J. R. (1986). Shifts in resource population in large marine ecosystems. *Variability and Management of Large Marine Ecosystems* (Ed. by K. Sherman & L. M. Alexander), pp. 9–18. Westview Press, Boulder, Colorado.

Beverton, R. J. H. & Holt, S. J. (1957). On the dynamics of exploited fish populations. *Fisheries Investigation, Ser.* 2, No. 19. Ministry of Agriculture, Fisheries and Food, London.

Brewer, P. G., Bruland, K., Eppley, R. W. & McCarthy, J. J. (1986). The joint global ocean flux study. *EOS,* **67**, 827–32.

Cheung, A. K. -T. (1985). *Network optimization in ecosystem development.* Ph. D. dissertation, Johns Hopkins University, Baltimore.

Chisholm, S. W., Olson, R. & Yentsch, C. M. (1988). Flow cytometry in oceanography: status and prospects. *EOS,* **69**, 562–72.

Cousins, S. (1983). *An alignment of diversity and energy models of ecosystems.* Ph. D. thesis, The Open University, Milton Keynes.

Cousins, S. H. (1985a). The trophic continuum in marine ecosystems: structure and equations for a predictive model. *Ecosystem Theory for Biological Oceanography* (Ed. by R. E. Ulanowicz & T. Platt), pp. 76–93. *Canadian Bulletin of Fisheries and Aquatic Sciences,* **213.**

Cousins, S. H. (1985b). Ecologists build pyramids again. *New Scientist,* **107**, 50–4.

Cushing, D. H. (1981). *Fisheries Biology: a Study in Population Dynamics,* 2nd edn. University of Wisconsin Press, Madison.

Cushing, D. H. (1988). The flow of energy in marine ecosystems, with special reference to the continental shelf. *Continental Shelves* (Ed. by H. Postma and J. J. Zijlstra), Ch. 7. Elsevier, Amsterdam.

Cushing, D. H. & Horwood, J. W. (1977). Development of a model of stock and recruitment. *Fisheries Mathematics* (Ed. by J. H. Steele), pp. 21–35. Academic Press, London.

Dame, R. F. & Patten, B. C. (1981). Analysis of energy flows in an intertidal oyster reef. *Marine Ecology Progress Series,* **5**, 115–24.

De Vries, I. & Hopstaken, C. F. (1984). Nutrient cycling and ecosystem behaviour in a salt-water lake. *Netherlands Journal of Sea Research,* **18**, 221–45.

Ducklow, H. W., Purdie, D. A., Williams, P. J. L. & Davies, J. M. (1986). Bacterioplankton: a sink for carbon in a coastal marine plankton community. *Science,* **232**, 865–7.

Ehrenberg, J. E., Carlson, T. J., Traynor, J. J. & Nelson, N. J. (1981). Indirect measurements of the mean acoustic scattering cross sections of fish. *Journal of the Acoustical Society of America,* **69**, 945–62.

Elton, C. S. (1927). *Animal Ecology.* Sidgwick & Jackson, London.

Fasham, M. J. R. (Ed.) **(1984a).** *Flows of Energy and Materials in Marine Ecosystems.* Plenum, New York.

Fasham, M. J. R. (1984b). Flow analysis of materials in the marine euphotic zone. *Ecosystem Theory for Biological Oceanography* (Ed. by R. E. Ulanowicz & T. Platt) pp. 139–62. *Canadian Bulletin of Fisheries and Aquatic Sciences,* **213.**

Fenchel, T. (1974). Intrinsic rate of natural increase: the relationship with body size. *Oecologia,* **14**, 317–26.

Field, J. G., Moloney, C. L. & Attwood, C. G. (in press). Network analysis of simulated succession after an upwelling event. *Flow Analysis of Marine Ecosystems* (Ed. by F. Wulff, J. G. Field & K. H. Mann). Springer, New York.

Frost, B. W. (1987). Grazing control of phytoplankton stock in the open subarctic Pacific Ocean: a model assessing the role of mesozooplankton, particularly the large calanoid copepods *Neocalanus* spp. *Marine Ecology Progress Series,* **39**, 49–68.

Giesy, J. P. (1980). *Microcosms in Ecological Research.* Department of Energy Symposium Series, 52, CONF-781101. National Technical Information Service, Springfield, Virginia, USA.

Goldman, J. (1984). Oceanic nutrient cycles. *Flows of Energy and Materials in Marine Ecosystems* (Ed. by M. J. R. Fasham), pp. 137–70. Plenum, London.

Gordon, D. C., Keizer, P. D., Daborn, G. R., Schwinghamer, P. & Silver, W. L. (1986). Adventures in holistic ecosystem modelling: the Cumberland Basin ecosystem model. *Netherlands Journal of Sea Research*, **20**, 325–35.

Gower, J. F. R., Denman, K. L. & Holyer, R. J. (1980). Phytoplankton patchiness indicates the fluctuation spectrum of mesoscale oceanic structure. *Nature*, **288**, 157–9.

Grice, G. D., Harris, G. P., Reeve, M. R., Heinbokel, J. F. & Davis, C. O. (1980). Large-scale enclosed water column ecosystems. *Journal of the Marine Biological Association, UK*, **60**, 401–14.

Grice, G. D. & Reeve, M. R. (1982). *Marine Mesocosms: Biological and Chemical Research in Experimental Ecosystems.* Springer, New York.

Haldane, J. B. S. (1928). On being the right size. *A Treasury of Science* (Ed. by H. Shapely, S. Raffort & H. Wright), pp. 321–5. Harper, New York.

Hannon, B. (1973). The structure of ecosystems. *Journal of Theoretical Biology*, **41**, 535–46.

Hannon, B. (1979). Total energy costs in ecosystems. *Journal of Theoretical Biology*, **80**, 271–93.

Harris, G. P. (1988). *Phytoplankton Ecology: Structure, Function and Fluctuation.* Chapman & Hall, London.

Hemmingsen, A. M. (1960). Energy metabolism as related to body size and respiratory surface, and its evolution. *Reports of the Steno Memorial Hospital, Copenhagen*, **9**, (2), 1–110.

Higashi, M. & Patten, B. C. (in press). Further aspects of the analysis of indirect effects in ecosystems. *Theoretical Studies of Ecosystems* (Ed. by M. Higashi & T. P. Burns). Cambridge University Press.

Hughes, R. G. (1985). A hypothesis concerning the influence of competition and stress on the structure of marine benthic communities. *Proceedings of the Nineteenth European Marine Biology Symposium: Plymouth, Devon, UK, 16–21 September 1984* (Ed. by P. E. Gibbs), pp. 319–400. Cambridge University Press.

Ivakhnenko, A. G., Krotov, G. I. & Visotsky, V. N. (1979). Identification of the mathematical model of a complex system by the self-organization method. *Theoretical Systems Ecology* (Ed. by E. A. Halfon), pp. 325–52. Academic Press, New York.

Jansson, B. -O. & Wulff, F. (1976). *Ecosystems Analysis of a Shallow Sound in the Northern Baltic.* Contribution No. 8., Asko Laboratory, University of Stockholm.

Jassby, A. D. & Platt, T. (1976). Mathematical formulation of the relationship between photosynthesis and light for phytoplankton. *Limnology and Oceanography*, **21**, 540–7.

JGOFS (1987). *The Joint Global Ocean Flux Study: Background, Goals, Organization, and Next Steps.* International Council of Scientific Unions, Paris.

Johnson, P. W. & Sieburth, J. M. (1979). Chroococcoid cyanobacteria at sea: a ubiquitous and diverse phototrophic biomass. *Limnology and Oceanography*, **24**, 928–35.

Kleiber, M. (1947). Body size and metabolic rate. *Physiological Researches*, **27**, 511–41.

Klir, G. (1981). On systems methodology and inductive reasoning. *General Systems Yearbook*, **26**, 29–38.

Klir, G. J. & Folger, T. A. (1988). *Fuzzy Sets, Uncertainty and Information.* Prentice Hall, Englewood Cliffs, New Jersey.

Kremer, J. N. & Nixon, S. W. (1978). *A Coastal Marine Ecosystem: Simulation and Analysis.* Springer, New York.

Lane, P. A. & Collins, T. M. (1985). Food web models of a marine plankton community network: an experimental mesocosm approach. *Journal of Experimental Marine Biology and Ecology*, **94**, 41–70.

Lehman, J. T. (1986). The goal of understanding in ecology. *Limnology and Oceanography*, **31**, 1160–6.

Leontief, W. (1951). *The Structure of the American Economy, 1919–1939*, 2nd edn. Oxford University Press.

Lindeman, R. L. (1942). The trophic-dynamic aspect of ecology. *Ecology*, **23**, 399–418.

Longhurst, A. R. (1978). Ecological models in estuarine management. *Ocean Management*, **4**, 287–302.

Longhurst, A. R. (1984). Importance of measuring rates and fluxes in marine erosystems. *Flow of Energy and Materials in Marine Ecosystems* (Ed. by M. J. R. Fasham), pp. 3–22. Plenum, New York.

Lotka, A. J. (1922). Contribution to the energetics of evolution. *Proceedings of the National Academy of Sciences*, **8**, 147–50.

Moloney, C. L., Bergh, M. O., Field, J. G. & Newell, R. C. (1986). The effect of sedimentation and microbial nitrogen regeneration in a plankton community: a simulation investigation. *Journal of Plankton Research*, **8**, 427–45.

Nihoul, C. J. (1986). *Marine Interfaces Ecohydrodynamics*. Elsevier, Amsterdam.

Nixon, S. W., Pilson, M. E. Q., Oviatt, C. A., Donaghay, P., Sullivan, B., Seitzinger, S., Rudnick, D. & Frithsen, J. (1984). Eutrophication of a coastal marine ecosystem—an experimental study using the MERL microcosms. *Flows of Energy and Materials in Marine Ecosystems* (Ed. by M. J. R. Fasham), pp. 105–35. Plenum, New York.

Odum, E. P. (1969). The strategy of ecosystem development. *Science*, **164**, 262–70.

Odum, H. T. (1971). *Environment, Power and Society*. Wiley, New York.

Okubo, A. (1974). *Diffusion-induced Instability in Model Ecosystems: Another Possible Explanation of Patchiness*. Technical Report, 86. The Chesapeake Bay Institute, Johns Hopkins University, Baltimore, Maryland.

Paloheimo, J. E. (1988). Estimation of marine production from size spectrum. *Ecological Modelling*, **42**, 33–44.

Paloheimo, J. E. & Dickie, L. M. (1966). Food and growth of fishes. III. Relations among food, body size and growth efficiency. *Journal of the Fisheries Research Board of Canada*, **23**, 1209–48.

Parsons, T. R., Lebrasseur, R. J. & Fulton, J. D. (1967). Some observations on the dependence of zooplankton grazing on the cell size and concentration of phytoplankton blooms. *Journal of the Oceanographical Society of Japan*, **23**, 10–17.

Patten, B. C., Bosserman, R. W., Finn, J. T. & Cale, W. (1976). Propagation of cause in ecosystems. *Systems Analysis and Simulation in Ecology* (Ed. by B. C. Patten), Vol. 4, pp. 457–79. Academic Press, New York.

Peters, R. H. (1983). *The Ecological Implications of Body Size*. Cambridge University Press.

Peters, R. H. (1986). The role of prediction in limnology. *Limnology and Oceanography*, **31**, 1143–59.

Peterson, B. J. & Fry, B. (1987). Stable isotopes in ecosystem studies. *Annual Review of Ecology and Systematics*, **18**, 293–320.

Pimm, S. L. & Lawton, J. H. (1977). Number of trophic levels in ecological communities. *Nature*, **268**, 329–31.

Platt, T. (1985). Structure of the marine ecosystem: its allometric basis. *Ecosystem Theory for Biological Oceanography* (Ed. by R. E. Ulanowicz & T. Platt), pp. 55–64. *Canadian Bulletin of Fisheries and Aquatic Sciences, 213.*

Platt, T. & Denman, K. L. (1975). Spectral analysis in ecology. *Annual Review of Ecology and Systematics*, **8**, 189–210.

Platt, T. & Denman, K. L. (1978). The structures of pelagic marine ecosystem. *Rapports et Procès-verbaux des Reunions du Conseil International pour l'Exploration de la Mer*, **173**, 60–5.

Platt, T. & Harrison, W. G. (1985). Biogenic fluxes of carbon and oxygen in the ocean. *Nature*, **318**, 55–8.

Platt, T., Lewis, M. & Geider, R. (1984). Thermodynamics of the pelagic ecosystem: elementary closure conditions for biological production in the open ocean. *Flows of Energy and Materials in Marine Ecosystems: Theory and Practice* (Ed. by M. J. R. Fasham), pp. 49–84. Plenum, New York.

Platt, T., Mann, K. H. & Ulanowicz, R. E. (1981). *Mathematical Models in Biological Oceanography.* Monographs on Oceanographic Methodology, 7. UNESCO, Paris.

Platt, T. & Sathyendranath, S. (1988). Oceanic primary production: estimation by remote sensing at local and regional scales. *Science,* **241**, 1613–20.

Platt, T. & Silvert, W. (1981). Ecology, physiology, allometry and dimensionality. *Journal of Theoretical Biology,* **93**, 855–60.

Platt, T., Subba Rao, D. V. & Irwin, B. (1983). Photosynthesis of picoplankton in the oligotrophic ocean. *Nature,* **300,** 702–4.

Pomeroy, L. R. (1974). The ocean's food webs, a changing paradigm. *BioScience,* **24**, 499–504.

Powell, T. M., Richerson, P. J., Dillon, T. M., Agee, B. A., Dozier, B. J., Godden, D. A. & Myrup, L. O. (1975). Spatial scales of current speed and phytoplankton biomass fluctuations in Lake Tahoe. *Science,* **189**, 1088–90.

Price, M. V. (1986). Approaches to the study of natural communities. *American Zoologist,* **26**, 3–4.

Ricker, W. E. (1975). Computation and interpretation of biological statistics of fish populations. *Bulletin of the Fisheries Research Board of Canada,* **191**, 382.

Robertson, B. R. & Button, D. K. (1987). Size and DNA distributions, populations and kinetic characteristics of bacterioplankton by flow cytometry. *EOS Transactions AGU,* **68**, 1677.

Rothschild, B. J. (1986). *Dynamics of Marine Fish Populations.* Harvard University Press, Cambridge, Massachusetts.

Rutledge, R. W., Basorre, B. L. & Mulholland, R. J. (1976). Ecological stability: an information theory viewpoint. *Journal of Theoretical Biology,* **57**, 355–71.

Schoener, T. W. (1986). Mechanistic approaches to community ecology: a new reductionism? *American Zoologist,* **26**, 81–106.

Shaffer, G. P. (1988). *K*-systems analysis for determining the factors influencing benthic microfloral productivity in a Louisiana estuary, USA. *Marine Ecology Progress Series,* **43**, 43–54.

Sheldon, R. W., Prakash, A. & Sutcliffe, W. H. (1972). The size distribution of particles in the ocean. *Limnology and Oceanography,* **17**, 327–40.

Sherr, E. B., Sherr, B. F. & Albright, L. J. (1987). Bacteria: link or sink? *Science,* **235**,88–9.

Silvert, W. & Platt, T. (1978). Energy flux in the pelagic ecosystem: a time-dependent equation. *Limnology and Oceanography,* **23**, 813–16.

Silvert, W. & Platt, T. (1980). Dynamic energy-flow model of particle size distribution in pelagic ecosystems. *Evolution and Ecology of Zooplankton Communities* (Ed. by W. C. Kerfoot), pp. 754–63. University Press of New England, Hannover, New Hampshire.

Simberloff, D. (1983). Competition theory, hypothesis testing, and other community-ecological buzz-words. *American Naturalist,* **122**, 626–35.

Sissenwine, M. P. (1986). Perturbation of a predator-controlled continental shelf ecosystem. *Variability and Management of Large Marine Ecosystems* (Ed. by K. Sherman & L. M. Alexander), pp. 55–85. Westview Press, Boulder, Colorado.

Slobodkin, L. B. (1961). *Growth and Regulation of Animal Populations.* Holt, Rinehart & Winston, New York.

Steele, J. H. (1974). *The Structure of Marine Ecosystems.* Harvard University Press, Cambridge, Massachusetts.

Sugimura, Y. & Suzuki, Y. (1988). A high-temperature catalytic oxidation method for the determination of non-volatile dissolved organic carbon in seawater by direct injection of a liquid sample. *Marine Chemistry*, **24**, 105–31.

Szyrmer, J. & Ulanowicz, R. E. (1987). Total flows in ecosystems. *Ecological Modelling*, **35**, 123–36.

Tansley, A. G. (1935). The use and abuse of vegetational concepts and terms. *Ecology*, **16**, 284–307.

Ulanowicz, R. E. (1983). Identifying the structure of cycling in ecosystems. *Mathematical Bioscience*, **65**, 219–37.

Ulanowicz, R. E. (1986). *Growth and Development: Ecosystems Phenomenology*. Springer, New York.

Ulanowicz, R. E. (in press). Formal agency in ecosystem development. *Search for Ecosystem Principles: a Network Perspective* (Ed. by M. Higashi & T. P. Burns). Cambridge University Press. In press.

Ulanowicz, R. E., Flemer, D. A., Heinle, D. R. & Huff, R. T. (1978). The empirical modelling of an ecosystem. *Ecological Modelling*, **4**, 29–40.

Ulanowicz, R. E. & Kemp, W. M. (1979). Toward canonical trophic aggregations. *American Naturalist*, **114**, 871–83.

Ulanowicz, R. E. & Platt, T. (Eds) **(1985).** *Ecosystem Theory for Biological Oceanography. Canadian Bulletin of Fisheries and Aquatic Sciences*, **213**.

Vezina, A. F. & Platt, T. (1988). Food web dynamics in the ocean. I. Best-estimates of flow networks using inverse methods. *Marine Ecology Progress Series*, **42**, 269–87.

Waring, R. H. (1989). Ecosystems: fluxes of matter and energy. *Ecological Concepts* (Ed. by J. M. Cherrett), pp. 17–41. British Ecological Society Symposium, 29. Blackwell Scientific Publications, Oxford.

Warwick, R. M., Joint, I. R. & Radford, P. J. (1979). Secondary production of the benthos in an estaurine environment. *Ecological Processes in Coastal Environments* (Ed. by R. L. Jefferies & A. J. Davy), pp. 429–50. British Ecological Society Symposium, 19. Blackwell Scientific Publications, Oxford.

Waterbury, J. B. S., Watson, S. W., Guillard, R. R. & Brand, L. E. (1979). Widespread occurrence of a unicellular marine plankton cyanobacterium. *Nature*, **277**, 293–4.

Wolff, W. F. (1988). Microinteractive predator-prey simulations. *Ecodynamics: Contributions to Theoretical Ecology* (Ed. by W. Wolff, C.-J. Soeder & F. R. Drepper), pp. 285–308. Springer, Berlin.

Wulff, F., Field, J. G. & Mann, K. H. (Eds) **(1989).** *Flow Analysis of Marine Ecosystems*. Springer, New York.

Yentsch, C. M. (1983). Flow cytometry and sorting: a powerful technique with potential applications in aquatic sciences. *Limnology and Oceanography*, **28**, 1275–80.

15. SOILS AS COMPONENTS AND CONTROLLERS OF ECOSYSTEM PROCESSES

E. A. PAUL
Department of Crop and Soil Sciences, Michigan State University, East Lansing, Michigan 48824, USA

INTRODUCTION

The recognition of soil as being at once an ecosystem component and a controller of ecosystem processes is based on both its physicochemical and its biological characteristics. Nearly everyone can think of a specific example where parent materials play a major role in ecosystem structure and function. The move from a clay to a sandy substratum, the effect of soil pH and that of drainage and aeration are especially noticeable in plant and soil animal components. The differential distribution of plant species is the characteristic that is most noticeable (Montagnini *et al.* 1986; Inouye *et al.* 1987).

It was long thought that soil characteristics such as the composition of parent materials, particle size, aggregation and bulk density together with chemical attributes such as pH, buffer capacity and the ability to absorb organic and inorganic constitutents were stable attributes. We now realize that all soils are sensitive to both long- and short-term perturbations that can cause major changes in ecosystem processess (Buyanovsky, Kuceru & Wagner 1987). Ecosystem–anthropogenic processes have over long periods of time generally (and over very short periods of time in a few cases) completely changed the characteristics of the soil and its ability to support plant growth under either native or agricultural systems. Examples include the extensive areas of the cerrado in Brazil where extensive weathering has caused extreme aluminium toxicity and phosphorus deficiency (Lathwell & Grove 1986). Examples of anthropogenic alterations include erosion and nutrient loss after cultivation or clear-cutting of forest (Vitousek & Matson 1985), the effects of acid rain, contamination with heavy metals, or, in the case of agricultural soils, amelioration with phosphate and lime. Man has an inordinate capability to alter the soil resource, and must realize that changing the steady state that took hundreds and even thousands of years to establish can lead to many unpredictable effects (Waring 1989).

The structure and nature of soil organic matter represents a long-term record of previous ecosystem production, and plant and animal community composition; it is also a continuing controller of the processes occurring within that ecosystem. At least 50% of the carbon dynamics and the vast majority of the other nutrient transformations occur underground (McGill & Cole 1981). The nutrient transformations in turn control plant growth directly, and many interactions between plants and animals and micro-organisms indirectly. Yet the soil was recently described as the only major 'black box' left in ecology (J. E. Cantlon 1988, private communication). There are divergent views concerning the need to unravel the complexities within this box. There is great diversity within the soil populations. Usually more than one type of organism is capable of carrying out a specific reaction. For example, there are many different cellulases. Is it important that we identify each class of enzyme or can we study the overall process? Can we look at the soil as somewhat akin to a complex biochemical entity where the overall processes are most important?

The general relationships governing plant decomposition are known and fairly well described mathematically (Paul & Van Veen 1978; Parton *et al.* 1987). There are a number of major environmental issues that will require a better knowledge of individual-organism controlled processes. These include the following: (i) degradation of pesticides, (ii) controls on organic matter and plant residue mineralization, (iii) nitrification and nitrogen leaching, (iv) a broad range of biological alternatives to the chemicals now used in weed and other pesticide management practices, and (v) the use of genetically engineered organisms in nature. These investigations have become much more feasible with the availability of rapid automated tracer techniques, powerful computers for modelling, high-resolution laser microscopes with image processing, automated organic analyses and molecular genetic techniques such as the gene probe and RNA sequencing.

The micro-organisms as a group dominate the underground biomass both in the size of the nutrient pools held therein and in the range of processes they carry out (Paul & Clark 1988). Feeding on these populations by the soil animals occurs in nearly all ecosystems. However, the significance of the animals varies from system to system. In some systems the mixing effect is extremely important (Coleman, Reid & Cole 1983). In other systems the animals play a major role in decomposition (Clarholm 1985; Hunt *et al.* 1987). In some, though animals are present and can benefit from the organic matter and microbial biomass as a source of food, they may not play a major role in controlling the nutrient cycling.

The products of biological activity are transformed by physical-chemical reactions to form the more stable soil organic matter (SOM) compounds through the processes collectively described as humification (Aiken *et al.* 1985). A better understanding of the SOM dynamics with more exact numbers is required for an improved understanding of past processes, and more importantly for the ability to manage these processes for greater ecosystem productivity and/or stability.

Carbon, nitrogen, phosphorus and sulphur are the major elements in organic matter transformations (McGill & Cole 1981). It has been recognized that the stoichiometry of the C, N and P observed by the Redfield in plankton can be linked to the protoplasmic contents of all life (Reiners 1986). Because soil phosphorus must be supplied by the weathering of minerals, or by its concentration in the surface through root feeding at greater depths, it is said to control the inputs of C and N into soils (Walker & Syers 1976; Tate & Salcedo 1988). The known controls of phosphorus on biological N_2 fixation (Hardy, Lamborg & Paul 1983) further emphasize the significance of this element. In the long term accumulation of bioactive phosphorus may be of the most significance to ecosystem functioning, but in the short term the many and varied transformations of nitrogen show the greatest changes (Parton *et al.* 1987). Accurate measurements of these transformations therefore are most important criteria in the advance toward a more exact ecology.

In this review I will concentrate on the carbon and nitrogen interactions through the soil organic matter and the microbial biomass. Anthropogenic effects on microbial populations are of great significance to many ecosystems, and the final section is concerned with measuring and understanding these changes.

STORAGE OF CARBON AND NITROGEN

The need for better information relative to the functioning of the global carbon cycle and ecosystem dynamics has lead to a series of intensive studies to establish good base-line data for soil organic matter contents. Information on carbon and nitrogen storage for 3600 profiles has been compiled by Zinke *et al.* (1984). These represent values where carbon and nitrogen contents have been measured to at least one metre depth, and where soil bulk density values are known so that calculations on an area basis can be made (Table 15.1).

It may seem somewhat strange to talk about global measurements in a review designed to lead toward a more exact ecology. I, however, have found that individual scientists are so enamoured by the relative numbers within their own system that they forget to ask the obvious

TABLE 15.1. Estimated amounts of carbon (kg m^{-3}) and nitrogen (g m^{-3}) in the soil profile to 1 m depth in various climatic zones (from Zinke *et al.* 1984)

Climatic zone	C	N	C:N
Dry tundra	3.1	500	6
Moist tundra	10.9	638	17
Wet tundra	21	1250	17
Rain tundra	37	1200	17
Boreal dry bush	10.2	630	16
Boreal wet forest	14.9	980	15
Boreal rain forest	32.0	1500	22
Cool temperate desert	9.9	500	20
Cool temperate steppe	13.3	1030	13
Cool temperate wet forest	12.0	626	19
Cool temperate rain forest	20.3	800	25
Warm temperate desert	1.4	106	13
Warm temperate moist forest	9.3	650	14
Warm temperate wet forest	27.0	1800	14
Subtropical desert bush	5.4	380	14
Subtropical moist bush	9.2	990	9
Tropical desert	1.0	50	20
Tropical dry forest	10.2	885	11
Tropical moist forest	11.4	803	14
Tropical wet forest	15.0	655	23
Tropical rain forest	18.0	600	30

questions, 'Do the numbers make any sense?' and 'What have other people found?' These questions are of most importance when making comparisons on a geographical scale between ecosystems differing in temperature and moisture supply.

The interactive effects of moisture and temperature on primary production and system respiration must be considered together with the effects of different inputs to the substratum, e.g. litter or roots (Aber & Melillo 1980; Berg 1986). The effects of the nature of the substratum on decomposition rates and humification processes, and the role of anaerobiosis, also have to be taken into account (Richards 1987). Table 15.1 shows that increased moisture results in higher carbon accumulation in each of the temperature regimes. It also shows that, although the tropical rain forest does not have the accumulation of C that is found in the temperate forests, there still is a very significant concentration of organic materials potentially available, especially when the deeper soil layers of these productive forests are taken into account. The N storage levels do not rise as much with moisture as those of C; thus, there is a general tendency for a rise in C : N quotient in more moist conditions. Whether this results from a reduction in decomposition of carbonaceous compounds, from an adaptation by the growing vegetation to lower N contents, from lower rates of N_2-fixation or possibly from higher rates of denitrification in the wetter systems has yet to be determined.

Lavelle (1984) investigated the effect of soil type on organic matter contents. However, organic matter accumulation in soils is so interwoven with moisture, temperature and, to some extent, the parent material that it proved difficult to differentiate the specific profile effects. Nevertheless, the analysis by Zinke *et al.* (1984) of 3600 profiles did show differences due to parent materials. The majority of soils, i.e. those derived from acid intrusive, ultra-basic and metamorphic parent materials, had mean carbon concentrations of 10–12 kg m^{-3}. The soils on basic extrusive materials (volcanic soils) known to contain minerals which have stabilizing effects on soil organic matter as well as having high inherent fertility showed an average of 22 kg m^{-3}. As was the case for all other studies, the diversity in the measurements for the various parent materials was high, showing that a great number of other factors also control soil organic matter content.

THE TURNOVER OF SOIL ORGANIC MATTER

It is now recognized that the majority of soil organic matter exists for extended periods with as much as 50% being at least 1000 years old (Stout, Goh & Rafter 1981; Van Veen & Paul 1981). The concept that a portion of the soil organic matter can be of physical importance in soil structure and erosion resistance, and yet be only very slowly decomposed has led to more exact numbers in nutrient turnover. Carbon dating has defined these old recalcitrant fractions. A second general fraction of organic matter, while not being as chemically resistant, is protected by inclusion within aggregates and by organo-mineral interactions (Jenkinson & Rayner 1977; Anderson & Paul 1984). This fraction persists for decades to as much as centuries. It is often depicted in soil organic matter models as 'protected' or 'stabilized' (Paul & Van Veen 1978; Parton, Stewart & Cole 1988).

The most important fraction, from a biological viewpoint, is that identified as the active or labile fraction. It is comprised of the microbial biomass itself, microbial detritus (especially cell walls), resistant plant materials and cytoplasmic constituents stabilized by adsorption to surfaces. Because of their adsorption to surfaces many otherwise decomposable organic compounds persist for periods ranging from years to decades. The chemical diversity of these labile components ensures that no one chemical or physical fractionation can separate them from the rest of the soil milieu. These components can, however, be operationally defined, and can be estimated using long-term incubation or calculations based on isotopic dilution. Isotope dilution experiments with ^{15}N can determine that proportion of soil N that is as equally available as the ^{15}N of constituents that have been allowed to equilibrate in the soil for periods long enough to enter the measurable fractions (Paul & Juma 1981).

The most common method of estimation of the size and mineralization kinetics of the labile pools is to incubate the soil in the laboratory for extended periods (up to 42 weeks) under conditions where moisture and temperature and even aggregate structure have been equalized (Jones, Ratcliff & Dyke 1982). The mineralized N is usually transformed to nitrate, and can be leached from the soil system. The maximum accumulation of mineral N, designated as N_m, is an estimate of the mineralizable N. The decomposition rate constants of individual components of this pool can be determined by computer analysis of the shape of the mineralization curves for either C (Freytag 1986, 1987) or N (Smith *et al.* 1980; Juma, Paul & Mary 1984).

Table 15.2 shows that the spodosols have the highest average organic matter concentration. These soils originated under the moist boreal and subtropical conditions shown by Zinke *et al.* (1984) to be those under which large concentrations of organic matter accumulate. These are followed by the mollisols of the dry temperate and subtropical grasslands, while the aridisols of dry, temperate and subtropical regions have the lowest organic C concentration. These findings are in agreement with those of Zinke *et al.* (1984), reached on the basis of an ecosystem classification. However, the mineral N accumulated under long periods of incubation (N_m) is not a consistent portion of the total N. Twenty-two per cent of the total N of ultisols — the soils from moist, subtropical and tropical regions — is in the labile pool, whereas only 10–11% of the N of mollisols and spodosols representing temperate climates is made available during this incubation period. These differences greatly affect ecosystem functioning.

A more exact ecology requires that a reasonable estimate of the fluxes of soil organic matter on an annual basis be available. The data from Zinke *et al.* (1984) can be utilized to determine the global N content of soils, i.e. 105×10^{15}g (Petagram, Pg). Reasonably good estimates of

TABLE 15.2. Characteristics of different types of cultivated soil, and the mineralization of nitrogen in them (recalculated from Jones, Ratcliff & Dyke 1982)

Soil type/climate	Organic carbon (mg g^{-1})	Total nitrogen (mg g^{-1})	C : N quotient	Mineral nitrogen accumulated ($\mu g\ g^{-1}$)	Mineral nitrogen accumlated as % of total nitrogen
Aridisols (dry, temperate–subtropical)	6.0	0.8	7.5	112	15
Ultisols (moist, subtropical–tropical)	6.7	0.5	13	114	22
Alfisols (moist, temperate–subtropical)	10	1.0	10	171	17
Mollisols (dry, temperate–subtropical)	19	1.7	11	170	10
Spodosols (moist, boreal–subtropical)	33	2.6	13	270	11

global C fluxes now exist (Whittaker & Likens 1973; Bolin *et al.* 1979; King, DeAngelis & Post 1987) and it is known that plant growth is fastest when N is in stoichiometric balance with other elements (Waring 1989). However, as far as I know, there is no published estimate of the global N mineralization capacity of soils. This could theoretically be calculated by summation of individual site data such as those brought together by Smith & Paul (1989), who calculated N mineralization rates on the basis of first-order kinetic analyses and environmental parameters. This approach yielded a range of values from a low of 26 kg ha^{-1} $year^{-1}$ for a eucalypt forest to a high of 1095 kg ha^{-1} $year^{-1}$ in an arable soil. Such incubation data, whether obtained in the laboratory or under field conditions, cannot take into account either the effects of plant uptake or the counteracting fluxes in mineralization-immobilization turnover. The alternative is to relate mineralization of N to uptake of N by plants, which in turn is related to dry matter production by plants.

The series on 'State of the Art Research on Global Carbon Dioxide' (Olson, Watts & Allison 1983; Zinke *et al.* 1984; Farell 1987) has produced a comprehensive compendium on pool sizes and fluxes involved in the global CO_2 cycles. The authors have developed site-specific as well as global models. Olson, Watts & Allison (1983), on the basis of a comprehensive summary of global primary production, estimated C production by plants to be 60 Pg per annum. This is only slightly higher than the estimates of Whittaker & Likens (1973). The transformation of data on plant primary production to plant N uptake requires a knowledge of plant N contents and of retranslocation of N within the plant. Annual plants such as wheat (Campbell, Nicholachuk & Warder 1977), because of root turnover and plant-N leaching during growth, take up more N than is found at harvest and can require an equivalent of 7% of their net primary C productivity as N uptake. Trees are known to retranslocate a considerable portion of their N content each year (Mellilo 1981) and there is similar evidence for perennial grasses (Clark 1977). While these translocations may lead to an overestimate of uptake by plants, there is a balancing tendency to underestimate uptake as a result of lack of information on rates of fine root turnover. This is known to be high in the few tree species that have been studied critically (Vogt, Grier & Vogt 1986). A stand of Douglas fir (*Pseudotsuga menziesii*) was found by Turner (1977) to redistribute 45% of its N and had a net N uptake requirement equal to 1% of C production.

Based on a value for N uptake of 1% of the net primary productivity of C in boreal forests, 2% for tropical forests, 1.5% for temperate forests, and 5% for grasslands, cultivated crops and tundra, one can use data such as those of Whittaker & Likens (1973) to calculate the global plant uptake of N. This is shown in Table 15.3 to be 1.1 Pg. The errors in such an estimate would come from uncertainties in the measurements of plant productivity and from the unknowns in the relative amounts of C and N accumulated. The latter would cause the greater error. While the estimate of 1.1 Pg probably has error levels of 25 to 50%, it provides a meaningful base for comparisons regarding other N fluxes. I hope it will also lead to more exact measurements with the aim of improving these preliminary calculations.

The global soil mineralization values can be estimated from the plant uptake values if the efficiency of soil N uptake is known. This number can be obtained from ^{15}N experiments. Numerous ^{15}N studies have shown that the proportion of added fertilizer N that is taken up by plants ranges from 0.1 to 0.7. The lower values come from forest studies (Heilman *et al.* 1982; Vitousek & Matson 1985) where there is a great deal of N immobilization. The higher values are from very highly fertilized agricultural fields with little N loss and negligible immobilization. The modal value for most field studies approximates 0.44.

TABLE 15.3. Estimates of total soil nitrogen and various fluxes of soil nitrogen for the earth (all in Pg)[a]

Soil N content	105[b]
Internal transformations	
Plant uptake of N	1.1[a]
Soil N mineralized	2.6[c]
Inputs	
Symbiotic fixation	0.12[d]
Asymbiotic fixation	0.060[d]
Fertilizer N applied	0.065[d]
Fertilizer N utilized	0.026[c]
Combustion and atmospheric	0.022[d]
Exports	
Denitrification	0.14[d]
Runoff and erosion	0.025[d]

[a]Pg = petagrams = 10^{15}g = 1 Gt.
[b]Zinke *et al.* (1984).
[c]Based on plant primary production, plant N translocation and efficiency of N uptake (see text).
[d]Knowles (1982).

Dividing the plant N uptake of 1.1 Pg by 0.44 results in an estimation of 2.5 Pg of soil N mineralized on a global basis annually. While the range in estimates has to be extensive for such a diversity of data, the data are very valuable in that they place the amount of soil N mineralization in perspective relative to the amounts of N_2 fixation, denitrification and N losses by erosion.

Biological N_2 fixation at 0.15 to 0.18 Pg (Knowles 1982) is small compared with the 2.6 Pg of soil N mineralized annually. However, the symbiotically fixed N can be assumed to be much more available to the plant, despite the fact that leakage and nodule turnover do occur. The N_2 fixed represents 14–16% of the 1.1 Pg of plant N uptake. The fertilizer N application of 0·065 Pg is subject to the partial uptake limitation discussed above, and, on a global basis, fertilizer accounts for only 2% of the global plant N uptake. Environmentalists who often point to fertilizers as major contributors to atmospheric pollution usually forget that soil N also is subject to low plant uptake, to losses, and to re-immobilization by soil micro-organisms as emphasized by Coleman, Reid & Cole (1983) and Vitousek & Matson (1985).

The values for denitrification and the other losses of N from the soil system have been calculated by techniques different from those used for N_2 fixation, yet the estimate of 0.14 Pg for denitrification is only slightly lower than that of 0.18 Pg for symbiotic and asymbiotic fixation combined. Runoff and erosion has been estimated to be 0.025 Pg, approximately equal to that of the input of N by combustion and atmospheric precipitation.

MICROBIAL BIOMASS

Reproducible, reasonably rapid techniques for the measurement of the microbial biomass have been available for approximately 10 years, and estimates for microbial biomass in many ecosystems are now available. These allowed Smith & Paul (1989) to relate the size of the microbial biomass and its turnover rate to C inputs (Table 15.4). There is a good relationship between soil C and N content, but none, on a global basis, between litter input and soil organic C or N. Similarly, the size of the microbial biomass depends on many factors and, in the analysis of Smith & Paul (1989), was shown not to be correlated with organic matter content. The scarcity of measurements in some of the major vegetation types precludes accurate comparisons of C to N ratios within the microbial biomass. If the C: N ratios are truly as divergent as the ranges shown in this table, they would be a major factor to be considered in studies on ecosystem functioning.

Field measurements of turnover rates of micro-organisms are not yet available. Microbial turnover therefore has been calculated on the basis of a model that relates microbial growth to available energy supplies (Smith & Paul 1989). The calculated microbial turnover time of 0.07 years in tropical forests shows that a high nutrient input together with optimum growing conditions can result in turnover times faster than have been previously published. However, turnover rates of 14 times per year are still very slow when compared with laboratory growth rates. The similar estimates for turnover time in the microbial biomass of temperate grasslands and forests (0.32 amd 0.30 years) are intriguing. It will be interesting to see whether these estimates continue to be found to be similar as more data become available.

Soils should be the best overall reflection of ecosystem processes. Detailed examination of the 3600 soil profiles in the analyses of Zinke *et al.* (1984) should yield good estimates of variability between and within major life-zones. The variability in samples differed with vegetation type. Forested areas showed the greatest variability, twice that of the mean. Semi-desert and grassland were the least variable.

The most exact numbers, if not always the best interpretations, are obtained from single sites, each with a simple vegetation type on a single soil type. Nevertheless the variability within one agricultural field is

TABLE 15.4. Plant biomass, input of litter to soil, soil organic matter and microbial biomass of major global vegetation types

	Tropical forest	Temperate forest	Boreal forest	Savannah	Temperate grassland	Tundra
Area (10^{12} m^2)	24.5	12.5	12.0	15	9	8
Carbon in plant biomass (kg m^{-2})	18	14	9	1.8	1.4	0.25
Carbon in above-ground litter fall (kg m^{-2})	1.71	0.37	0.25	0.36	0.67	0.075
Soil carbon (kg m^{-2})	13	9	15	5.4	23	22
Soil nitrogen (kg m^{-2})	0.82	0.64	1.1	0.33	2.1	1.1
Carbon in microbial biomass (g m^{-2})	50	110	35	60	215	20
Nitrogen in microbial biomass (g m^{-2})	2	14	2.5	8.7	51	1
Microbial turnover (years)	0.07	0.30	0.14	0.17	0.32	0.27

Plant biomass and net primary productivity from Whittaker & Likens (1973). Litter from Atjay, Ketney & Duvigneaud (1979). Soil C and N from Zinke *et al.* (1984). Microbial biomass from Smith & Paul (1989).

often as great as that obtained across a transect of a climatic zone. Differences in slope, aspect and drainage cause differences in major soil type, plant growth and microbial content and activity. Not as easily recognized are the differences found in areas that, although relatively uniform in slope, have differences in parent material and internal drainage.

Geostatistic techniques are making it possible to relate adequately parameters such as plant community structure and nutrient availability to spatial patterns (Montagini *et al.* 1986; Inouye *et al.* 1987). Robertson *et al.* (1988) measured spatial characteristics of N mineralization, nitrification and denitrification at a resolution of 1 m over 0.5 ha of an old field, and showed a high degree of spatial dependence among points sampled within 1 to 40 m of one another. Most of the variation within the sample population was attributed to spatial auto-correlation at a scale > 1 m.

The grassland component of an oak–annual grass savannah on an alfisolic soil in the mediterranean-type climate of central California (Table 15.5) gives an example of the relative sizes of N pools and fluxes found on one site. The site, in the year of analysis, produced 400 g m^{-2} $year^{-1}$ litter C in grasses and 7 g litter N. Although other studies were con-

TABLE 15.5. Carbon and nitrogen transfers in an area occupied by annual grasses within an oak savannah in California; data from Brooks, Jackson & Paul (1986) and Jackson, Strauss & Firestone (1985)

Litter C (g m^{-2} $year^{-1}$)	400 ± 60[a]
Litter N (g m^{-2} $year^{-1}$)	7 ± 1
Microbial biomass C in top 10 cm (g m^{-2})	78 ± 8
Microbial biomass N in top 10 cm (g m^{-2})	19 ± 3
Microbial turnover (years)	0.21
^{15}N uptake	
Plants (%)	29
Soil (%)	45
Unaccounted for (%)	26
Microbial (%)	24
N mineralized in 100 days	3.5
Biomass N/soil N (%)	9.3

[a]SE (n = 5).

ducted on the tree–grass interactions (Jackson, Strauss & Firestone 1985), the data in Table 15.5 are for areas that were not under a tree canopy. The microbial biomass represented 78 g C and 19 g of N. The competitive demand by the microbial biomass is illustrated by the fact that plant uptake accounted for 29% of the added ^{15}N, whereas the microbial biomass accounted for 24% of the N added at a time when the plants were at their most rapid rate of growth. A significant amount of N moved through the microbial biomass into the soil, i.e. 21%, so that the microbial biomass plus soil organic matter constituted the largest pool at 45% of the ^{15}N added. The 24% of the ^{15}N not accounted for was attributed to leaching, denitrification and some herbivory by small animals.

The above figures for a semi-natural system are not greatly different from the averages shown in Table 15.4 for plant and microbial uptake and soil residual N and losses on a much broader geographic basis. However, the uncultivated alfisolic system showed a mineralization rate during a 100-day incubation that accounted for only 3.5% of the total soil N. The cultivated alfisols shown in Table 15.2 mineralized 17% of their N. The microbial biomass of the 0–10 cm depth accounted for 9% of the total N in this fraction, and the turnover time of 0.21 years for the microbial C is supported by the fact that only 25% of the ^{15}N in the microbial biomass accumulated as mineral N in a subsequent 100-day laboratory incubation. Re-immobilization of the N in the presence of available C in this semi-natural system and the great stability of the microbial population are therefore major ecosystem controllers in this savannah.

Table 15.5 shows that the amount of microbial N in the top 10 cm of soil is more than twice the amount in a year's litter fall. The biomass can be either a source or a sink of nutrients. It represents 9% of the N in the surface 10 cm. The range for most values is 5–15% of the total (Smith & Paul 1989). Much of the ^{15}N added to soils remains in the biomass for extended periods after immobilization (Paul & Juma 1981; Schimel, Coleman & Horton 1985). This competition for available N by the microbes and its incorporation into soil organic matter is the reason for the low plant uptake values when ^{15}N is utilized in uptake studies. During immobilization of the ^{15}N there is, however, concurrent release of ^{14}N by microbial turnover. The actual amount of N accumulated by plants comprises both ^{15}N and the released ^{14}N, and thus is greater than that calculated from the ^{15}N per cent excess in the plant. The difference is attributed to the mineralization–immobilization turnover, and further study of this process (Jansson & Persson 1982) will lead to a more exact knowledge of the nutrient dynamics in ecosystems.

THE FORMATION OF SOIL ORGANIC MATTER

A considerable proportion (30–50%) of the organic N of most soils is present in forms which cannot be identified by present-day techniques (Stevenson 1982); these forms are thought to consist of amino N in organo-metal complexes, and N in ring structures as well as in pyrolidines and pyridines (Schnitzer 1982). The latter compounds do not occur to a significant extent in plant residues, but similar forms are found throughout the world on terrestrial sites as well as in sediments. The present theories on the formation of soil organic matter involve the condensation of amino acids, sugars and phenolic substrates to produce dark-coloured, recalcitrant humates high in N (Liu *et al.* 1985; Paul & Clark 1988). The phenolic content can vary with soil type, and NMR studies have shown that the aromatic constituents are not present in as high concentrations as was previously thought (Preston *et al.* 1982; Aiken *et al.* 1985; Wershaw 1985).

The type and extent of decomposition controls both the eventual accumulation and the mineralization of organic matter and nutrients within ecosystems and in different agricultural toposequences (Schimel, Coleman & Horton 1985). People are now asking whether new agricultural managment systems, such as sustainable agriculture, can alter the size of the labile fraction. Continuing research shows that the very large size and diversity of the soil microbial population can supply the needed organisms and enzymes, and that inoculation of foreign organisms into soil seldom has any long-term effect. Development of models that allow the prediction of litter decay rates and nutrient dynamics have shown that the effects of climate and the chemical composition of macro-litter can be reasonably predicted (Ågren & Bosatta 1987; Parton *et al.* 1987) but internal cycling is still largely looked upon as a 'black box'. The enzyme activity and the ratios of the constituents involved are also being delineated.

Cellulose, a major component of plant litter, disappears more rapidly than lignin during early stages of decay, and overall decay rates are generally higher in litter that has a low initial lignin concentration or a high initial N concentration (Aber & Melillo 1980). Work on kinetics of organic matter has led to a reassessment of the role of lignin in soil organic matter formation. We know that polyphenols, which are degradation products of lignin, can lead to humate formation. Microbial products produced from other substrates can, however, also be of major importance. Voroney, Paul & Anderson (1989), in studying the relative rates of straw and glucose decomposition under field conditions, found that after seven years in the field ^{14}C originally added as glucose was present in higher concentrations and had slower decomposition rates than ^{14}C added as straw. This can be explained by the fact that the addition of the highly available substrate led to the rapid production of a large microbial biomass which on its death and decay resulted in the stabilization of a higher concentration of ^{14}C than did the growth on the straw ^{14}C.

Degradation of lignin ^{14}C by micro-organisms has been found to result in a negligible uptake of the ^{14}C into the microbial bodies themselves (Scott *et al.* 1983). A large proportion of ^{14}C remains in the soil. Whether this remains as humified constituents or as partial degradation products which cannot be separated from the humified constituents by present-day techniques is not known. Other questions concerning the role of lignin in soil organic matter stabilization have come from analyses using ^{18}O (Dunbar & Wilson 1983). The ^{18}O content of lignin is different from that of cellulose. The ^{18}O content of soil humic acids was found to be similar to that of cellulose, but not to that of lignin, indicating that the major source of oxygen in the organic matter may be cellulose rather than lignin.

ANTHROPOGENIC EFFECTS ON THE SOIL ENVIRONMENT

Anthropogenic chemicals such as toxic wastes, sewage sludge, acid rain, pesticides and fertilizers have the capacity to influence the soil ecosystem and associated ground waters even at some distance from points of addition. Specific impurities can be monitored in the soil and in the ground water with great sensitivity. However, we have little knowledge of the effects of long-term exposures of very low concentrations or the interactive effects of different compounds. More exact numbers in this regard are essential. There is a tendency to rely on large safety factors to ensure the protection of both humans and the environment from poorly defined risks, especially where the effects of breakdown products are not known. More measurements as well as mathematical models of the fate and persistence of chemicals are needed to assist in development of monitoring, movement and management strategies. The precision and accuracy of the models describing these reactions need to be defined for a much broader range of ecosystems. We cannot measure these factors empirically in each system, and must be able to rely with confidence on predictive models.

The problems associated with movement and toxicity of chemicals have resulted in an increased interest in the development of biological control agents. The techniques of molecular biology were developed with micro-organisms, and many of the genes to be inserted in plants will come from micro-organisms or will be introduced via microbial vectors (Moses 1987). It is considered that alteration of microbial populations and activities in the rhizosphere can contribute the following: (i) modification of plant growth and development, (ii) control of pathogens, insects, nematodes or weeds, (iii) enhanced nutrient availability, and (iv) degradation of toxic substances. The role of native and of genetically altered organisms, including bacteria, fungi, protozoa and viruses, is now being defined in a number of areas. This work is needed to provide the background information required for further advances in this field.

The release of genetically engineered organisms into the environment will continue to depend upon the decisions of appropriate regulatory agencies and society at large. To realize the potential of biological control mechanisms, and at the same time control the potential risks associated with the introduction of new organisms into nature, much more information is required on the factors that control their population dynamics and their function in the soil and rhizosphere. Specific information is required on the following: (i) development of methodology to measure the potential effects of engineered organisms on other susceptible life, (ii) the ability of genetically altered organisms to grow and survive under field conditions, (iii) the effects of the new organisms on natural environmental processes, and (iv) the ability of introduced organisms to transfer genetic information to native strains in the environment.

Because of the above uncertainties, the modification of plant growth and development through the use of inserted genes will probably occur earlier than the introduction of genetically engineered micro-organisms. Genes inserted for resistance to specific herbicides are being tested by a number of agencies. A gene for resistance to a herbicide that is reasonably safe environmentally, such as glyphosphate, can now be engineered into crop plants. This could allow the control of all other plants in the field by the application of one, supposedly innocuous, herbicide at very low levels. The possibility of a gene such as this moving from a plant to non-target plant is considered more remote than if a micro-organism had been used as a host. Another example of alteration of plants is the incorporation directly into plants of the broad range of proteinaceous toxins from *Bacillus thuringiensis* (Board on Agriculture 1987). These are activated only after ingestion into insect guts, but questions concerning the safety of the plant for human consumption and the effect on the ecology of non-target crops and soil-associated animals still remain to be answered.

Enhancement of nutrient availability through molecular genetics can readily be imagined in the case of increased symbiotic N_2 fixation or altered phosphatase levels on roots. A less direct approach but one still with great potential impact is the possibility of managing the vast store of nutrients in the microbial biomass. This would require a proper synchronization of microbial growth with plant growth such that nutrients are mineralized during periods of active plant growth and immobilized when plant uptake is not occurring. The active fraction of the soil organic matter (SOM) consists of the microbial biomass and microbial metabolites. These account for the majority of nutrients released and offer some hope for management. Cultivation is one such management tool as it provides new surfaces for microbial activity, and incorporates plant residues. The older fractions of SOM will continue to degrade slowly when abiotic factors make microbial growth feasible. Nutrients released during this time could be taken up by cover crops which would later be incorporated at the time of planting of the major crop. This scenario already occurs to some extent in agro-ecosystems through timing of application of fertilizers and plant residues as well as cultivation. The continued protection of soil by interseeded and cover crops is a major aspect of many of the sustainable agricultural techniques now being tested.

The potential for tinkering with the environment is immense. We, however, do not have the necessary background in basic microbial physiology, plant biology or ecological interactions to be able to predict adequately the outcome of anthropogenic influences. There are also many ethical, economic and social questions that need to be answered. The basic questions in microbial ecology that need to be answered before widespread introduction of genetically engineered micro-organisms into the environment include the following.

1 What factors control competitiveness in nature? Organisms cannot act in nature unless they find a niche for themselves. It is an axiom of microbial ecology that foreign organisms are very difficult to introduce unless given a special advantage (Paul & Clark 1988). Symbiotic N_2 fixers will establish themselves if the host legume is grown, but many years of inoculation studies have shown the difficulty in trying to introduce more efficient N_2-fixing symbionts if they are not more competitive in the soil itself or in the multifaceted absorption–infection process (Havelka, Boyle & Hardy 1982). Conversely, in genetic engineering for pest or weed control the need for an introduced organism to do its job and then die is a prerequisite, not only from an environmental standpoint but also from an economic one. The company introducing the organism wants to be able to sell the inoculum year after year, thus requiring microbial die-back. Suicidal genes for these organisms are thus being investigated.

2 What is the diversity of the soil microbial community? Only a very small percentage (0.1–10%) of the microbial populations identified in soils and sediments can be grown on laboratory media (Klug & Reddy 1984). Does this mean that there is a great deal of untapped diversity in the soil population that may be easier to address for scientific purposes than new organisms that must be produced by modifying known ones? Novel organisms isolated from one environment to be used in another would probably pass regulatory examinations more easily than genetically engineered organisms. It is, however, possible that we already know much of the diversity in soil organisms but that the environmental 'stress' under which soil organisms normally grow keeps them from developing under laboratory conditions.

Techniques of molecular genetics have made available to us the analytical methodology required to answer the diversity question. The use of gene probe methodology (Holben & Tiedje 1988) and the analysis of the higher-order structure of ribosomal RNA sequences make it possible to analyse phylogenetic and quantitative aspects of naturally occurring mixed microbial populations (Pace *et al.* 1986). The need to understand the microbial ecology, diversity and functioning of organisms in nature before genetically engineered organisms are introduced into the environment will also provide some of the funds necessary for the basic studies, now that tools for these studies are available.

3 What is the effect of the physico-chemical environment on function?

The effects of 'stress' on microbial activity and morphology is starting to be understood (Atlas 1984). Other relevant effects include the known controls played by interaction with the extensive organic and inorganic surfaces in nature (Fletcher & McEldowney 1984). The inability to isolate vesicular-arbuscular mycorrhiza, the study of the role of fungistasis in plant diseases, and the role of anaerobiosis in decomposition are examples of problems where biochemical activity cannot be studied until we obtain a more quantitative understanding of the interaction of genetic and environmental controls on microbial activity such that we can grow the organisms adequately in the laboratory.

4 What is the accessibility of the gene pool of soil organisms? What we know about gene exchange has come from laboratory studies with a restricted list of bacteria such as *Agrobacterium, Pseudomonas* and, to some extent, *Rhizobium* and *Bradyrhizobium*. Bacteriophages are known to be able to transfer genetic information, but their mechanism of crossing barriers between bacterial species is not known. The ability of bacteria to overcome substances such as antibiotics, heavy metals, pesticides and anthropogenic wastes is often carried on plasmid genes (Barkay & Olson 1986). The movement of these genes in nature and the effects of the extra plasmids on the host micro-organisms in the competitive natural environment must be studied. Most of the genes controlling the bacterial contribution to symbiotic N_2 fixation are also plasmid-borne. The fates of plasmids in natural ecosystems, therefore, are of great importance to the ecology of both managed and native systems.

It is of paramount importance that the basic questions in microbial ecology be answered before there is a widespread introduction of new microbial genetic material into nature. The analytical power associated with genetic research makes much of this research feasible (Berger & Kimmel 1987). The recognition that research centres, such as one in microbial ecology, are required to develop the background knowledge required before genes are introduced into nature now makes exact knowledge in this field of ecology possible.

Microbial ecology and 'macro-ecology' have tended to operate in separate fields, and the concepts developed for 'macro-ecology' have not played a major role in microbial ecology (Check 1988). At the same time the huge population sizes, and now the funding in microbial ecology, should make it possible to utilize this field to answer many questions in 'macro-ecology'. This can come about only through the application of a more exact biology, using the power of the new analytical techniques and the capabilities of mathematical analyses available to us now and in the near future.

REFERENCES

Aber, J. D. & Melillo J. M. (1980). Litter decomposition: measuring relative contributions of organic matter and nitrogen to forest soils. *Canadian Journal of Botany,* **58**, 416–21.

Ågren, G. A. & Bosatta, E. (1987). Theoretical analysis of the long term dynamics of carbon and nitrogen in soils. *Ecology,* **68**, 1181–4.

Aiken, G. R., McKnight, D. M., Wershaw, R. L. & MacCarthy, P. (1985). *Humic Substances in Soil, Sediment and Water.* Wiley, New York.

Anderson, D. W. & Paul, E. A. (1984). Organo-mineral complexes and study by radiocarbon dating. *Soil Science Society of America Journal,* **48**, 298–301.

Atjay, G. L., Ketney, P. & Duvigneaud, P. (1979). Terrestrial primary production and phytomass. *The Global Carbon Cycle* (Ed. by B. Bolin, E. Degens, S. Kempe & D. P. Ketinnea), pp. 129–81. Wiley, New York.

Atlas, R. M. (1984). Use of microbial diversity to assess environmental stress. *Current Perspectives in Microbial Ecology,* (Ed. by M. J. Klug & C. A. Reddy), pp. 540–5. ASM, Washington, DC.

Barkay, T. & Olson, B. H. (1986). Phenotypic and genotypic adaptation of aerobic heterotrophic sediment bacterial communities to mercury stress. *Applied and Environmental Microbiology,* **52**, 403–6.

Berg, B. (1986). Nutrient release from litter and humus in coniferous forest soils—a mini review. *Scandinavian Journal of Forest Research,* **1**, 359–69.

Berger, S. L. & Kimmel, A. R. (Eds) **(1987).** Guide to molecular cloning techniques. *Methods in Enzymology,* **152.** Academic Press, New York.

Board on Agriculture (1987). *Agricultural Biotechnology: Strategies for National Competitiveness.* National Academy Press, Washington, DC.

Bolin, B., Degens, E. T., Kempe, S. & Ketner, P. (Eds) **(1979).** *The Global Carbon Cycle.* Wiley, New York.

Brooks, P. D., Jackson, L. E. & Paul, E.A. (1986). Grassland nitrogen transformations. *Process Controls and Nitrogen Transformation in Terrestrial Ecosystems 1985–86* (Ed. by J. L. Smith & M. K. Firestone), pp. 79–115. University of California Press, Berkeley.

Buyanovsky, G. A., Kucera, C. L. & Wagner, G. H. (1987). Comparative analysis of carbon dynamics in native and cultivated ecosystems. *Ecology,* **68**, 2023–31.

Campbell, C. A., Nicholachuk, W. & Warder, F. G. (1977). Effects of fertilizer N and soil moisture on growth N control and moisture use of spring wheat. *Canadian Journal of Soil Science,* **57**, 289–310.

Check, W. (1988). The reemergence of microbial ecology. *Mosaic,* **24**, 24–35.

Clarholm, M. (1985). Interaction of bacteria, protozoa and plants leading to mineralization of soil nitrogen. *Soil Biology and Biochemistry,* **17**, 181–7.

Clark, F. E. (1977). Internal cycling of nitrogen in shortgrass prairie. *Ecology,* **58**, 1322–33.

Coleman, D. C., Reid, C. P. P. & Cole, C. V. (1983). Biological strategies in nutrient cycling in soil systems. *Advances in Ecological Research,* **13**, 1–55.

Dunbar, J. & Wilson, A. T. (1983). The origin of oxygen in soil humic substances. *Journal of Soil Science,* **34**, 99–103.

Farell, M. P. (1987). Master index for the carbon dioxide research, state of the art report series. *US Department of Energy Report* DOE/ER0316, 1–253. Washington DC.

Fletcher, M. & McEldowney, S. (1984). Microbial attachment to non biological surfaces. *Current Perspectives in Microbial Ecology* (Ed. by M. J. Klug & C. A. Reddy), pp. 121–4. ASM, Washington, DC.

Freytag, H. E. (1986). Derivation of a general mineralization function for soil organic matter. *Archiv für Acker- und Pflanzenbau und Bodenkunde,* **30**, 201–9.

Freytag, H. E. (1987). Simultaneous determination of the parameters for the carbon mineralization function on the basis of carbon dioxide measurements under constant conditions. *Archiv für Acker- und Pflanzenbau und Bodenkunde*, **31**, 23–32.

Hardy, R. W. F., Lamborg, M. R. & Paul, E. A. (1983). Microbial effects. *CO_2 and Plants: the Response of Plants to Rising Levels of Atmospheric Carbon Dioxide* (Ed. by E.R. Lemon), pp. 131–76. American Association for the Advancement of Science Selected Symposium, 84.

Havelka, U. D., Boyle, M. G. & Hardy, R. W. F. (1982). Biological nitrogen fixation. *Nitrogen in Agricultural Soils* (Ed. by F. J. Stevenson), pp. 365–413. *Agronomy, Madison*, **22**.

Heilman, P. E., Dao, T., Cheng, H. H., Webster, S. R. & Harper, S. S. (1982). Comparison of fall and spring applications of ^{15}N-labelled urea to Douglas-fir: I. Growth response and nitrogen levels in foliage and soil. *Soil Science Society of America Journal*, **46**, 1293–9.

Holben, W. A. & Tiedje, J. M. (1988). Application of nucleic acid hybridization in microbial ecology. *Ecology*, **69**, 561–4.

Hunt, H. M., Coleman, D. C., Ingham, E. R., Ingham, R. E., Elliot, E. T., Moore, J. C., Rose, S. L., Reid, C. P. P. & Morley, C. R. (1987). The detrital food web in a short grass prairie. *Biology and Fertility of Soil*, **3**, 57–68.

Inouye, R. S., Huntly, N. J., Tilman, D. & Tester, J. R. (1987). Pocket gophers (*Geomys bursarius*), vegetation and soil nitrogen along a successional sere in east central Minnesota. *Oecologia*, **72**, 178–84.

Jackson, L. E., Strauss, R. & Firestone, M. K. (1985). Seasonal changes in plant productivity and nitrogen pools in oak savannah. *Process Controls and Nitrogen Transformations in Terrestrial Ecosystems 1984–85* (Ed. by J. L. Smith & E. A. Paul), pp. 1–22. University of California, Berkeley.

Jansson, S. L. & Persson, J. (1982). Mineralization and immobilization of soil nitrogen. *Nitrogen in Agricultural Soils* (Ed. by F. J. Stevenson), pp. 229–52. American Society of Agronomy, Madison, Wisconsin.

Jenkinson, D. S. & Rayner, J. H. (1977). The turnover of soil organic matter in some of the Rothamsted classical experiments. *Soil Science*, **123**, 298–305.

Jones, C. A., Ratcliff, L. F. & Dyke, P. T. (1982). Estimation of potentially mineralizable soil nitrogen from chemical and taxonomic criteria. *Communications in Soil Science and Plant Analysis*, **13**, 75–86.

Juma, N. G., Paul, E. A. & Mary, B. (1984). Kinetic analyses of net nitrogen mineralization in soil. *Soil Science Society of America Journal*, **48**, 753–7.

King, W. W., DeAngelis, D. L. & Post, W. M. (1987). *The Seasonal Changes of Carbon Dioxide between the Atmosphere and the Terrestrial Biosphere: Extrapolation from Site Specific Models to Regional Models.* Oak Ridge National Laboratory Environmental Sciences Division Publication, 2988. Oak Ridge, Tennessee.

Klug, M. J. & Reddy, C. A. (1984). *Current Perspectives in Microbial Ecology*, ASM Press, Washington, DC.

Knowles, R. (1982). Denitrification in soils. *Advances in Agricultural Microbiology* (Ed. by N. S. Subba Rao), pp. 243–66. Butterworths, London.

Lathwell, D. J. & Grove, T. L. (1986). Soil plant relations in the tropics. *Annual Review of Ecology and Systematics*, **17**, 1–16.

Lavelle, P. (1984). The soil system in the humid tropics. *Biologie Internationale*, **9**, 2–19.

Liu, S.-Y., Freyer, A. J., Minard, R. D. & Bollag, J.-M. (1985). Enzyme-catalyzed complex-formation of amino acid esters and phenolic humus constituents. *Soil Science Society of America Journal*, **49**, 337–42.

McGill, W. B. & Cole, C. V. (1981). Comparative aspects of organic C, N, S and P cycling through soil organic matter during pedogenesis. *Geoderma*, **26**, 267–86.

Melillo, J. M. (1981). Nitrogen cycling in deciduous forests. *Terrestrial Nitrogen Cycles*

(Ed. by F. E. Clark & T. Rosswall), pp. 427–42. Ecological Bulletins, 33. Swedish Natural Science Research Council, Stockholm.

Montagnini, F., Haines, B., Boring, L. R. & Swank, W. (1986). Nitrification potentials in early successional black locust and in mixed hardwood forest stands in the southern Appalachians, USA. *Biogeochemistry*, **2**, 197–210.

Moses, P. B. (1987). Gene transfer methods applicable to agricultural organisms. *Agricultural Biotechnology* (Ed. by Board on Agriculture), pp. 149–92. National Academy Press, Washington, DC.

Olson, J. S., Watts, J. A. & Allison, L. J. (1983). *Carbon in Live Vegetation of Major World Ecosystems.* Oak Ridge National Laboratory Environmental Sciences Division Publication, 1997. Oak Ridge, Tennessee.

Pace, N. R., Stahl, D. A., Lane, D. J. & Olsen, G. J. (1986). The analyses of natural microbial populations by ribosomal RNA sequence. *Advances in Microbial Ecology*, **9**, 1–56.

Parton, W. J., Schimel, D. S., Cole, C. V. & Ojima, D. S. (1987). Analysis of factors controlling soil organic matter levels in Great Plains grasslands. *Soil Science Society of America Journal*, **5**, 1173–9.

Parton, W. J., Stewart, J. W. B. & Cole, C. V. (1988). Dynamics of carbon, nitrogen, phosphorus and sulfur in grassland soils—a model. *Biogeochemistry*, **5**, 109–32.

Paul, E. A. & Clark, F. C. (1988). *Soil Microbiology and Biochemistry.* Academic Press, New York.

Paul, E. A. & Juma, N. G. (1981). Mineralization and immobilization of soil nitrogen by plants. *Terrestrial Nitrogen Cycles* (Ed. by F. E. Clark & T. Rosswall), pp. 179–95. Ecological Bulletins, 33. Swedish Natural Science Research Council, Stockholm.

Paul, E. A. & Van Veen, J. A. (1978). The use of tracers to determine the dynamic nature of organic matter. *Transactions of the 11th International Congress of Soil Science*, **3**, 61–102.

Preston, C. M., Rauthan, B. S., Rodger, C. & Ripmeester, J. A. (1982). A hydrogen-1, carbon-13, and nitrogen-15 nuclear magnetic resonance study of *p*-benzoquinone polymers incorporating amino nitrogen compounds ('synthetic humic acids'). *Soil Science*, **134**, 277–93.

Reiners, W. A. (1986). Complementary models for ecosystems. *American Naturalist*, **127**, 59–73.

Richards, B. N. (1987). *The Microbiology of Terrestrial Ecosystems.* Longman, Harlow.

Robertson, G. P., Huson, M. A., Evans, F. C. & Tiedje, J. M. (1988). Spatial variability in a successional plant community: patterns of nitrogen availability. *Ecology*, **69**, 1517–24.

Rosenberg, M. & Kjelleberg, S. (1986). Hydrophobic interactions: role in bacterial adhesion. *Advances in Microbial Ecology*, **9**, 353–94.

Schimel, D. S., Coleman, D. C. & Horton, K. A. (1985). Soil organic matter dynamics in paired rangeland and cropland toposequences in North Dakota. *Geoderma*, **36**, 201–14.

Schnitzer, M. (1982). Quo vadis soil organic matter research. *Transactions 12th International Congress of Soil Science*, **5**. *Whither Soil Research*, 67–78.

Scott, D. E., Kassim, G., Jarrell, W. M., Martin, J. P. & Haider, K. (1983). Stabilization and incorporation into biomass of specific plant carbons during biodegradation in soil. *Plant and Soil*, **70**, 15–26.

Smith, J. L. & Paul, E. A. (1989). The significance of soil microbial biomass estimation in soil. *Soil Biochemistry*, Vol. 6 (Ed. by G. Stotsky & J. M. Bollag) Marcel Dekker, New York.

Smith, J. L., Schnabel, R. R., McNeal, B. L. & Campbell, G. S. (1980). Potential errors in the first order model for estimating soil nitrogen mineralization potentials. *Soil Science*

Society of America Journal, **44**, 996–1000.

Stevenson, F. J. (1982). *Humus Chemistry*. Wiley, New York.

Stout, J. D., Goh, K. M. & Rafter, T. A. (1981). Chemistry and turnover of naturally occurring resistant organic compounds in soil. *Soil Biochemistry*, Vol. 5 (Ed. by E. A. Paul & J. N. Ladd), pp. 1–74. Marcel Dekker, New York.

Tate, K. R. & Salcedo, I. (1988). Phosphorus control of soil organic matter accumulation and cycling. *Biogeochemistry*, **5**, 99–108.

Turner, J. (1977). Effect of nitrogen availability on nitrogen cycling in a Douglas fir stand. *Forest Science*, **23**, 307–16.

Van Veen, J. A. & Paul, E. A. (1981). Organic carbon dynamics in grassland soils. 1. Background information and computer simulation. *Canadian Journal of Soil Science*, **61**, 185–201.

Vitousek, P. M. & Matson, P. A. (1985). Disturbance, nitrogen availability and nitrogen losses in an intensively managed loblolly pine plantation. *Ecology*, **66**, 1360–76.

Vogt, K. C., Grier, C. C. & Vogt, D. J. (1986). Production turnover and nutritional dynamics of above and below ground detritus. *Advances in Ecological Research*, **15**, 303–77.

Voroney, R. P. Paul, E. A. & Anderson, D. W. (1989). Decomposition of wheat straw and stabilization of microbial products. *Canadian Journal of Soil Science*, **69**, 63–7.

Walker, T. W. & Syers, J. K. (1976). The fate of phosphorus during pedogenesis. *Geoderma*, **15**, 1–19.

Waring, R. H. (1989). Ecosystems: fluxes of matter and energy. *Ecological Concepts* (Ed. by J. M. Cherrett), pp. 17–41. Symposia of the British Ecological Society, 29. Blackwell Scientific Publications, Oxford.

Wershaw, R. L. (1985). Application of nuclear magnetic resonance spectroscopy for determining functionality in humic substances. *Humic Substances in Soil, Sediment and Water* (Ed. by G. R. Aiken, D. M. McKnight, R. L. Wershaw & P. MacCarthy), pp. 561–82. Wiley, New York.

Whittaker, R. H. & Likens, G. E. (1973). Carbon in the biota. *Carbon and the Biosphere*. (Ed. by G. M. Woodwell & E. L. Power), pp. 281–302. *Proceedings of the Brookhaven Symposia, 24*.

Zinke, P. J., Stangenbeiber, A. P., Post, W. M., Emanuel, W. R. & Olson, J. S. (1984). *Worldwide Organic Soil Carbon and Nitrogen Data*. Oak Ridge National Laboratory Environmental Sciences Division Publication, 2217. Oak Ridge, Tennessee.

VII. APPLIED ECOLOGY

Long before the Society launched its *Journal of Applied Ecology* in 1963, there was a strong tradition of papers with an overt or implied practical application amongst the work of ecologists in many fields. These were, however, often the work of individual scientists who could see how the results of their fundamental studies might be applied. Just as the last 20 years have seen an increasing tendency for ecologists to work in interdisciplinary groups, it has become apparent that if some of the major ecological problems of the world in the late twentieth century are to be solved, there is a need for collaborative efforts by many kinds of scientists. There are many such problems, but we have chosen three to illustrate the theme.

Krause reviews that most complex intertwining of biological and environmental changes which have become known in continental Europe as 'forest decline', but are in reality the latest and potentially most damaging stage in man's centuries-old destruction of forests. Understanding the multiple causes of the decline, and their interactions, can be achieved only by the co-operation of scientists with as varied backgrounds as meteorology, microbiology, soil chemistry, plant physiology and biochemistry, and entomology as well as forestry. Integrating and synthesizing the information gathered so far is the classic role of the ecologist.

On a more intimate scale is the study of the management problems of parts of the Norfolk Broads in eastern England, reported and discussed by Moss. It is a fascinating account, not viewed through the rosy glass of hindsight, but presented rather as an evolving approach to an interdisciplinary project destined to force the author into unexpected channels of enquiry. It illustrates most clearly our intention of looking at exactness in ecology in a much more flexible way than the reductionism which has characterized exactness in so many other sciences. It is a very human study in its honest illustration of how progress in science is strongly dependent upon the idiosyncracies of its practitioners. At the same time it reveals much about the ways in which human endeavours have created the problem in the first place.

Ecology as a whole, and applied in particular, cannot be divorced from human and social elements of the real world. The truth of this is borne out by the fact that many of the major ecological crises of the present day are man-made. But it is important also to recognize that the developing exactness of mainstream ecological thinking, traditionally applied to non-human situations, is adaptable enough to offer insight into global issues of economics and politics. Slesser modestly eschews claim to be an ecologist but his contribution, the last in this volume, shows how traditional ecological concepts can enlighten other disciplines, and how its arguments and models may be extended to embrace much of human endeavour.

16. FOREST DECLINE IN CENTRAL EUROPE: THE UNRAVELLING OF MULTIPLE CAUSES

G. H. M. KRAUSE
Landesanstalt für Immissionsschutz des Landes Nordrhein-Westfalen, 4300 Essen 1, Federal Republic of Germany

INTRODUCTION

The productivity of European forests has been deleteriously affected by man's activities for thousands of years. Before the industrial revolution damage was done by removal of the mineral nutrient stock, not only in woody parts used for fuel or building but also in litter used for fertilizing fields (Ellenberg 1988). Moreover, many present-day forests grow on soils that were degraded or even truncated after removal of the original forests.

Rettstadt (1845) was one of the first to deal with damage to forests by chemicals produced in industrial processes. He recognized that 'sulphur acids' and heavy metals (from the refining of Ag and Pb) were the cause of damage to plants and animals in the Harz Mountains around the refineries there. The need to remove the pollutants from the flue gases, and not merely distribute them further by increasing stack height, was emphasized by von Schroeder & Reuss (1883). Much modern research has been concerned with the impact of SO_2 and heavy metals (often Cd, Cu and Zn as well as Pb), and these pollutants are believed to be the cause of acute forest damage in localized areas of Czechoslovakia and Poland at the present time.

This paper tackles the question of what has caused the 'novel forest decline', recognized in France, Germany, Austria, Switzerland and northern Italy in the 1980s. It is widely accepted that the potential causes include not only SO_2 but also NO_x, O_3, acidic rain (and associated soil changes) and possibly organic compounds too. The pattern of damage in space and time suggests strongly that there are predisposing factors reflected in site position, and that the natural phenomena of drought and exceptional frost may aggravate or even trigger the development of the condition. In the context of a symposium entitled *Toward a More Exact*

Ecology, this paper illustrates the real difficulties in unravelling the interactive effects of the many partial causes of an environmental problem which has grave economic consequences.

First, the nature of the novel forest decline is defined: symptoms, temporal development and spatial distribution. Secondly, the hypothesis that the decline could have resulted from natural causes is considered. Thirdly, the effects of single types of pollutant are reviewed, and compared with the symptoms of the novel forest decline. Fourthly, the same is done for the effects of various combinations of pollutants. Finally, the conclusions that may be drawn so far about the causes are summarized, and an indication is given of the kinds of new work needed.

THE NATURE OF THE NOVEL FOREST DECLINE

Symptoms

The symptoms considered specific to the novel forest decline are given below for the tree species most affected.

In silver fir (*Abies alba*) injury is characterized by (i) chlorosis of needles (mainly on the upper sides of older needles), (ii) casting of older needles (usually starting from the base of the crown and from inner parts of branches), and (iii) premature reduction in height increment so that the uppermost branches form a 'stork's nest' even in relatively young trees. Several episodes of 'fir die-back' have been reported over the last two centuries (Seitscheck 1981) with some similarities in the symptoms, making it difficult to differentiate the 'new' and the 'old' die-back.

In Norway spruce (*Picea abies*) injury is like that in the fir, and characterized by (i) chlorosis of older needles (which also go partly reddish-brown), and (ii) casting of older needles (which is delayed when compared with fir). The yellowing of the needles is associated with deficiency of Mg, Ca and Zn (Krause, Prinz & Jung 1983; Zöttl & Hüttl 1985). It has been shown experimentally that the decline in concentration of Mg in fully grown leaves in the spring, and induction of chlorosis, is a result of translocation to the newly growing buds (Lange *et al.* 1987). There are also clear shading effects. When yellowing of the needles occurs, it is nearly always more intense on the upper side of each needle than on the lower side, and where one branch lies over another the needles on the upper branch become yellowed while those on the lower remain green. However, yellow needles are not the predominant symptom of novel forest decline in all European countries.

Symptoms on other tree species are not so clear-cut, but beech (*Fagus sylvatica)* seems to have been affected for more than 10 years, showing reduced shoot growth and die-back of lateral branches (Roloff 1985a). Premature leaf-drop, seen as a symptom of novel forest decline by Schütt & Summerer (1983), may be more closely related to biotic factors (Roloff 1985b). The overall symptomatology in beech is too unspecific to be related to particular causes (Forschungsbeirat Waldschäden 1986).

Temporal development

At the beginning of the present decade the vitality of silver fir and Norway spruce in southern Germany deteriorated rapidly for no apparent reason, and in 1982 the German Forest Services began to conduct detailed annual surveys throughout the country. Neighbouring countries started surveys in 1983 or 1984, but there are problems in comparing the results of these different studies. The UNECE now co-ordinates an annual survey in many European countries using standard methods. By 1986 a region in Europe between 53 and 46 °N appeared to be affected. Information for thirteen countries was summarized by Nilsson & Duinker (1987). While in some countries (Denmark, France, Federal Republic of Germany and Sweden) the extent of affected coniferous forest was markedly reduced between 1985 and 1986, there occurred at the same time an increase in the number of trees affected in other countries (Austria, the Netherlands, Switzerland and the United Kingdom). In nearly all countries a marked increase in injury was recorded during this period for deciduous trees, particularly beech and oak. Overall 15–20% of the coniferous trees and 30–40% of the deciduous trees are now affected in countries with large forested areas; a more detailed analysis is given by PCC-W (1987).

Spatial distribution

Novel forest decline is found mostly in elevated areas, starting above some altitude in the range 200 to 2000 m (Ammer *et al.* 1988). The details of distribution on a local scale are very important for interpreting the cause of the condition. Schöpfer & Hradetzky (1984), in a thorough statistical study in Baden-Württemberg, based partly on permanent observation plots, showed for both fir and spruce that trees were most frequently ded as affected if in stands of heterogeneous composition or with low canopy density, or at edges or with open-site exposure. Slope and aspect were also shown to be important. Intermixture of conifers and deciduous trees was associated with increased damage, and likewise multi-storeyed stands of conifers were more susceptible than single-storeyed ones.

It has also been shown that throughfall and bulk deposition of anions and cations are substantially greater at the edge of a forest stand than just 50 m into the stand (Hasselrot & Grennfelt 1987). Similarly the forest edge shows a greater filtering effect for gaseous pollutants such as sulphur dioxide or ozone (Thiele *et al.* 1988). Schöpfer & Hradetzky (1984) argued that these circumstantial proofs lead to the conclusion that air pollution is the major causative factor. They could not find a correlation between soil properties and incidence of decline in their study area. However, in the Harz Mountains in north-eastern Germany, where the same general patterns are seen in the local distribution of damage (Hartman, Uebel & Stock 1988), an association has been found between the novel forest decline and soil nutrient status (Blank *et al.* 1988). The association is with low concentrations of Ca and Mg in the soil solution, and with low Ca/Al and Mg/Al ratios, all indicators of acidification processes in the soil (Rost-Siebert 1985).

NATURAL FACTORS AS POSSIBLE CAUSES OF NOVEL FOREST DECLINE

Under this heading we have to consider weather events, soil factors and pathogenic micro-organisms. It is convenient to consider silvicultural practices also at this point. We are interested in the idea that natural factors may have predisposed the forest trees to damage, or may even have triggered the start of decline, as much as in the idea that natural factors are the chief cause.

In the experience of foresters in recent centuries, extreme weather events, particularly droughts, have had the most significant effects on tree vitality and stand development (Krause *et al.* 1986). All three tree species of major economic interest in the mountainous areas of central Europe (beech, fir and spruce) are relatively drought-sensitive, and these are the species which show the most extensive development of novel forest decline. The last important dry period in central and northern Europe was in the 1970s and early 1980s (Asthalter 1984). Long-lasting dry periods are believed to be the most important predisposing factors for many kinds of diseases, and for epidemic spreading of bark beetles and other pests and pathogens (Schwerdtfeger 1981). Indirect effects of drought that involve physical factors have also been recorded, for example, increased acidification of the soil due to so-called 'acid pulses' (Matzner & Cassens-Sasse 1984) and increased impact of photochemical oxidants on plants during hot periods with high values of solar radiation (Arndt & Lindner 1981; Prinz, Krause & Stratmann 1982). Furthermore, high levels of solar radiant flux, together with water shortage and nutrient imbalance, can lead to photo-inhibition and reduction in assimilate production (Lichtenthaler & Buschmann 1984). The triggering of damage to beech in north-eastern Germany has been specifically assigned by Roloff (1985b) to droughts of 1976, 1982 and 1983.

In southern Germany there were no significant droughts in the period 1980–7, when the novel forest decline developed, and Bosch & Rehfuess (1988) favour the idea of other critical weather events as synchronizing factors: early frosts in autumn or late frosts in spring, frost desiccation in late winter, extreme low temperatures in midwinter, or abrupt temperature changes after warm periods in winter. Periods of severe frost, with rapid and extreme temperature changes, were observed in Germany and neighbouring countries during the winter of 1978–9 and the spring of 1981, and these were associated with increased susceptibility of trees to various biotic and abiotic 'stresses' (Materna 1985; Cramer & Cramer-Middendorf 1984), considerable needle loss (Friedland *et al.* 1984) and retarded development of leaves (Cramer & Cramer-Middendorf 1984). The susceptibility of individuals was found to be very much determined by genetic composition and mineral nutrient status (Bosch & Rehfuess 1988).

The finding that novel forest decline in southern Germany is linked with nutrient status is in agreement with conclusions reached in other areas. The key symptom of chlorotic needles is known to be related to a 'sudden' deficiency of Mg, Ca or K (Prinz, Krause & Stratmann 1982; Zech & Popp 1983). The fact that the novel forest decline appears on many different soil types, including the calcareous soils of the Schwabische Alb, does not favour the idea that a soil factor is the one primarily responsible, but the decline does seem to be related to poor soil conditions and low nutrient availability in nearly all affected areas (Pollanschütz 1985; Blank *et al.* 1988). It is reasonable to suppose that soil conditions are very significant predisposing factors.

Viruses or virus-like organisms have been considered as a possible cause of novel forest decline, because of the apparent spreading of the disease from rather specific areas of origin. Recent findings have ruled out such organisms as the primary cause, but have not excluded them from the list of possible predisposing factors (Hüttl & Mehne 1988).

Silvicultural practice can be discussed here only briefly. As methods have been developed for producing more timber on a sustained-yield basis, the nature of the forest environment has been changed dramatically in many parts of Europe, and stands of deciduous trees have decreased in area while those of the economically more important conifers, especially Norway spruce, have increased greatly (Hasel 1985) — sometimes without regard to the requirements of the tree for site type and climate. Other problems have arisen from the provenances of the trees used; the fact is that the trees have not always been genetically suited to the environment in which they are grown. Forest management has certainly been considered a predisposing factor for novel forest decline (Manion 1987), but it cannot have been the major cause. Injury is not limited to unsuitable sites or particular forest types, but is also found on optimal sites and in well-managed stands (Bosch & Rehfuess 1988).

Some people may be tempted to accept the hypothesis of Kandler (1987) that 'Nature is again showing her capriciousness' — with the novel forest decline being a result of natural factors, as has happened many times before. However, it is apparent from studies on the permanent observation plots of the Forest Research Institute of Baden-Württemberg, where the health of particular trees has been evaluated over many years for the development of specific symptoms of decline, such as yellowing of needles, that the decline presently observed is not due to natural and silvicultural factors alone (Schröter, Achstetter & Holzapfel 1985).

AERIAL POLLUTANTS AS POSSIBLE CAUSES OF NOVEL FOREST DECLINE

Oxides of sulphur

Concentrations of SO_2 in remote areas, away from natural volcanic influence, are well below 5 $\mu g\ m^{-3}$, and annual means in European areas with clean air range between 5 and 25 $\mu g\ m^{-3}$. In areas close to industrial agglomerations they exceed 25 $\mu g\ m^{-3}$. Annual mean concentrations in areas of Nordrhein-Westfalen with severe forest decline are in the range 30–40 $\mu g\ m^{-3}$, and are about one-third lower than in the Ruhr area. Half-hour means (maximum per month) are also lower in remote areas, but in the winter, with easterly winds and high atmospheric stability, peak concentrations of up to 1200 $\mu g\ m^{-3}$ (comparable with those in the Ruhr area) have been recorded (Pfeffer & Buck 1985). On the other hand, in other high-elevation areas where forests are severely affected, such as the southern Black Forest, southern Bavaria and parts of Switzerland, the annual means are well below 10 $\mu g\ m^{-3}$ (Bucher 1987).

Sulphur dioxide is probably the best-known phytotoxic pollutant and its deleterious effects on forest trees have been appreciated for more than 100 years (Wislicenus 1908). Several excellent reviews of cause and effect have been published, for example, those of Guderian (1977) and Roberts, Darrall & Lane (1983). In most cases exposure to SO_2 was related to point sources, and identification of the cause of injury was possible because of the specific symptoms, variation between species in tolerance, and a clear gradient in foliar S concentration away from the source.

In a major experimental study near to a point source of SO_2 near Biersdorf in West Germany various tree species were grown at increasing distances from the source (Guderian & Stratmann 1968). In general, growth was not affected over a 4-year period at concentrations below 50 $\mu g\ m^{-3}$. Other studies near point sources revealed similar results (Linzon 1971).

In contrast, in the more extreme climate of western Canada, a study near a sour-gas processing plant identified changes in a conifer forest at concentrations of SO_2 currently encountered in Europe (Lester, Rhodes & Legge 1986). Similarly, in Finland where field surveys were carried out to assess the effects of low concentrations of SO_2, and concentrations were in the range 20–40 $\mu g\ m^{-3}$, chronic injury was detected in the leaf tissues of both conifers (*Picea abies* and *Pinus sylvestris*) and deciduous trees (*Betula pubescens*); the results were summarized by World Health Organisation (1987). These effects were frequently observed in the spring, and were especially pronounced after harsh winters (Havas & Huttunen 1980). It seems that resistance to frost was lowered by increased leakage of electrolytes and sugars from the protoplasts (Feiler 1985). In Czechoslovakia also SO_2 concentrations of 20–40 $\mu g\ m^{-3}$ were blamed for degradation of large areas of *Picea abies* forest (Materna 1981). However, according to Materna (1985), the process of forest decline there is slow, and is very much influenced by other factors such as soil nutrient imbalances, the climate in general, unusually harsh winters, rapid changes in temperature or drought, all of which can trigger or intensify SO_2 effects. The lack of ecologically comparable areas free from SO_2 makes it very difficult to attribute the damage to trees unequivocally to SO_2.

The role of incidents when the SO_2 concentration is very high for a very short time has been frequently discussed, but so far experimental results have led to the conclusion that the main effects of SO_2 can be best related to the medium-term means (Garsed & Rutter 1984). However, this issue has not been definitively resolved because peaks tend to be seasonal, occurring mainly in winter, when the trees are dormant, rather than in the summer growing period.

Oxides of nitrogen

In remote areas the concentrations of NO_2 and NO are still usually very much lower than those of SO_2 despite the fact that emission of nitrogen oxides has increased considerably more than that of SO_2 in the last 15 years (Prinz, Krause & Stratmann 1982). At the monitoring sites in Nordrhein-Westfalen the annual mean concentrations of NO_2 are in the range 20–25 $\mu g\ m^{-3}$ (with maximum daily means about 60–90 $\mu g\ m^{-3}$), and those of NO rarely exceed 5 $\mu g\ m^{-3}$ (Pfeffer & Buck 1985). These values are about half those measured in the Ruhr area. Concentrations of NO_2 at higher-altitude sites, such as Brotjacklriegel in Bavaria and Schauinsland in the Black Forest, are generally below 20 $\mu g\ m^{-3}$, even during winter periods, as shown in the Umweltbundesamt Monatsberichte for 1985 and 1986. Concentrations of NO_2 in a forested area in Sweden were below 2 $\mu g\ m^{-3}$ in the summer months, and below 5 $\mu g\ m^{-3}$ during the winter, with a maximum daily concentration of 26 $\mu g\ m^{-3}$ (Sjödin & Grennfelt 1984).

In terms of direct deleterious effects on plants, NO_2 is more important than NO not only because it occurs in higher concentrations, but also because its effects at a given concentration are greater (Rowland, Murray & Wellburn 1985). The phytotoxicity of both gases is comparatively low (Guderian & Tingey 1987), and therefore their importance is likely to be based more on their function as precursors for ozone formation, or when co-occurring with SO_2 (see below).

Undoubtedly both dry and wet input of nitrogen in forested areas has increased considerably during the last decade and has become a cause for serious concern. G. Kenk (Forstliche Versuchsanstalt, Freiburg, personal communication 1986) has reported N-deposition rates of 40 kg ha^{-1} $year^{-1}$ in certain areas of the Black Forest. In the agricultural areas of northern Germany the loads reach 80 kg ha^{-1} $year^{-1}$, and in parts of the Netherlands even >100 kg ha^{-1} $year^{-1}$ (H. Werner, Universität Bonn, personal communication 1987; W. de Vries, Universiteit Wageningen, personal communication 1988).

Higher N input leads to increased N concentrations in the needles and roots, and greater growth increment in the shoots (Röhrle 1985; G. Kenk, cited by Krause *et al.* 1987). On the other hand, frost-hardiness of trees may be lowered (Nihlgard 1985) when total nitrogen levels in needles rise to 1.8–2.0 % dry weight of total needle N (A. Aronsson, cited by Andersen 1986). Similar observations have been reported by Davison, Barnes & Renner (1987), the Gesellschaft für Strahlen und Umweltforschung (1987) and Bosch & Rehfuess (1988). There is also evidence for an increased risk of infection by micro-organisms (Nihlgard 1985; Glatzel *et al.* 1987).

Photochemical oxidants

Effects of photochemical oxidants on vegetation were first observed more than 30 years ago in the area around Los Angeles (Middleton, Kendrick & Schwalm 1950). Under the influence of ultraviolet light, reactions involving NO_x and certain hydrocarbons lead to a mixture of oxidizing pollutants of which ozone is the major component. Unlike the pollutants discussed so far, O_3 is found at higher concentrations at greater altitude, remote from urban and industrial sources (Ashmore, Bell & Rutter 1985). In some southern parts of West Germany and in East Germany the concentration of O_3 increased between 1967 and 1982 by 50–80%, and by as much between 1981 and 1982 as in the two preceding 7-year periods (Warmbt 1979; Attmannspacher, Hartmannsgruber & Lang 1984). In the past 25 years even the winter concentrations of O_3 have been found to increase (Feistner & Warmbt 1984). In mountainous areas of Nordrhein-Westfalen severely affected by the novel forest decline the O_3 concentrations are generally higher than those in the Ruhr area by a factor of 2.0–2.4, and reach monthly means of 100 $\mu g\ m^{-3}$ during the summer (Pfeffer & Buck 1985). Concentrations up to 220 $\mu g\ m^{-3}$ have been reported for subalpine and alpine regions of Bavaria (Paffrath & Peters 1988).

The ozone concentrations in higher-altitude regions of West Germany and other European countries are high enough to injure sensitive plant species (Arndt 1985; Ashmore, Bell & Rutter 1985; Guderian, Küppers & Six 1985). However, the concentrations observed are not likely to cause acute visible injury. Native European trees, such as *Picea abies*, have turned out to be relatively ozone-tolerant. Even fumigation with 150 $\mu g\ m^{-3}$ for nearly a month produced only slight visible injury, changing after 48 days of continuing fumigation to an unspecific chlorosis, mainly on the youngest, light-exposed needles, and on the upper parts of the crown. No visible ozone-induced injury was observed in *Abies alba*, even after 6 weeks of continuous fumigation with 500 $\mu g\ m^{-3}$. Nevertheless concentrations of 150 $\mu g\ m^{-3}$ did inhibit photosynthesis, respiration and transpiration in *Picea abies* (Krause & Prinz

1988). The primary sites for the action of ozone are the cell membranes; changes in permeability in the plasmalemma and chloroplast are involved, as shown by fluxes in organic and inorganic metabolites (Tingey & Taylor 1982; Guderian, Tingey & Rabe 1985). One result may be leaching of essential nutrients from the plant (Krause, Prinz & Jung 1983; Krause 1988), accompanied and perhaps enhanced by ozone-induced weathering of the cuticle (Krause, Jung & Prinz 1985). However, Brown & Roberts (1988) have shown that the usual method of ozonating air for experiments on this issue introduces substantial quantities of anhydrous nitric acid, and they argue that the latter is responsible for accelerating leaching. Roberts, Skeffington & Blank (1989) further argue that even the total leaching by O_3 plus HNO_3 does not contribute substantially to the cause of novel forest decline.

The effects of ozone are much more difficult to evaluate than those of SO_2 because there are no point sources to allow observations along concentration gradients. In areas with chronic ozone exposure (and I think many areas in Europe should be ranked in this category) decline in vigour of forest trees is a commonly observed response (Guderian, Tingey & Rabe 1985). Symptoms of chronic decline differ markedly from acute visible injury, and according to McLaughlin *et al.* (1982) include (i) premature senescence with cast of older needles in autumn, (ii) reduced storage of assimilates in roots at the end of the growing season and reduced resupply capacity in spring, (iii) increased reliance of new needles on self-support during growth, (iv) shorter new needles and thus reduced assimilate production, and (v) reduced availability of photosynthates for homeostasis.

Such symptoms are not very specific, and are therefore difficult to differentiate from the effects of the other causal factors already considered. However, Paffrath & Peters (1988) have been able to show that the altitudinal distribution of ozone concentration in the subalpine in Bavaria matches rather well the altitudinal distribution of injury on coniferous trees. As a result of local photochemical sources and long-distance transport, high concentrations build up during the morning under the inversion layer, with a peak at around 1200 m (Fig. 16.1a). In the afternoon the inversion layer rises, and so does the peak in ozone concentration (Fig. 16.1b). The maximum damage to spruce on south-facing slopes is at around 1200 m, corresponding to the morning peak in O_3. The maximum damage on west-facing slopes is at around 1600 m, corresponding with the late afternoon peak. It is speculated that on warm, dry, south-facing slopes the stomata close during the time of peak ozone concentrations, and that uptake of ozone is reduced considerably.

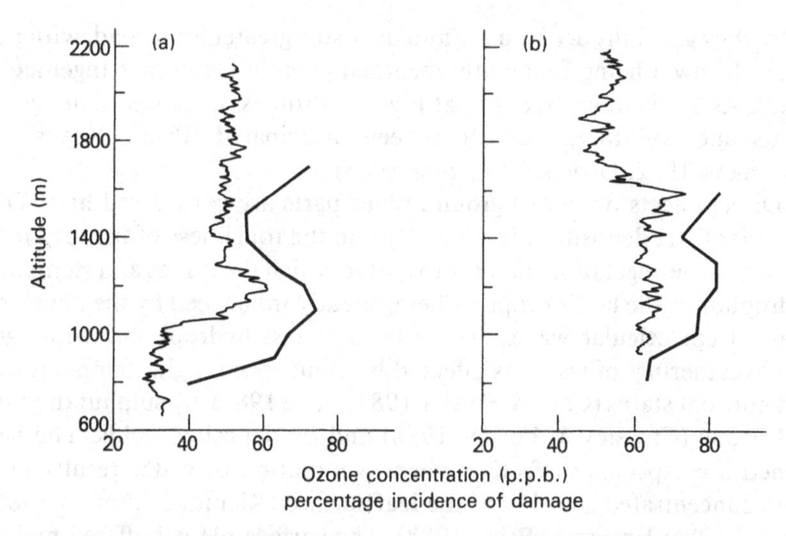

FIG 16.1 Altitudinal profiles of ozone concentration in the air at different times of day (very wavy lines), and their relation to the incidence of damage to spruce (injury classes 2 + 3), in the subalpine zone in Bavaria (after Paffrath & Peters 1988): (a) profile at 1020 hours and incidence of damage on south-facing slopes; (b) profile at 1620 hours and incidence of damage on west-facing slopes. Note common scale for ozone concentration and percentage incidence of tree damage.

The wetter and cooler west-facing slopes, where the stomata are supposed to remain open, receive sunshine in the afternoon, just when ozone concentrations are at their highest in the range 1200–1600 m. It is at least plausible that in this study area ozone is a major causal agent.

Effects of acidic deposition

Both wet and dry deposition occur. Wet deposition happens through rain, snow, dew and interception of cloud (fog). The solutes are numerous but we are here concerned primarily with the anions of three strong acids. Commonly the relative contributions are about 60% SO_4, 30% NO_3 and 10% Cl (Beilke 1983). However, the ratios vary, and the deposition of nitrate seems to increase with distance from source while that of sulphate does not. The acidity of rain-water in industrialized areas has not increased appreciably since 1940, when a point of 'acid saturation' at a pH of about 4.2 was reached (Winkler 1982). Since that time the affected area has enlarged greatly, and there are now few spatial variations in the pH of rainfall in central Europe (Winkler 1983). Because fog droplets have a smaller volume than raindrops, and may be efficient at scavenging gases, they generally accumulate and deposit a greater ionic load, with pH values below 3 being frequently recorded (Verein Deutscher Ingenieure 1987). As fog is more frequent at higher altitudes, increased damage to plants and soil there has often been anticipated (Prinz, Krause & Stratmann 1982; Ulrich & Matzner 1983).

Direct effects on above-ground plant parts are considered first. The amount of wet deposition is dependent on the roughness of the response surface of the vegetation, and the extent to which the leaves and stems are hydrophobic, the latter property being greatly influenced by the development of epicuticular waxes. These become less hydrophobic with age. The 'weathering' of waxes is affected by wind speed, light, temperature, and nutrient status (Cape & Fowler 1981; Cape 1982), by sulphur dioxide and ozone (Crossley & Fowler 1986) and by deposited acids. The last named are especially effective when evaporation of water results in a more concentrated solution at the leaf surface (Klemm & Frevert 1985; Wenzel 1985; Krause & Prinz 1988). The surface pH is buffered by ion exchange (Skiba, Peirson-Smith & Cresser 1986) but this process is associated with the loss of not only inorganic ions but also amino acids, proteins and carbohydrates from the cells of the leaf (Tukey 1970; Scherbatskoy & Klein 1983). Thus deposition of acids may reduce plant growth through leaching of metabolites, and the plant may be unable to compensate for loss of mineral nutrients from the shoots by increased uptake from the soil (Mecklenburg & Tukey 1964). Of course, such effects may be partly compensated by increased input of nutrient ions in the rain (Reich *et al.* 1986).

It has yet to be established whether or not deposition of acids in rain of the present-day pH is by itself deleterious to above-ground parts of forest trees. However, in assessing the combined effects of acidic deposition and other pollutants, it must be remembered that the amounts of nutrients leached increase as trees lose their vitality (Krause & Prinz 1986).

Indirect effects of acidic deposition via the soil must now be considered. Because of the high filtering capacity of tree canopies, the impact of acidic deposition is especially great on forest soils (Matzner & Ulrich 1984). As the buffering capacity of many forest soils is limited, high rates of acidic deposition can lead to soil acidification. This process involves release to the soil solution of ions which are potentially toxic to the roots, notably H^+, Al_3^+ and certain heavy metals. Acidification of forest soils during recent decades has been reported in numerous papers (summarized by Verein Deutscher Ingenieure 1987). Silvicultural practices are important because the form of the canopy so greatly affects the amount of deposition.

Hydrogen-ion budgets have revealed that atmospheric deposition represents the major load of strong acidity to forest soils (Matzner & Ulrich 1984). Nevertheless the effects of acidic deposition are enhanced by natural release of HNO_3 in soils during warm, dry seasons. The magnitude of such pulses has been measured by Matzner & Cassens-Sasse (1984). Their effects depend on the buffering capacity of the soil. Acidic deposition continuously exhausts this capacity, and is especially serious on soils with low natural base-saturation capacity.

Damage to roots has been observed frequently in declining forest stands in Germany (Schütt *et al.* 1983), and has been attributed to toxic concentrations of H^+, Al_3^+ and other ions in solution. Laboratory experiments have shown that low concentrations of Al can reduce uptake of Mg and Ca (Verein Deutscher Ingenieure 1987), and hence symptoms of deficiency in Mg and Ca in needles might be explained. A ratio of <1 Ca : 1 Al, said by Rost-Siebert (1985) to induce root damage and reduce uptake of Ca and Mg, has been shown to occur in the soil solution of some declining stands, e.g. in the Solling Forest (Ulrich, Murach & Pirouzpanah 1984). However, the hypothesis that high concentrations of H^+ and Al_3^+ are responsible for the root damage is controversial (Verein Deutscher Ingenieure 1987), and the fact that the novel forest decline is found on some soils where excessive mobilization of toxic metal ions is not to be expected leads to the conclusion that Al toxicity cannot explain the forest decline syndrome at every site. Nevertheless it must surely be of importance on naturally acidic soils.

Organic compounds

Little information is available on the effects of organic air pollutants on vegetation, although the worldwide emissions are enormous: 3×10^6 t $year^{-1}$ according to Frank & Frank (1985). Particular attention has been paid to hydrocarbons (HCs) and halogenated hydrocarbons (HHCs). Part of the HC 'burden' arises naturally, as large quantities are emitted by forests — particularly terpenoids. These can be consumed rapidly by reactions with OH radicals, and are thus linked with the formation of photochemical oxidants under appropriate weather conditions. Discontinuous measurements of polycyclic aromatic hydrocarbons over one year in the Ruhr region and in the Egge Mountains about 150 km away showed no significant difference between the two areas (Pfeffer & Buck

1985). Measurements of HHCs have been carried out in the Schwäbische Alb by Frank (1985), who found concentrations of 3.2–7.0 $\mu g\ m^{-3}$ for freons, and 0.4–1.3 $\mu g\ m^{-3}$ for chloroethene, trichloroethene and tetrachloroethene.

Polycyclic aromatic hydrocarbons and HHCs tend to become concentrated in the waxy surface layers of leaves. Perhaps more importantly they can also react with the lipids in cell membranes. When activated by u.v. light, HHCs produce reactive triplet states or free radicals, and may decompose to form highly toxic chemical species such as phosgene, dichloroethene and chloroacetylchloride, which can interfere with plant metabolism (Frank 1985).

It is believed that the concentrations of HCs and HHCs in the rural areas of industrialized states have increased, but further measurements are needed. The phytotoxic potentials of the compounds also need to be established, using realistic concentrations and long-term exposures.

COMBINED EFFECTS OF AIR POLLUTANTS

Few studies are available on combined effects, but it is established that they may be additive or antagonistic or show positive interaction (Guderian & Tingey 1987; Krause 1986). The combined effects of the oxides of sulphur and nitrogen are considered first. When *Populus nigra* was fumigated in summer with SO_2 and NO_2 together (110 ppb) growth was reduced more than by either pollutant alone (Freer-Smith 1984); no effect was found on dormant trees. Fumigation with chronic doses of the two gases (62 ppb) over 2 years resulted in greater needle loss in *Picea sitchensis* and *Pinus sylvestris* than with SO_2 alone (Freer-Smith 1985); NO_2 alone initially enhanced needle retention in the latter species. Effects took nearly 2 years to develop, and no seasonal response pattern was observed.

Several experimenters have exposed plants to both SO_2 (and/or NO_2) and O_3. When *Abies alba* and *Picea abies* were exposed in open-top chambers for 2 years at SO_2 concentrations of 30–50 and 15–30 $\mu g\ m^{-3}$ during winter and summer respectively, and O_3 was applied during sunny days only for 12 h per day (80–100 $\mu g\ m^{-3}$), it was found that formation of the anti-oxidants glutathione and vitamin E was greater than when plants were subjected to only one of the pollutants (Mehlhorn *et al.* 1986). In another experiment with *Picea abies* Kettrup *et al.* (1987) found that the concentrations of phenolic compounds and tannins in the needles of plants exposed to both SO_2 and O_3 were markedly different from those in needles of plants exposed to only one of the pollutants.

Treatment with the single pollutants did not result in morphological changes, but treatment with the two together did.

A comparison of treatments with ambient concentrations of SO_2 + O_3, NO_2 + O_3, and SO_2 + NO_2 + O_3 was made by Kress, Skelly & Kinkelmann (1982) with *Pinus taeda*. Height growth was reduced to a similar extent in all three treatments. Thus the interaction of O_3 with SO_2 or NO_2 was more important than that between SO_2 and NO_2. In similar experiments with *Pinus strobus* growth was reduced more by O_3, SO_2 or O_3 + SO_2 than by NO_2, O_3 + NO_2 or SO_2 + NO_2 (Yang, Skelly & Chevone 1983). Mooi (1984) found in addition a greater decrease in growth of *Populus nigra* when treating plants with NO_2 + O_3 or NO_2 + O_3 + SO_2 than NO_2 + SO_2.

In *Pinus taeda* a combination of NO_2 and O_3 was found to induce chlorotic mottling of needles, and occasional tip-burn (Kress, Skelly & Kinkelmann 1982). In *Picea abies* the effect of a combination of SO_2, NO_2 and O_3 was yellowing of the needles, and, when there was also a deficiency of nutrients (especially Mg), the symptoms were remarkably close to those of the novel forest decline (Guderian, Küppers & Six 1985).

Experiments with two or three air pollutants plus acidic rain have concentrated on the degree of leaching of nutrients. Treatment of *Abies alba, Fagus sylvatica* and *Picea abies* for 2 years in open-top chambers with near-ambient concentrations of SO_2, O_3 and acidic rain caused a marked increase in the leaching of cations, while treatment with O_3 and acidic rain without SO_2 did not have such a marked effect (Seufert & Arndt 1986). On the other hand, in other experiments leaching of cations (Mg, Ca, Cu and Zn) from *Picea abies* has generally been found to be dependent on O_3 dose (Krause, Prinz & Jung 1983; Krause & Prinz 1988). In the San Gabriel Mountains of California leaching of ions from pine needles into cloud water was found to be closely related to previous episodes of high ozone concentration (Waldman & Hoffmann 1989). Similar results have been reported for European forests (Fabig *et al.* 1987). The position is complicated by the fact that leaching is generally greater in plants of low vitality, and by the additional enhancement of leaching by frost episodes (Bosch *et al.* 1986).

CONCLUSIONS

Most of the factors which have been thought to play some part in the development of forest decline are included in the summary diagram in Fig. 16.2. This review has shown that it is most improbable that either 'natural' factors (i.e. weather, diseases and pests) or forest management

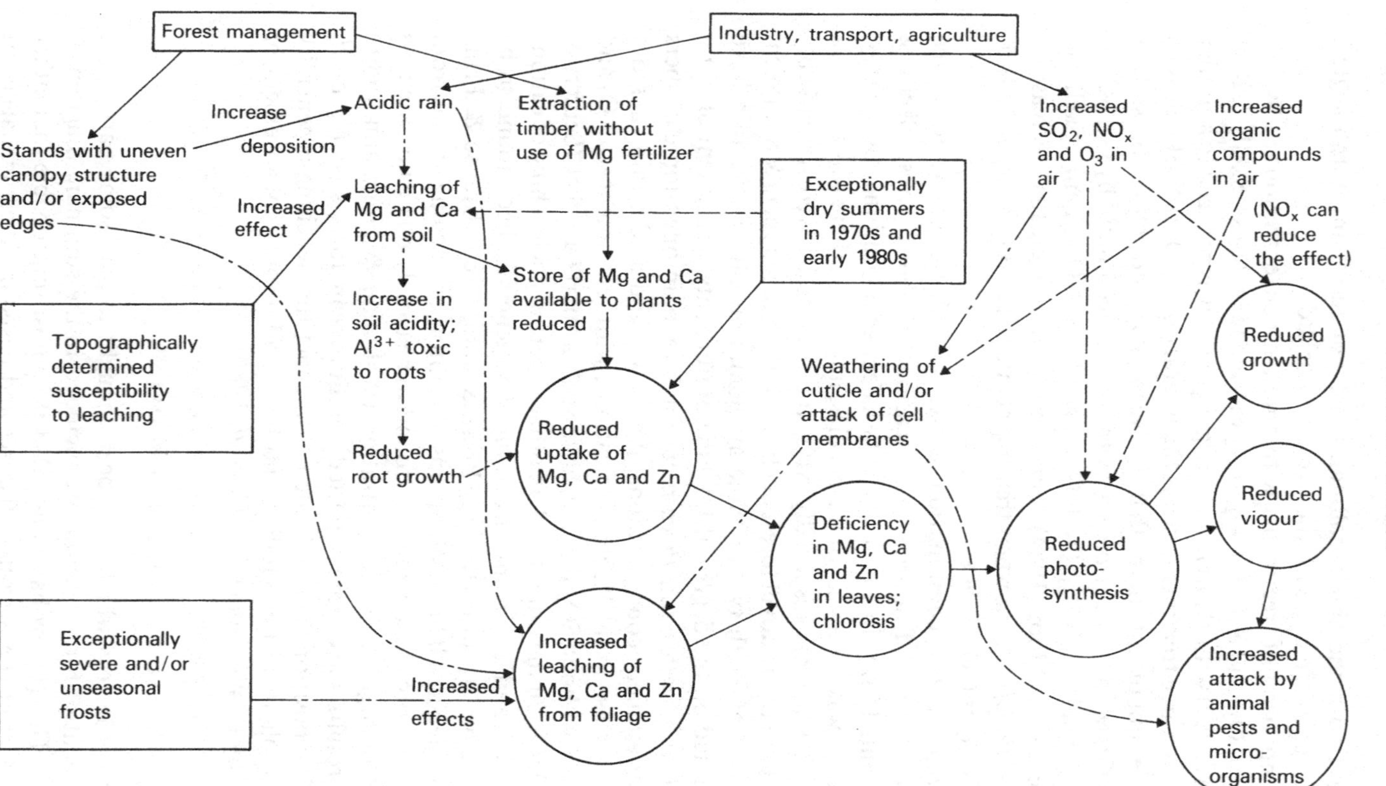

FIG 16.2 A diagram summarizing the influences apparently involved in the novel forest decline. The continuous lines indicate the influences believed at the moment to be the major ones; the pecked lines indicate what are probably lesser influences, and the lines with alternate dashes and dots indicate influences that are in contention.

practices can wholly account for the phenomenon. On the other hand the exceptional droughts of the 1970s and early 1980s do seem to have been important triggering factors in some areas, and perhaps exceptionally cold spells in the same period had a parallel effect elsewhere. Also forest management, by depleting the soil without addition of Mg fertilizer, must play an important part. The pattern of incidence in the field suggests that forest management also plays a part through the creation of exposed edges to stands, and the encouragement of stands with an uneven canopy. Observations on the topographical distribution of incidence also support the view that impaction of aerial pollutants is an important part of the phenomenon. The specific symptoms of needle-yellowing do not arise from simple exposure to any of the major aerial pollutants (SO_2, NO_x and O_3), either alone or in mixture, even when experiments are carried on over long periods. Critical experiments are still needed on the possible influence of organic pollutants, but it is now fairly certain that the part played by aerial pollution involves acidic rain, and *either* simply involves the leaching of basic cations from the soil (as advocated by some researchers) *or* involves not only that effect but also accelerated leaching of basic cations from the foliage under the influence of acidic rain combined with high concentrations of ozone (as argued in this chapter).

Future research should be designed to clarify this last issue, to explore the many interactions between factors (such as direct effects of drought, pulses of HNO_3 in dried soils, and soil Mg status) and to understand the timing and pattern of spread of the decline (cf. Prinz 1987) — not least so that future problems, for example related to the 'greenhouse effect', may be anticipated.

REFERENCES

Ammer, U., Burgis, M., Koch, B. & Martin, K. (1988). Untersuchungen über den Zusammenhang zwischen Schädigungsgrad und Meereshöhe im Rahmen des Schwerpunktprogramms zur Erforschung der Wechselwirkungen von Klima und Waldschäden. *Forstwissenschaftliches Centralblatt,* **107**, 145–51.

Andersen, B. (1986). Impact of nitrogen deposition. *Critical Loads for Sulfur and Nitrogen* (Ed. by J. Nilsson), pp. 159–97. Milijo Rapport 1986, II. Nordisk Ministerrad, Stockholm.

Arndt, U. (1985). Ozon und seine Bedeutung für das Waldsterben in Mitteleuropa. *Was wir über das Waldsterben wissen* (Ed. by E. Nieslein & G. Voss), pp. 160–74. Deutscher Instituts Verlag, Cologne.

Arndt, U. & Lindner, G. (1981). Zur Problematik phytotoxischer Ozonkonzentrationen im südwestdeutschen Raum. *Staub-Reinhaltung Luft,* **41**, 349–52.

Ashmore, M. R., Bell, N. J. B. & Rutter, A. J. (1985). The role of ozone in forest damage in W. Germany. *Ambio,* **14**, 81–7.

Asthalter, K. (1984). Trockenperioden und Waldschäden aus forstgeschichtlicher und standortkundlicher Sicht. *Allgemeine Forstzeitung*, **39**, 549–51.

Attmannspacher, W., Hartmannsgruber, R. & Lang, P. (1984). Langzeittendenzen des Ozons der Atmosphäre aufgrund der 1967 begonnenen Ozonmeβreihen am Meteorologischen Observatorium Hohenpeiβenberg. *Meteorologische Rundschau*, **37**, 193–9.

Beilke, S. (1983). Acidic deposition: the present situation in Europe. *Proceedings of CEC Workshop on Physico-chemical Behavior of Atmospheric Pollutants* (Ed. by S. Beilke & A. Elshout), pp. 235–41. Reidel, Dordrecht.

Blank, K., Matzner, E., Stock, R. & Hartmann, G. (1988). Der Einfluβkleinststandörtlicher bodenchemischer Unterschiede auf die Ausprägung von Vergilbungssymptomen an Fichten im Harz. *Forst und Holz*, **12**, 288–92.

Bosch, C., Pfannkuch, E., Rehfuess, K. E., Runkel, K. M., Schramel, P. & Senser, M. (1986). Einfluβ einer Düngung mit Magnesium und Calcium, von Ozon und saurem Nebel auf Frosthärte, Ernährungszustand und Biomasseproduktion junger Fichten (*Picea abies* [L.] Karst.). *Forstwissenschaftliches Centralblatt*, **105**, 218–42.

Bosch, C. & Rehfuess, K. E. (1988). Über die Rolle von Frostereigissen bei den 'neuartigen Waldschäden'. *Forstwissenschaftliches Centralblatt*, **107**, 123–30.

Brown, K. A. & Roberts, T. M. (1988). Effects of ozone on foliar leaching in Norway spruce (*Picea abies* [L.] Karst.): confounding factors due to NO_x production during ozone generation. *Environmental Pollution*, **55**, 55–73.

Bucher, J. B. (1987). Forest damage in Switzerland, Austria and adjacent parts of France and Italy in 1984. *Effects of Acid Deposition and Air Pollution on Forests, Wetland and Agricultural Ecosystems* (Ed. by T. C. Hutchinson & K. M. Meeme), pp. 43–58. NATO Advanced Study Institute Series G: Ecological Sciences, 16. Springer, Berlin.

Cape, J. N. (1982). Contact angles of water droplets on needles of Scots pine (*Pinus sylvestris*) growing in polluted atmospheres. *New Phytologist*, **93**, 293–9.

Cape, J. N. & Fowler, D. (1981). Changes in epicuticular wax of *Pinus sylvestris* exposed to polluted air. *Silva Fennica*, **15**, 457–8.

Cramer, H. H. & Cramer-Middendorf, M. (1984). Untersuchungen über Zusammenhänge zwischen Schadensperioden und Klimafaktoren in mitteleuropäischen Forsten seit 1851. *Pflanzenschutz-Nachricht Bayer*, **37**, 208–334.

Crossley, A. & Fowler, D. (1986). The weathering of Scots pine epicuticular wax in polluted and clean air. *New Phytologist*, **103**, 207–18.

Davison, A. W., Barnes, J. D. & Renner, C. J. (1987). Interactions between air pollutants and cold stress. *Air Pollution and Plant Metabolism* (Ed. by S. Schulte-Hostede, N. M. Darrall & L. W. Blank), pp. 307–28. Elsevier Applied Science Publishers, London.

Ellenberg, H. (1988). *Vegetation Ecology of Central Europe*, 4th edn. Cambridge University Press.

Fabig, W., Boose, C., Fritsche, U., Grundmann, B., Hochrainer, D., Klöppel, H., Mönig, F.-J. & Oldiges, H. (1987). Anthropogene Immissionen als Belastungsfaktoren in terrestrischen Ökosystemen und Wege zur Minderung ihrer Schadwirkung. *Umweltforschungsplan des Bundesministers des Inneren Forschungsvorhaben* 106070446/01. Fraunhofer Institut für Umweltchemie und Ökotoxikologie, Schmallenberg-Grafschaft.

Feiler, S. (1985). Influence of SO_2 on membrane permeability and consequences on frost sensitivity of spruce (*Picea abies* L.). *Flora, Jena*, **177**, 217–26.

Feistner, U. & Warmbt, W. (1984). Long-term surface ozone increase at Arkona (54° 68′ N, 13° 43′ E). *Atmospheric Ozone* (Ed. by C. S. Zerefos & A. E. Ghazi), pp. 782–7. Reidel, Dordrecht.

Forschungsbeirat Waldschäden (1986). *Luftverunreinigungen der Bundesregierung und der Länder. 2. Bericht*, Forschungsbeirat Waldschäden.

Frank, H. (1985). The new forest decline. Destruction of photosynthetic pigments by UV-activation of chloroethenes: a probable cause of forest decline. *Environmental Protection Agency Newsletter,* 6 April, 7–13.

Frank, H. & Frank, W. (1985). Chlorophyll bleaching by atmospheric pollutants and sunlight. *Naturwissenschaften,* **72**, 139–41.

Freer-Smith, P. H. (1984). The responses of six broadleaved trees during long-term exposure to SO_2 and NO_2. *New Phytologist,* **97**, 49–61.

Freer-Smith, P. H. (1985). The influence of gaseous SO_2 and NO_2 and their mixture on the growth and physiology of conifers. *Air Pollution and the Stability of Coniferous Forest Ecosystems* (Ed. by E. Klimo & R. Saly), pp. 235–47. University of Agriculture, Brno.

Friedland, A. J., Gregory, R. A., Kärenlampi, L. & Johnson, A.H. (1984). Winter damage to foliage as a factor in red spruce decline. *Canadian Journal of Forest Research,* **14**, 963–5.

Garsed, S. G. & Rutter, A. J. (1984). The effects of fluctuating concentrations of sulfur dioxide on the growth of *Pinus sylvestris L.* and *Picea sitchensis* (Bong.) Carr. *New Phytologist,* **97**, 175–95.

Gesellschaft für Strahlen und Umweltforschung (1987). *Klima und Witterung im Zusammenhang mit den neuartigen Waldschäden.* GSF-Bericht, 10. München-Neuherberg.

Glatzel, G., Kazda, M., Grill, D., Halbwachs, G. & Katzensteiner, W. (1987). Ernährungsstörungen bei Fichte als Komplexwirkung von Nadelschäden und erhöhter Stickstoffdeposition — ein Wirkungsmechanismus des Waldsterbens? *Allgemeine Forst- und Jagdzeitung,* **158**, 91–7.

Guderian, R. (1977). *Air Pollution: Phytotoxicity of Acid Gases and its Significance in Air Pollution Control.* Springer, Berlin.

Guderian, R., Küppers, K. & Six, R. (1985). Wirkungen von Ozon, Schwefeldioxid und Stickstoffdioxid auf Fichte und Pappel bei unterschiedlicher Versorgung auf Magnesium und Kalzium sowie auf die Blattflechte *Hypogymnia physodes. Verein Deutscher Ingenieure-Berichte,* **560**, 657–701.

Guderian, R. & Stratmann, H. (1968). Freilandversuche zur Ermittlung von Schwefeldioxidwirkungen auf die Vegetation. III. Grenzwerte schädlicher SO_2 -Immissionen für Obst- und Forstkulturen sowie für landwirtschaftliche und gärtnerische Pflanzenarten. *Forschungsberichte des Landes Nordrhein-Westfalen,* 1920, pp. 1–113. Westdeutscher Verlag, Cologne.

Guderian, R. & Tingey, D. T. (1987). Notwendigkeit und Ableitung von Grenzwerten für Stickstoffoxid. *Umweltbundesamt-Berichte, Berlin,* **1987 (1)**, 1–170.

Guderian, R., Tingey, D. T. & Rabe, R. (1985). Effects of photochemical oxidants on plants. *Air Pollution by Photochemical Oxidants* (Ed. by R. Guderian), pp. 129–333. Springer, Berlin.

Hartman, G., Uebel, R. & Stock, R. (1988). Zur Verbreitung der Nadelvergilbung an Fichte im Harz — Vorläufige Mitteilung von Luftauswertungsergebnissen. *Forst- und Holzwirt,* **10**, 286–92.

Hasel, K. (1985). *Forstgeschichte — Ein Grundriß für Studium und Praxis.* Paul Parey, Hamburg.

Hasselrot, B. & Grennfelt, P. (1987). Deposition of air pollutants in a wind exposed forest edge. *Water, Air and Soil Pollution,* **34**, 135–43.

Havas, P. J. & Huttunen, S. (1980). Some special features of the ecophysiological effects of air pollution on coniferous forests during winter. *Effects of Acid Precipitation in Terrestrial Ecosystems* (Ed. by T. C. Hutchinson & P. J. Havas), pp. 236–41. Plenum, New York.

Hüttl, R. E. & Mehne, B. (1988). Diagnostische Düngungsversuche zur Revitalisierung geschädigter Fichtenkulturen (*Picea abies* Karst.) in Südwestdeutschland. *Forstwissenschaftliches Centralblatt,* **107**, 173–83.

Kandler, O. (1987). Klima und Baumkrankheiten. *Gesellschaft für Strahlen und Umweltsforschung München-Neuherberg-Bericht*, **10**, 269–75.

Kettrup, A., Masuch, G., Kicinski, H. G. & Boos, K. S. (1987). Untersuchungen zur Schädigung von Jung- und Altfichten (*Picea abies* L.) und Jung- und Altbuchen (*Fagus sylvatica* L.) durch Ozon als Luftverunreinigung. *Annual Report for* 1986 *of Bundesminister für Forschung und Technologie Project* 03–7399–0. University of Paderborn.

Klemm, O. & Frevert, T. (1985). Säure- und redoxchemisches Verhalten von Niederschlagswasser beim Abdampfen von Oberflächen. *Verein Deutscher Ingenieure-Berichte*, **560**, 457–63.

Krause, G. H. M. (1988). Ozone induced nitrate formation in needles and leaves of *Picea abies, Fagus sylvatica* and *Quercus robur*. *Environmental Pollution*, **52**, 117–30.

Krause, G. H. M., Arndt, U., Brandt, C. J., Bucher, J. B., Kenk, G. & Matzner, E. (1987). Forest decline in Europe: development and possible causes. *Water, Air, and Soil Pollution*, **31**, 647–68.

Krause, G. H. M., Jrng, K.-D. & Prinz, B. (1985). Experimentelle Untersuchungen zur Aufklärung der neuartigen Waldschäden in der BRD. *Verein Deutscher Ingenieure-Berichte*, **560**, 627–56.

Krause, G. H. M. & Prinz, B. (1986). Zur Wirkung von Ozon und saurem Nebel auf phänomenologische und physiologische Parameter an Nadel-und Laubgehölzen im kombinierten Begasungsexperiment. *Spezielle Berichte der Kernforschungsanlage Jülich*, **369**, 208–21.

Krause, G. H. M. & Prinz, B. (1988). Neuere Untersuchungen zur Ursachenforschung der neuartigen Waldschäden. *Landesanstalt für Immissionsschutz der NRW Berichte* Nr. 80.

Krause, G. H. M. & Prinz, B. (1986). Zur Wirkung von Ozon und saurem Nebel auf phänomenologische und physiologische Parameter an Nadel-und Laubgehölzen im kombinierten Begasungsexperiment. *Spezielle Berichte der Kernforschungsanlage*

Kress, L. W., Skelly, J. M. & Kinkelmann, K. H. (1982). Growth impact of O_3, NO_2, and/or SO_2 on *Pinus taeda*. *Environmental Monitoring Assessment*, **1**, 229–39.

Lange, O. L., Zellner, H., Gebel, J., Schramel, P., Kostner, B. & Czygan, F. C. (1987). Photosynthetic capacity, chloroplast pigments, and mineral content of the previous year's spruce needles with and without the new flush: analysis of the forest decline phenomenon of needle bleaching. *Oecologia*, **73**, 351–7.

Lester, P. F., Rhodes, E. C. & Legge, A. H. (1986). Sulphur gas emissions in the boreal forest: the West Whitecourt case study. IV: Air quality and the meteorological environment. *Water, Air, and Soil Pollution*, **27**, 85–108.

Lichtenthaler, H. K. & Buschmann, C. (1984). Beziehungen zwischen Photosynthese und Baumsterben. *Allgemeine Forstzeitung*, **39**, 12–16.

Linzon, S. N. (1971). Economic effects of sulfur dioxide on forest growth. *Journal of the Air Pollution Control Association*, **21**, 81–6.

McLaughlin, S. B., McConathy, R. K., Duvick, D. & Mann, L. K. (1982). Effects of chronic air pollution stress on photosynthesis, carbon allocation and growth of white pine trees. *Forest Science*, **28**, 60–70.

Manion, P. D. (1987). Decline as a phenomenon in forests: pathological and ecological considerations. *Effects of Atmospheric Pollutants on Forest Wetlands and Agricultural Ecosystems* (Ed. by T. C. Hutchinson & K. M. Meeme), pp. 267–75. Springer, Berlin.

Materna, J. (1985). Luftverunreinigungen und Waldschäden. *Symposium von Umweltschutz — eine internationale Aufgabe* (Ed. by Verein Deutscher Ingenieure), pp. 19.1–19.13. VDI, Düsseldorf.

Matzner, E. & Cassens-Sasse, E. (1984). Chemische Veränderungen der Bodenlösung als Folge saisonaler Versauerungsschübe in verschiedenen Waldökosystemen. *Berichte des Forschungszentrums Waldökosysteme/Waldsterben*, **2**, 50–60.

Matzner, E. & Ulrich, B. (1984). Raten der Deposition, der internen Produktion und des Umsatzes von Protonen in Waldökosystemen. *Zeitschrift der Pflanzenernährung und Bodenkunde,* **147**, 290–308.

Mecklenburg, R. A. & Tukey, H. B. (1964). Influence of foliar leaching on root uptake and translocation of calcium[45] to the stems and foliage of *Phaseolus vulgaris. Plant Physiology,* **39**, 533–6.

Mehlhorn, H., Seufert, G., Schmidt, A. & Kunert, K. J. (1986). Effect of SO_2 and O_3 on production of antioxidants in conifers. *Plant Physiology,* **82**, 336–8.

Middleton, J. T., Kendrick, J. B. & Schwalm, H. W. (1950). Injury to herbaceous plants by smog or air pollution. *Plant Diseases Reporter,* **34**, 245–52.

Mooi, J. (1984). Wirkungen von SO_2, NO_2, O_3 und ihrer Mischungen auf Pappeln und einige andere Pflanzenarten. *Forst- und Holzwirt,* **39**, 438–44.

Nihlgard, B. (1985). The ammonium hypothesis — an additional explanation to the forest dieback in Europe. *Ambio,* **14**, 4–16.

Nilsson, S. & Duinker, P. (1987). A synthesis of survey results — the extent of forest decline in Europe. *Environment,* **29**, 4–9.

Paffrath, D. & Peters, W. (1988). Betrachtung der Ozonvertikalverteilung im Zusammenhang mit den neuartigen Waldschäden. *Forstwissenschaftliches Centralblatt,* **107**, 152–9.

PCC-W (1987). *Forest Damage and Air Pollution — Report of the 1986 Forest Damage Survey in Europe.* International Co-operative Programme on Assessment and Monitoring of Air Pollution Effects on Forest (PCC-W), Hamburg.

Pfeffer, H. U. & Buck, M. (1985). Meβtechnik und Ergebnisse von Immissionsmessungen in Waldgebieten. *Verein Deutscher Ingenieure-Berichte,* **560**, 127–55.

Pollanschütz, J. (1985). Waldzustandsinventur 1984: Ziele-Inventurverfahren-Ergebnisse. *Schriftenreihe Forstliche Bundesversuchsanstalt, Wien,* **8**, 1–29.

Prinz, B. (1987). Causes of forest damage in Europe — major hypotheses and factors. *Environment,* **29**, 11–37.

Prinz, B., Krause, G. H. M. & Stratmann, H. (1982). Waldschäden in der Bundesrepublik Deutschland. *Landesanstalt für Immissionsschutz des NRW-Berichte,* **28**, 1–146.

Reich, P. B., Schoettle, A. W., Stroo, H. F. & Amundson, R. G. (1986). Acid rain and ozone influence mycorrhizal infection in tree seedlings. *Journal of the Air Pollution Control Association,* **36**, 724–6.

Rettstadt, G. (1845). Über die Einkwirkungen des Rauches der Silberhütten auf die Waldbäume und den Forstbetrieb. *Allgemeine Forst- und Jagdzeitung,* **14**, 132–40.

Roberts, T. M., Darrall, N. M. & Lane, P. (1983). Effects of gaseous air pollutants on agriculture and forestry in the UK. *Advances in Applied Biology,* **9**, 1–142.

Roberts, T. M., Skeffington, R. A. & Blank, L. W. (1989). Causes of type 1 spruce decline in Europe. *Forestry,* **62**, 179–222.

Röhrle, H. (1985). Ertragskundliche Aspekte der Wald-Erkrankungen. *Forstwissenschaftliches Centralblatt,* **104**, 225–45.

Roloff, A. (1985a). *Morphologie der Kronenentwicklung von* Fagus sylvatica *L. (Rotbuche) unter besonderer Berücksichtigung möglicherweise neuartiger Veränderungen.* Dissertation, Forstwissenschaftlicher Fachbereich, Göttingen University.

Roloff, A. (1985b). Untersuchungen zum vorzeitigen Laubfall und zur Diagnose von Trockenschäden in Buchenbeständen. *Allgemeine Forstzeitung,* **40**, 157–60.

Rost-Siebert, K. (1985). Untersuchungen zur H- und Aluminium-Ionentoxizität an Keimpflanzen von Fichte (*Picea abies* [L.] Karst.) und Buche (*Fagus sylvatica* L.) in Lösungskultur. *Berichte des Forschungszentrums Waldökosysteme/Waldsterben,* **12**, 1–319.

Rowland, A., Murray, A. J. S. & Wellburn, A. R. (1985). Oxides of nitrogen and their impact upon vegetation. *Reviews of Environmental Health,* 7, 253–88.

Scherbatskoy, T. & Klein, R. M. (1983). Response of spruce and birch foliage to leaching by acidic mists. *Journal of Environmental Quality,* **12**, 189–95.

Schöpfer, W. & Hradetzky, J. (1984). Der Indizienbeweis: Luftveschmutzung maβgebliche Ursache der Walderkrankung. *Forstwissenschaftliches Centralbatt,* **103**, 241–8.

Schröter, H., Achstetter, L. & Holzapfel, W. (1985). Gesundheitszustand von Tannen, Fichten und Buchen auf Dauerbeobachtungsflächen der FVA in Baden-Württemberg. *Allgemeine Forst- und Jagdzeitung,* **156**, 123–32.

Schütt, P., Blaschke, H., Hoque, E., Koch, W., Lang, K. J. & Schuck, H. J. (1983). Erste Ergebnisse einer botanischen Inventur des 'Fichtensterbens'. *Forstwissenschaftliches Centralblatt,* **102**, 158–66.

Schütt, P. & Summerer, H. (1983). Waldsterben-Symptome an Buche. *Forstwissenschaftliches Centralblatt,* **102**, 201–6.

Schwerdtfeger, F. (1981). *Die Waldkrankheiten.* Paul Parey, Hamburg.

Seitscheck, O. (1981). Verbreitung und Bedeutung der Tannenerkrankung in Bayern. *Forstwissenschaftliches Centralblatt,* **100**, 138–48.

Seufert, G. & Arndt, U. (1986). Beobachtungen in definiert belasteten Modellökosystemen mit jungen Bäumen. *Allgemeine Forstzeitung,* **41**, 545–9.

Sjödin, A. & Grennfelt, P. (1984). Regional background concentrations of NO_2 in Sweden. *3rd European Community Symposium on Physico-chemical Behaviour of Atmospheric Pollutants,* Varese 10–12 April 1984, pp. 346–63.

Skiba, U., Peirson-Smith, T. J. & Cresser, M. S. (1986). Effects of simulated precipitation acidified with sulfuric and/or nitric acid on the throughfall chemistry of sitka spruce (*Picea sitchensis*) and heather (*Calluna vulgaris*). *Environmental Pollution,* **B, 11**, 255–70.

Thiele, V., Specovius, J., Metzger, F. & Prinz, B. (1988). Meβ- und Experimentierstation im Forst (MEXFO)—Errichtung und Betrieb einer Station zur Messung von Luftverunreinigungen im Wald. *Forschungsberichte zum Forschungsprogramm des Landes NRW 'Luftverunreinigungen und Waldschäden'* (Ed. by Ministerium für Umwelt, Raumordnung und Landwirtschaft des Landes, Nordrhein-Westfalen, Düsseldorf), Vol. 1, pp. 111–31. Landesanstalt für Immissionsschutz des Landes NRW, Essen.

Tingey, D. T. & Taylor, G. E. (1982). Variation in plant response to ozone: a conceptual model of physiological events. *Effects of Gaseous Air Pollution in Agriculture and Horticulture* (Ed. by M. H. Unsworth & D. P. Ormrod), pp. 113–38. Butterworth Scientific, London.

Tukey, H. B. (1970). The leaching of substances from plants. *Annual Review of Plant Physiology,* **21**, 305–29.

Ulrich, B., Matzner, E. (1983). Raten der ökosystem-internen H^+-Produktion und der sauren Deposition und ihre Wirkung auf Stabilität, Elastizität von Waldökosystemen. *Verein Deutscher Ingenieure-Berichte,* **500**, 289–300.

Ulrich, B., Murach, D. & Pirouzpanah, D. (1984). Beziehungen zwischen Bodenversauerung und Wurzelentwicklung von Fichten mit unterschiedlich starken Schadsymptomen. *Forstarchiv,* **55**, 127–34.

Verein Deutscher Ingenieure (1987). *Acidic Precipitation — Formation and Impact on Terrestrial Ecosystems.* VDI, Kommission Reinhaltung der Luft, Düsseldorf.

Von Schroeder, J. & Reuss, C. (1883). *Die Beschädigung der Vegetation durch Rauch und die Oberharzer Hüttenrauchschäden.* Paul Parey, Hamburg.

Waldman, J. M. & Hoffmann, M. R. (in press). Nutrient leaching from pine needles impacted by acidic cloud water. *Water, Air, and Soil Pollution.*

Warmbt, W. (1979). Ergebnisse langjähriger Messungen des bodennahen Ozons in der DDR. *Zeitschrift für Meteorologie,* **29**, 24–31.

Wenzel, K. F. (1985). Hypothesen und Theorien zum Waldsterben. *Forstarchiv,* **56**, 51–6.

Winkler, P. (1982). Zur Trendentwicklung des pH-Wertes des Niederschlags in Mitteleuropa. *Zeitschrift für Pflanzenernährung und Bodenkunde,* **145**, 576–85.

Winkler, P. (1983). Der Säuregehalt von Aerosol, Nebel und Niederschlägen. *Verein Deutscher Ingenieure-Berichte,* **500**, 141–7.

Wislicenus, H. (1908). Sammlung von Abhandlungen über Abgase und Rauchschäden. Reprinted in 1985 in *Waldsterben im 19. Jahrhundert.* Verein Deutscher Ingenieure-Verlag, Düsseldorf.

World Health Organisation (1987). *The Effects of Sulfur Oxides on Vegetation.* WHO Regional Publications, European Series No. 23, Copenhagen.

Yang, Y. S., Skelly, J. M. & Chevone, B. I. (1983). Effects of pollutant combinations at low doses on growth of forest trees. *Aquilo, Series Botany,* **19**, 406–18.

Zech, W. & Popp, E. (1983). Magnesiummangel einer der Gründe für das Fichten- und Tannensterben in NO-Bayern. *Forstwissenschaftliches Centralblatt,* **102**, 50–5.

Zöttl, H. W. & Hüttl, R. (1985). Schadsymptome und Ernährungszustand von Fichtenbeständen im südwestdeutschen Alpenvorland. *Allgemeine Forstzeitung,* **40**, 197–9.

17. WATER POLLUTION AND THE MANAGEMENT OF ECOSYSTEMS: A CASE STUDY OF SCIENCE AND SCIENTIST

B. MOSS*
School of Environmental Sciences, University of East Anglia, Norwich NR4 7TJ, UK

INTRODUCTION

The desirability for increased 'exactness' in ecology is seemingly self-evident and nowhere more so than in applied ecology where advice given by ecologists may influence the future of a crop, a national park or, potentially, the biosphere. A problem arises in what is meant by exactness. It might mean, for example, a precise measure of the effect of a substance on an organism under defined conditions, a prediction (with a high degree of success) of the course of change in numbers in a population, or a correct forecast of the application of a particular technique of management to an ecosystem.

The first example is very familiar to pollution biologists working with freshwaters. The standard way to assess the effects of a new chemical or to determine how much of a substance might be discharged to a river is to conduct an LC_{50} or similar test (Mason 1981). A large literature exists on the LC_{50} values for some domesticated laboratory organisms like *Daphnia pulex* and rainbow trout (*Salmo gairdneri*) and values are given to two or even three significant figures (Commission of the European Communities 1978). The test results are precise and carry the credibility which biologists have sought in the face of prejudice in the past from the more 'exact' physical sciences. However, these values are almost useless in predicting the effects of the substance in a real ecosystem, where not only is there a complex and often changing community but also the physical conditions change. The reactions of even the test organism may vary by several orders of magnitude depending on its age, its sex, its

*Present address: Department of Environmental & Evolutionary Biology, The University, Liverpool L69 3BX, UK.

degree of acclimation and the composition of the ambient water (Clark 1986). In such a case any exactness in the LC_{50} data is spurious.

The second example (of population prediction) may seem closer to the reality with which applied ecologists must deal. Fish in temperate waters produce growth rings on scales and various bones which allow reliable ageing of individuals in a population. Because of the commercial importance of many marine fish, the relationship between population size and catch is also well understood (Cushing 1975; Pitcher & Hart 1982). This combination of data on age and population size allows construction of models which predict the effect of fishing intensity with particular gear. The models are increasingly exact. Yet most commercial fish stocks are over-fished (McHugh 1984) because again the exactness is confined to only part of the problem. The human element is ignored and, no matter how good the model *per se*, it has little value if it cannot help sustain the fishery.

The third example, of predicting the effects of a particular style of management on an ecosystem, might encompass a low degree of exactness for an orthodox, increasingly reductionist ecologist. Such experiments often can have only limited controls or none at all; they are usually done under environmental conditions that are changing because of weather cycles or the economic climate surrounding agriculture; and they must involve a great deal of judgement. The scale and speed of change in the biosphere, however, is such that painstaking synthesis of the results of innumerable small controlled experiments to predict the behaviour of the whole, even if they could do this, is impracticable. In the *Realpolitik* of applied ecology, exactness may owe little to the precise measurement of the efficiency of an enzyme or the gene frequency in a population, but rests on a shrewd assessment of what may be practicable in the political and financial climate of the time. It may owe as much to intuition as to experiment, and it cannot ignore the fact that most 'ecological' problems are in fact the problems created by human societies and their aspirations.

Frequently the ecological goals at which the management is aimed will be set by ecologists, for lay political bodies rely upon experts to tell them what might be achieved and why it should be undertaken. This particular advice depends much on the personality and values of the ecological adviser. The suggestion that such subjectivity is inevitable in the role of the professional ecologist will be unacceptable to some (Medawar 1982, 1986). But it seems to me that the idea that scientists proffer advice which is value-free is not only naive but also dangerously arrogant. It is more objective (and enlightening) to take the prejudices and conditionings of the adviser into account (Russell 1987) than to pretend that scientists are *tabulae rasae* whose minds are unimpeded by considerations other than the (often unconsciously selected) data of the current experiments. A much-cluttered palimpsest is by far the better analogy.

I have been involved in the problems caused by eutrophication in the Norfolk Broadland for 16 years. In that time my colleagues and I have been attempting to give bodies which have responsibilities for the management of the area increasingly accurate advice on the ways to restore the waterway from eutrophication. An account of progress toward this end may illustrate steps 'toward a more exact ecology' in the context of a very complex problem seen against its political background, and may also show the limits to which exactness may be taken.

THE NORFOLK BROADLAND

The Broadland in the counties of Norfolk and Suffolk in eastern England specifically includes the rivers Waveney, Yare, Bure, Ant, Thurne and Chet in their lower, flood-plain reaches, and their valley floors. The valley floors were covered originally with a freshwater wetland grading to brackish marshes and mud flats as the three main rivers converged in a common estuary, Breydon Water, near Great Yarmouth.

Probably around 900 AD peat cutting in the valley wetlands began a process by which large pits were eventually excavated (Lambert *et al.*1960). The pits were worked until sometime between about 1250 and 1450 AD when, as a result probably of wetter weather for some decades, they were flooded as water tables rose on the valley floors. Channels were then dug for navigation, and meant that the pits became permanent lakes, the Broads, surrounded by a wetland which was extensively managed for production of thatching reed, alder poles, marsh hay and other wetland products. In recent years the communities of all of the major habitat types of Broadland have become generally less diverse. The undrained wetlands have suffered from a loss of the traditional management which maintained a varied tapestry of successional vegetation types (Nature Conservancy 1965; George 1976). And the waterway itself has been altered by eutrophication, to some extent abetted by recreational boating (Moss 1983). The development of our understanding of the eutrophication problems is the subject of this paper.

Former state of the waterway and aims of restoration

Systematic studies of the Broads and rivers began only in the 1960s, but a wealth of photographic, anecdotal and natural history information, backed by palaeolimnological work, allows reconstruction of the ecosystem prior to the recent changes. In the River Bure these changes began in the mid-nineteenth century as local towns expanded after the Napoleonic wars and developed sewerage systems, but they have intensified since the Second World War (Moss 1988). Many Broads had charophyte swards, and very clear water (as is characteristic of pristine marl lakes in chalk or limestone catchments). Few planktonic diatoms are preserved from this period, which we call Phase 1 in sediment cores.

Phase 1 merged into Phase 2 at various times for different Broads. Phase 2 had ranker-growing plants (e.g. *Ceratophyllum demersum, Hippuris vulgaris, Myriophyllum spicatum, Potamogeton pectinatus, Stratiotes aloides* and nymphaeids) and may have had significant phytoplankton development though not necessarily so. Nomenclature for vascular plants follows Tutin *et al.* (1964–80).

Most of the waterway is now in Phase 3, with very large phytoplankton populations and no submerged plants. With the loss of plants there has also been a decline in invertebrate diversity (Mason & Bryant 1975; Mason 1977), in fish stock and diversity (Anglian Water Authority 1979) and very obviously in water clarity. Restoration of submerged plant beds and clear water has been a main aim for conservation bodies in the area for some time and is enshrined in the formal policies of the Broads Authority (1987).

Initial work — a dominance of boats, chemistry and botany

The *Broadland Study and Plan* (Broads Consortium 1971) was produced by a group of local official bodies. It reflected a concern with the state of the Broads, but there was little mention in it of eutrophication problems, sewage effluent or agricultural drainage. The main problems were assumed to be those of boats, with crowding, bank erosion and latrine discharge causing concern (Duffey 1964). Turbidity in the water was taken to be the result of bottom-disturbance by boat propellers. The author of a 'New Naturalist' book (Ellis 1965) published 6 years previously had also seen no reason to be much concerned about eutrophication — encroachment of the open water by reedswamp was the item of concern. Ellis, however, had hinted at a problem in a previous newspaper article (1964).

Among the ranks of Nature Conservancy officers there was suspicion that there was a eutrophication problem and a concern about loss of submerged plants (Morgan 1972) but the committee-written *Report on Broadland* of the Nature Conservancy (1965) did not raise this as a major problem. In the early 1970s Mason & Bryant (1975) noted the widespread losses of submerged plants and a lack of diversity in the bottom invertebrate communities, and associated this with the increasing intensification of agriculture in the Broadland catchments. They also placed emphasis on boat activity, finding correlations between phytoplankton chlorophyll *a* concentrations and light penetration in the Broads closed to boats but not in those open to navigation.

Nonetheless, the symptoms of eutrophication were readily to be seen in the early 1970s. There was green turbid water in the lower Broads of the River Bure and in Barton Broad (Moss 1977). A mapping of sewage treatment works in the Broadland would have shown that almost every side stream as well as every main river was receiving effluent. And a recent controversy in North America over the roles of phosphorus and carbon as factors limiting algal growth had given especial prominence to the effects of sewage effluent (Likens 1972). This had not been given much attention in Britain, however, and a report of the Water Research Centre (Collingwood 1977) had dismissed eutrophication as not a real problem. In this context it took several years of monitoring to establish eutrophication — not boats — as the key problem in Broadland, and to elicit an official recognition of this by the Water Authority in 1977 (Skinner 1978). Even then the Countryside Commission in its early discussion papers about the possibility of a Broads National Park (Countryside Commission 1977) was still seduced by the prominence of the boats.

Our approach to the eutrophication problems in the 1970s and early 1980s was essentially a chemical one. Osborne (1981) established a nutrient budget for Barton Broad showing that the bulk of the phosphorus loading came from sewage effluent, whilst most of the nitrogen supply was from catchment drainage. This general pattern was also found for the River Bure System (Moss *et al.* 1984), whilst for Hickling Broad the main source of phosphorus was still ultimately of excretory origin, though in this case from a flock of roosting blackheaded gulls (*Larus marinus*), which had risen in size from 1000 birds in 1911 to perhaps 250 000 by the early 1970s (Moss & Leah 1982). The sediments of the Broads were shown to be phosphorus-rich and capable of releasing large quantities of phosphate in summer (Osborne & Phillips 1978) when the surface oxidized-microzone broke down as a result of bacterial activity, despite a well-aerated water column (Wilkinson 1985). Osborne (1980) successfully accounted for the concentration of phosphorus in Barton Broad using Vollenweider's (1975) model, which predicts the concentration in terms of loading, lake depth, flushing rate and exchange with the sediments. We saw increasingly the solution to the Broads' problems as a reduction in the supply of phosphorus from the sewage treatment works either by diversion of effluent to the sea or by precipitation ('stripping', 'tertiary treatment') from the effluent (Osborne & Moss 1977). The literature was dominated by the dramatically successful restoration of Lake Washington by effluent diversion (Edmondson 1970), and legislation had been enacted, or was soon to be so, in North America and mainland Europe, particularly Scandinavia, requiring phosphate precipitation at sewage-treatment works.

Examination of our data from Broads in Phases 1 and 2 led us to associate each phase with particular upper mean total phosphorus concentrations — 50 $\mu g\ l^{-1}$ for Phase 1, 100 $\mu g\ l^{-1}$ for Phase 2 (Moss, Leah & Clough 1979). A target of 100 $\mu g\ l^{-1}$ became widely accepted, almost totemically, as a negotiating concentration by the conservation lobbies (and myself). We were trying to persuade a reticent water authority to install phosphate-stripping at treatment works on the River Ant, for which we had a reasonably accurate nutrient budget.

As a target, 100 $\mu g\ l^{-1}$ is probably too high and this simple Liebigian concept proved to be flawed. However, it was necessary to determine a particular target value because we were dealing with engineers whose approach was based on meeting specifications for a particular project. Although a request to reduce phosphorus by as much as possible would have been wiser ecologically (though I see this in hindsight), an apparent exactness was necessary to stimulate political action. My then confidence in this chemical engineering approach to lake restoration can be traced in a traditional training as a botany student. A predominance of physiological and biochemical attitudes had conditioned me to think in terms of explaining plant (or algal) growth largely in terms of physical and chemical environmental conditions.

Isolation experiments

Phosphate-stripping is the only widely available method of reducing nutrient concentrations in Broadland. Nitrate control is less practicable because of the high solubility and diffuse sources from which the nitrate comes. However, isolation of Broads from nutrient-rich rivers offers a possibility of severe reduction of both nitrate and phosphate. But there is a problem with this technique because it can be used only where there is no conflict with navigation, and the Broads in the main waterway are all tidal, with ancient rights of navigation. In the 1960s the Eastern Regional Officer of the Nature Conservancy, Martin George, had attempted restoration of the submerged plant communities of the dykes in the

Woodbastwick Marshes within the Bure Marshes National Nature Reserve by having the accumulated mud sucked out and encouraging vigorous flushing with river water. This was intuitively a sensible piece of management, although it was unsuccessful until the cleaned dykes had been closed off from the river by wooden dams. By then it was realized that the river water quality was the problem. Diverse communities of Phase 2 plants then grew in the dykes, and encouraged George to propose similar isolation of suitable Broads (discussion on pp. 37–8 of Simmonds 1978).

The same approach was taken by the Norfolk Naturalists' Trust at Alderfen Broad in 1979, where the inflow stream carrying effluent (Moss, Forrest & Phillips 1979) was dammed. It was diverted into a network of pre-existing dykes in the fens surrounding the Broad, to discharge into the sluiced-off former outlet stream. The cost of this was small (less than £1000 at the time). A more ambitious scheme (1982) at Cockshoot Broad not only dammed the Broad from the River Bure, leaving it supplied with a small stream of low phosphorus content, but also removed about 1-m depth of sediment from two-thirds of the Broad's area. This was necessary in any case because the Broad had so filled with sediment that only 10–20 cm of water remained. However, it also greatly reduced any possibility of internal phosphorus loading from the sediments. Results of these experiments up to 1985 have been reported in Moss *et al.* (1986). At Alderfen Broad, a predicted reduction in the internal release of phosphate was found and phytoplankton crops dwindled. A dense cover of *Ceratophyllum demersum* had re-established by 1982 after 3 years of isolation. It persisted in 1983, though it was less dense in 1984. In 1985 submerged plants were scarce and in 1986 not detected (Fig. 17.1). Spring diatom growth had returned in 1984 and the summers of 1985 and 1986 saw blooms of a nitrogen-fixing cyanophyte, *Anabaena planctonica.*

Organic matter from the aquatic plant growths had resulted in resumed release of phosphate from the sediments as it decayed at the sediment surface, and had probably caused release of ammonium ions also. The isolation of the Broad means that there is now very little throughflow, water entering in winter doing little more than replacing evaporation losses from the previous summer. Phosphate and ammonium, released from the sediments and decay of the plants, thus accumulated in the water over winter and supported algal crops in the spring. Phosphate remained abundant even after the spring diatom crops were limited by nitrogen shortage, and so could support cyanophyte nitrogen-fixers in summer. It appeared that the original nutrient sources for the catchment had simply been replaced in Alderfen Broad by sediment and atmospheric sources and that the Broad had returned to a Phase 3 state and might permanently remain so. Alternatively an alternation of plant-dominated and algal-dominated systems might arise within a period of several years. Results from 1987 support this. Algal crops were small and plant biomass had significantly increased (Fig. 17.1).

Alderfen Broad thus had two lessons for us. Isolation without removal of sediment was a feasible technique, but success was likely to be partial because of interactions with the sediment. Our interpretation was still in terms of chemistry and plant growth alone.

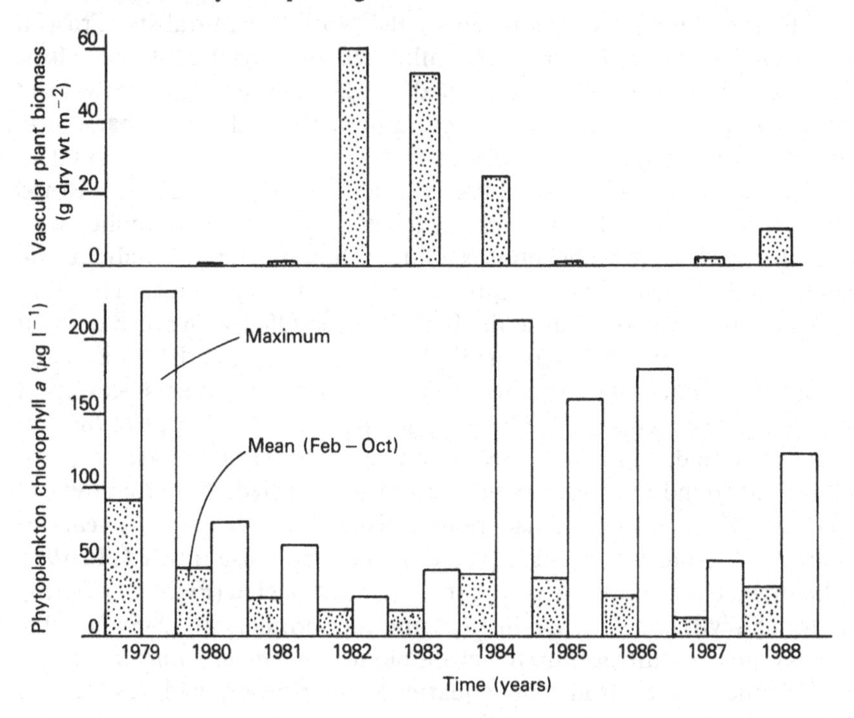

FIG 17.1 Changes in the maximum density of biomass of aquatic plants (largely *Ceratophyllum demersum*) and in the mean and maximum concentrations of chlorophyll *a* in Alderfen Broad since 1979.

Cockshoot Broad was a more expensive experiment (£90 000 in 1982) because of the costs of sediment removal and disposal, but has been highly successful in restoring submerged aquatic plants. The Broad comprises a main basin and a long thin arm which used to be the dyke (Cockshoot Dyke) connecting it with the River Bure. Colonization was very rapid in Cockshoot Dyke, but has been slower in the main basin, perhaps due to movement northwards by wind of potential inocula into the more sheltered dyke area. The deepened part of the main basin had acquired a sparse but extensive cover of *Ceratophyllum demersum* by 1987. During the slow colonization of the main basin between 1982 and 1984 there was a steady fall in phytoplankton crop densities, measured as chlorophyll *a*. However, since 1985 there has been a renewed small increase in chlorophyll *a* concentration though the quality of the inflow stream water has not changed significantly. This recent change is not apparently explicable in terms of nutrients and algal growth alone. It has resulted from a reduction in the population of efficiently grazing zooplankton after predatory fish populations recovered following the disturbance when the Broad was mud-pumped.

Phosphate-stripping in the River Ant

Since 1980 Anglian Water, the regional water authority, has progressively diverted sewage effluent from the River Ant or treated the effluent with ferrous sulphate to precipitate phosphate. Effluent from a food-processing factory has also been treated. The phosphorus loading to the river and hence to Barton Broad, which lies astride it, has been reduced considerably. Concentrations of phytoplankton chlorophyll *a* have decreased, but to a much lower extent, and the Broad remains phytoplankton-dominated (Fig. 17.2). In some of the early years this seems to have been because release of phosphate from the sediments increased the phosphate concentrations in summer, but concentrations in spring and early summer are at or below the 100 $\mu g\ l^{-1}$ target which we had previously determined as representing the upper limit for submerged plant growth.

In retrospect our suggestion that phosphate-stripping would be sufficient to restore Barton Broad was strongly influenced by the success of effluent diversion at Lake Washington and by its widespread adoption elsewhere (Welch 1980; Lack 1981). We had not perhaps appreciated that in the Broads we were attempting to restore a much more eutrophicated system with higher total phosphorus concentrations than those in the systems selected for restoration elsewhere. Perhaps more importantly we were attempting to change the nature of the system from algal to plant dominance, and not merely attempting to change the established state only by degree. A large literature, with sophisticated treatments of the relationship between chlorophyll *a* and total phosphorus concentrations, gives a strong quantitative backing where the latter (see Vollenweider & Kerekes 1981) is the intention, but misled us into oversimplification about the restoration of the Broads. Observations on zooplankton behaviour and experiments in ponds were to widen our view.

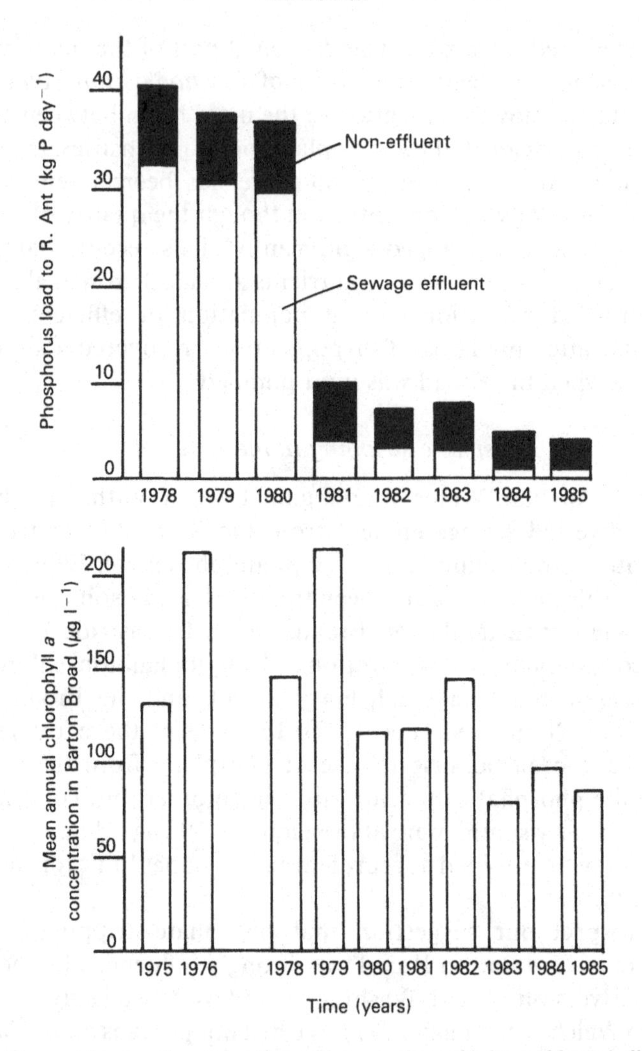

FIG 17.2 Changes in phosphorus loading and mean phytoplankton chlorophyll *a* concentration in Barton Broad as a result of sewage effluent diversion and phosphate-stripping of the remaing effluent. Data from University of East Anglia and Anglian Water sources.

Pond experiments

The mechanism by which phytoplankton competes successfully with aquatic plants as lakes become eutrophicated is not necessarily the obvious one of shading out of the underlying by the overlying. Plants may disappear when adequate light still reaches the bottom. Also vigorously growing species, or those with large energy stores in their rhizomes, can survive in very turbid water in parts of Broadland. Phillips, Eminson & Moss (1978) proposed that epiphytic algae might be more important than planktonic algae in reducing the available energy reaching the plants, and that phytoplankton might develop subsequently to loss of the plants. They suggested that the plants limit phytoplankton growth by secretion of inhibiting substances but that epiphytes, growing intimately on the plant surface, might not be so sensitive to these substances and hence respond earlier to increased nutrient loading than the phytoplankton. Subsequent studies have shown the importance of epiphytes (Sand-Jensen & Borum 1984) and added further credence to the role of allelopathy (Wium-Andersen *et al.* 1982; Van Vierssen & Prins 1985). A competitive advantage might be given to faster- and taller-growing Phase 2 plants over the shorter Phase 1 plants by this mechanism. However, epiphyte burdens seem not to be a problem for the vigorous growth of *Hippuris vulgaris* and *Myriophyllum spicatum* in Hickling Broad (Moss 1981), whose phytoplankton populations also develop to levels consistent with the available nutrient concentrations, despite any potential allelopathy. Other studies (Maberley & Spence 1983) have suggested a possible competition for available CO_2 between phytoplankton and plants, with the former having the advantage of short diffusion pathways when, at high pH, the concentrations of CO_2 and HCO_3 are reduced.

In 1982 we decided to study experimentally the effects of eutrophicating the water supply on a submerged plant community in Broadland (Balls, Moss & Irvine, in press; Irvine, Moss & Balls, in press). In the undrained fens there are nearly parallel-sided dykes originally constructed for the removal by boat of marsh products like reed and hay. Along such a dyke in the Bure Marshes National Nature Reserve we created a series of nineteen 3.5–4 × 10 × 0.7 m ponds with wooden-board dams. Such dams had been used to protect parts of the reserve from River Bure water, and consequently the dyke had a dense community of *Ceratophyllum demersum, Hydrocharis morsus-ranae, Lemna minor* and *L. trisulca* to which we added *Stratiotes aloides.* All of the ponds were heavily fertilized with ammonium nitrate, and were divided into two series. The ponds of one series were regularly raked clear of plants (the 'open ponds'), whilst in the 'plant ponds' the community was left intact. A graded series of regular phosphate additions was made during the summer to the ponds in each series. This ranged from zero addition to about 2.6 g m^{-2}, a value which, with the negligible summer flushing of the ponds, gave effective loadings as high as those in the Broads. In 1982 we removed fish from the ponds by electrofishing so as to eliminate one variable that we assumed to be irrelevant to the botanical–chemical hypothesis we were testing.

We expected that large phytoplankton populations would develop at the highest phosphate loadings in both series, and that the plant growth would be reduced in the plant ponds. Neither happened. The plant community remained, phytoplankton populations were negligible and the concentrations of total phosphorus established in the water were much lower than anticipated from the amounts of phosphate added.

Large populations of *Daphnia* and other large-bodied Cladocera were present in the ponds, and a combination of uptake of nutrients by the plants, denitrification, and grazing of any developing phytoplankton by the Cladocera was acting to remove the added nutrients as gaseous nitrogen, as plant biomass or as rapidly sedimenting zooplankton faeces. The development of the large cladoceran populations, again in retrospect, can be seen to reflect the removal of fish, their main predators (Hrbacek *et al.* 1961: Brooks & Dodson 1965; Hall *et al.* 1976). It became apparent that there could be zoological dimensions to a problem that we had hitherto tackled only in botanical and chemical terms.

There had been earlier indications of an important role for the zooplankton which should have alerted us. An isolation experiment at Brundall Broads (Leah, Moss & Forrest 1980) had 'gone wrong' when effluent-rich water from the River Yare penetrated at high tide through leaks caused by tree roots in the embankments intended to isolate the experimental Broad from such influence. Large phytoplankton populations thus developed in the 'isolated' Broad as well as in a control Broad kept open to the river. In the subsequent year, however, major differences developed. The control Broad retained high phytoplankton populations whilst the 'isolated' one, despite very high phosphate and nitrate concentrations, had low algal populations and very clear water after the end of May. This was associated with large populations of *Daphnia hyalina* and negligible fish stocks, in contrast to a rotifer-copepod zooplankton community in the control Broad which had a substantial population of roach (*Rutilus rutilus*). The predatory effects of small fish on zooplankton communities are well understood; see, for example, Kerfoot & Sih (1987). They remove large Cladocera in preference to small organisms like rotifers and small Cladocera and fast-moving ones like copepods. Development of the *Daphnia* population and its high grazing potential clearly reflected the loss of the fish stock in the 'isolated' broad, perhaps as a result of predation by cormorants (*Phalacrocorax carbo*) in a situation where replenishment of the stock was not possible. Large stands of *Nitella flexilis* developed in the 'isolated' broad, and it was thus shown that phytoplankton could be reduced and plant growth encouraged in the Broads, despite high nutrient loadings and concentrations.

Loss of the fish stock is not a normal feature of the Broads. A second set of observations, however, showed that if the zooplankton community could find refuge against fish predation amongst beds of submerged plants, they could effectively control the phytoplankton by grazing, despite the presence of planktivorous fish (Timms & Moss 1984). Hoveton Great Broad is phytoplankton-dominated but connected by open channels with Hudsons Bay and the effluent-rich River Bure. Both Broads receive nutrient-enriched water, but Hudsons Bay, for reasons that are not clear, retains stands of white water-lilies (*Nymphaea alba*). In open water, adjacent to the stands, chlorophyll *a* concentrations fall to negligible values in the lily season, compared with high values in the almost lily-free Hoveton Great Broad. The lily beds support large communities of Cladocera, which probably move out of the beds more or less continually. By day they are probably removed by fish, but at night their numbers build up in the open water, and their grazing on phytoplankton is very effective.

With these new realizations, we repeated in 1983 our 1982 experiments in the ponds after we had added uniform stocks of small zooplanktivorous fish (mainly roach) to the ponds. There was a differential survival of fish and complex effects on the zooplankton community, but the chemical results were clear-cut. The plant ponds again retained low total phosphorus concentrations despite the high phosphate loadings given, whilst a graded increase in total phosphorus concentrations reflected the loadings to the cleared ponds. And, where fish survival was high in the cleared ponds, cladoceran populations were reduced and phytoplankton crops, similar in chlorophyll *a* concentrations to those of the Broads, were able to develop. With a somewhat rueful regret at not having included any formal zoological courses in my past education, I mentally rearranged the Broads' ecological theatre for a complex play more like Ibsen than a medieval mystery.

Alternative states and eutrophication

Aquatic plant communities clearly could survive and there could be clear water with low phytoplankton concentrations despite high nutrient loadings (ponds experiments) or concentrations (Hudsons Bay, Brundall Broad). The communities could buffer the concentrations and prevent phytoplankton build-up by harbouring grazers and perhaps by allelopathy.

In turn a phytoplankton community might remain despite decreased nutrient loadings (Barton Broad), perhaps because in the open water in shallow lakes there are limited refuges for grazing Cladocera against fish predation and little possibility of control by grazing. Support for this also comes from experiments done in butyl rubber reservoirs (Lund tubes) in Hickling Broad (Moss & Leah 1982). Despite isolation from catchment nutrient sources and bird droppings, a dense phytoplankton community, dominated by *Aphanothece* sp., persisted for the 3 years of the experiment. The zooplankton populations in the water were negligible.

It thus seems that there might be two alternative states for the aquatic ecosystem in the Broads. Once established, either can persist, stabilized by various buffering mechanisms (Fig. 17.3), over a range of nutrient loadings or concentrations. At loadings or concentrations lower than this range it is possible to re-establish an aquatic plant community by nutrient control alone, as has been done in the isolation experiments at Alderfen and Cockshoot Broads. There may also be an extreme upper range of nutrient loadings and concentrations uniquely occupied by phytoplankton dominance. This is predicted by mathematical modelling (Bondi 1985). The present problem is that the available method of nutrient control, phosphate-stripping, though it is needed to bring concentrations to within the middle range which can support alternative stable systems, may be insufficient alone to reduce nutrient availability enough to support plant dominance as a unique state. Our original 100 $\mu g\ l^{-1}$ target seems to fall within this middle range.

Forward and backward switches between the two states

Our original working hypothesis for the effects of eutrophication in Broadland was of a steady change from plant to algal dominance as nutrient loadings increased. This was familiar to many limnologists, to judge by the emphasis given to phosphorus–chlorophyll regressions in the planning of strategies for eutrophication control (Vollenweider & Kerekes 1981). It was a convenient model, readily communicable to the planners and engineers of government bodies who have the responsibility for implementation of such strategies. By 1985, however, we were forced to accept that matters were not simple. (I use the verbs with particular care because many scientists will recognize the reticence with which long-cherished hypotheses are dropped despite mounting evidence against them.)

Not only was our target of 100 $\mu g\ P\ l^{-1}$ too simple a concept but we now had to explain what caused the forward switch from plant domi-

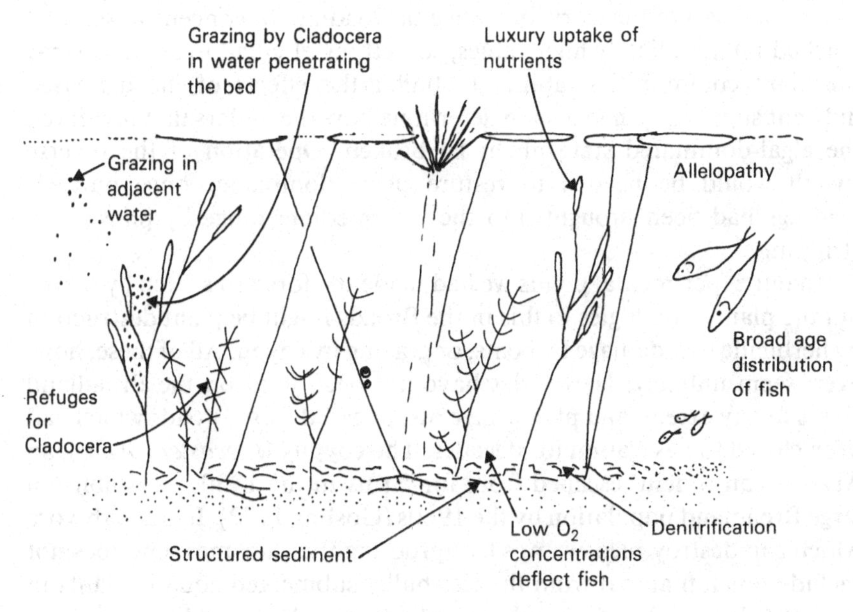

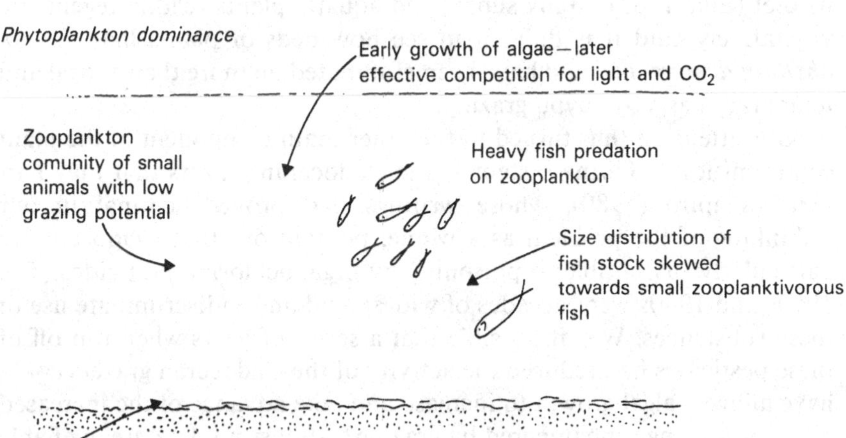

FIG 17.3 Mechanisms likely to contribute to the buffering of submerged plant communities against the effects of increasing nutrient loading and to the buffering of phytoplankton communities against decreasing nutrient loading.

nance to algal dominance in the 1950s and 1960s. Nutrient increase was clearly one part of the story, but, once the loadings or concentrations had reached intermediately high values, something else had to act to destroy the plant communities' ability to buffer the effects of the increased nutrient supply. We had also to determine how the buffers that preserved the algal-dominated state might be broken. Operation of the reverse switch would be needed to restore plant dominance once nutrient loadings had been brought into the intermediate range by phosphate-stripping.

In our experimental ponds we had made the forward switch by raking out the plants. Analogies to this in the Broads might be plant destruction by herbicide use, damage by boats, or grazing by coypu. All of these, however, seem unlikely. Herbicides have not been used on the Broadland rivers to any extent, and plants have disappeared from Broads which had been closed to navigation for decades. The coypu (*Myocaster coypus*), an Argentinian rodent, escaped from fur-farms in the 1930s to establish a large Broadland population by the 1960s (Gosling 1972). It is a herbivore which can destroy reedswamps by uprooting the rhizomes, but does not include much material from the less bulky submerged aquatic plants in its diet (Ellis 1963). Many submerged aquatic plants readily regenerate vegetatively, and it is difficult to see how beds of plants like *Ceratophyllum demersum* could have been eliminated on more than a local and temporary basis by coypu grazing.

Our attention thus turned to the other main component of the plant communities' buffering system — the cladoceran grazers that they harbour. Shapiro (1980), whose writings had proved seminal in our rethinking of the problem as a whole, pointed out that Cladocera are particularly susceptible to poisoning by organochlorine pesticides. The 1950s and 1960s were decades of widespread and indiscriminate use of these substances. Was it possible that a series of years when run off of these pesticides had reduced the activity of the cladoceran grazers could have allowed algal growth to increase and take advance of the increased nutrient loading, unhampered by grazing? Almost no data are available on dissolved pesticide concentrations for the critical period, and those that are available are for the end of the period when the dangers of organochlorines were being realized and greater care was being exercised in their use (Croll 1969; Lowden, Saunders & Edwards 1969). No quantitative data are available for the zooplankton communities at the time either.

Well-layered sediments are, however, preserved in many of the

Broads and have been used to establish the changes that have occurred as eutrophication has proceeded (Osborne & Moss 1977; Moss 1980, 1988). We have examined sediment cores from Hoveton Great Broad, from which submerged plants have been lost, and Martham Broad, which has a continuous history of plant dominance, isolated as it is from sewage effluent and intensive agriculture. At Hoveton Great Broad, but not at Martham Broad, peaks of DDT derivatives have been found (Stansfield, Moss & Irvine, in press), coincident with the change from Phase 2 to Phase 3. Relating a concentration of a pesticide in sediment to its concentration in the water at the time the sediment was laid down is very difficult. There are several pathways by which the pesticide may reach the sediment (direct absorption, burial of animal corpses which have accumulated the pesticide, movement into the lake of soil particles to which the pesticide is absorbed). By using the most conservative data on partition between water and sediment and on half-life of the pesticide, we calculate that concentrations as great as those killing *Daphnia* and *Simocephalus* species in toxicity tests (Sanders & Cope 1968) could have been present. Sub-lethal effects which could influence population growth and grazing efficiency would occur at much lower concentrations (Gliwicz & Sieniawska 1986). The hypothesis of a role of organochlorine pesticides in effecting the forward switch in at least parts of Broadland is thus worthy of further examination.

Understanding the causes of the loss of submerged plants from Broadland is obviously necessary, not least to prevent it happening again once the plants have been restored. What to do to bring about the latter is the second problem created by the failure, so far, of limited nutrient reduction to do it alone.

Again our thinking has turned to the Cladocera. Plants originally provided refuges for their populations to build up. Provision of artificial refuges to husband the Cladocera might then increase populations sufficiently to reduce algal populations and allow submerged plants to re-establish. We have tested a variety of materials (polypropylene rope, plastic fruit-cage netting, bundles of alder twigs) and found that the twig-bundles may offer a cheap and workable option (Moss, Irvine & Stansfield 1988), though a quite large proportion of a Broad might need to be covered with such refuges. A further option for biomanipulation as a restorative measure would be removal of zooplanktivorous fish. In a popular angling area like Broadland which is also extensive, with interconnected rivers and lakes, this is politically and technically difficult.

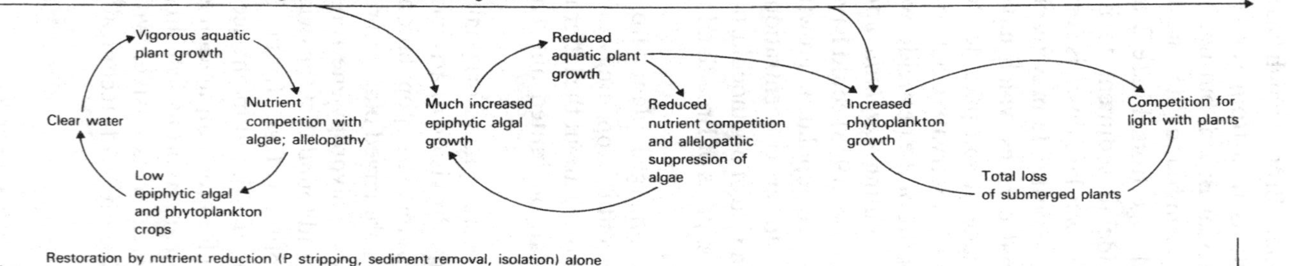
FORMER MODEL
(See Phillips et al. 1978). Influences on thinking: 1. Classic studies (e.g. L. Washington). 2. Limited experience of investigators. 3. Need to simplify situation for discussions with engineers and planners. 4. Success of isolation experiments.
Steadily increasing nutrient loading (P from effluent, N from agriculture)
Vigorous aquatic plant growth
Nutrient competition with algae; allelopathy
Low epiphytic algal and phytoplankton crops
Clear water
Much increased epiphytic algal growth
Reduced aquatic plant growth
Reduced nutrient competition and allelopathic suppression of algae
Increased phytoplankton growth
Competition for light with plants
Total loss of submerged plants
Restoration by nutrient reduction (P stripping, sediment removal, isolation) alone

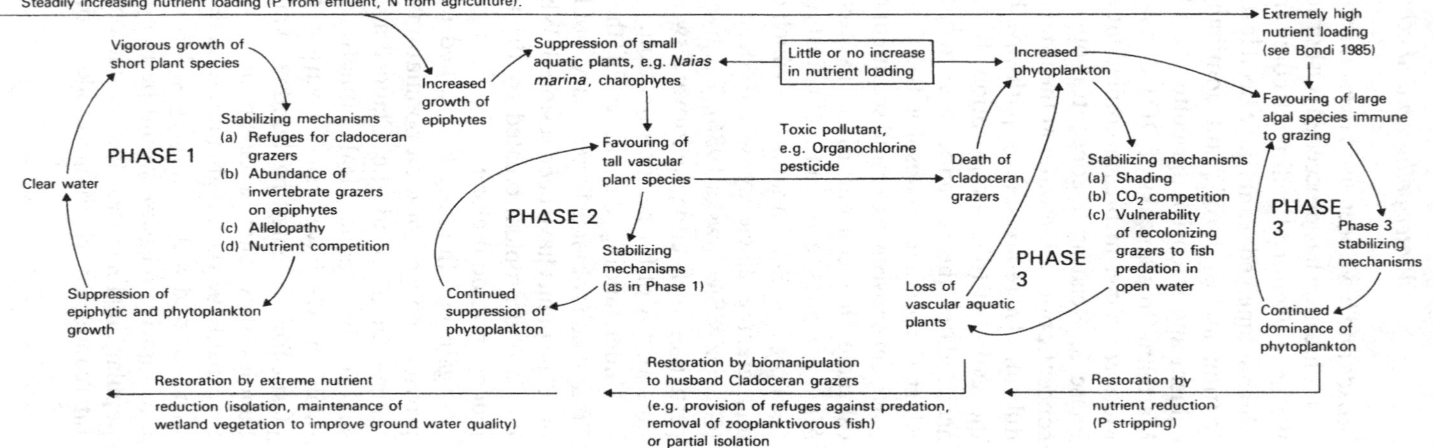
CURRENT MODEL
Influences on thinking: 1. New influences e.g. Shapiro (1980). 2. Widening experience of investigators. 3. Failure of phosphorus stripping at Barton Broad. 4. Field experiments and observations on role of cladoceran grazers.
Steadily increasing nutrient loading (P from effluent, N from agriculture).
Extremely high nutrient loading (see Bondi 1985)
Vigorous growth of short plant species
Stabilizing mechanisms
(a) Refuges for cladoceran grazers
(b) Abundance of invertebrate grazers on epiphytes
(c) Allelopathy
(d) Nutrient competition
PHASE 1
Clear water
Suppression of epiphytic and phytoplankton growth
Increased growth of epiphytes
Suppression of small aquatic plants, e.g. Naias marina, charophytes
Little or no increase in nutrient loading
Favouring of tall vascular plant species
PHASE 2
Stabilizing mechanisms (as in Phase 1)
Continued suppression of phytoplankton
Toxic pollutant, e.g. Organochlorine pesticide
Death of cladoceran grazers
Increased phytoplankton
Stabilizing mechanisms
(a) Shading
(b) CO_2 competition
(c) Vulnerability of recolonizing grazers to fish predation in open water
PHASE 3
Loss of vascular aquatic plants
Favouring of large algal species immune to grazing
PHASE 3
Phase 3 stabilizing mechanisms
Continued dominance of phytoplankton
Restoration by extreme nutrient reduction (isolation, maintenance of wetland vegetation to improve ground water quality)
Restoration by biomanipulation to husband Cladoceran grazers (e.g. provision of refuges against predation, removal of zooplanktivorous fish) or partial isolation
Restoration by nutrient reduction (P stripping)

THE IMPLICATIONS OF EXACTNESS

Are there lessons to be learned from this story, as summarized in Fig. 17.4? Exactness at the scale of solving practical problems need not mean the elaborate and increasingly precise measurement of isolated components but an increasingly accurate understanding of how the components fit together. This needs a wide view; individuals like myself are often hampered by the narrow prisons of their prior training but the view may be expanded by groups of people of diverse backgrounds. Our understanding of Broadland has benefited immensely from the combined contributions of those who knew about the properties of phosphorus but had never heard of furcal rami, and vice versa. Frequently the understanding has come from 'we' and the hold-ups from me!

The wide view is important because solution of the earth's ecological (*sensu lato*) problems requires it (Vallentyne & Hamilton 1987). They are the problems of the only ecosystem that is definable, because it is bounded. One report of the British Ecological Society's first jubilee symposium (North 1988) suggests, on the basis of what he heard, that the modern ecologist 'distinctly does not believe in a "balance of nature"', and 'dislikes the greatest single invention of the breed: "ecosystems"'. Yet the biosphere is a system far departed from chemical equilibrium (Lovelock 1979) and has been maintained by living organisms within the rather narrow limits which will support a living system based on carbon compounds and liquid water for at least 2×10^9 years. This must imply some validity for a holistic ecosystem concept. It is an approach being taken over by the chemists and other environmental scientists whilst the ecologists — *sensu stricto* — as they increasingly seem to want to be, could be left exploring minutiae. If this is the value of exactness, there are great dangers in it. No matter how well we appreciate the construction and performance of the fiddle, it is readily combustible, along with Rome.

REFERENCES

Anglian Water Authority (1979). *Report of Preliminary Fisheries Surveys carried out in Broadland during the period July 1977–June 1978.* Norwich.

Balls, H., Moss, B. & Irvine, K. (in press). The loss of aquatic plants on eutrophication. I. Experimental design, water chemistry, aquatic plant and phytoplankton biomass in experiment carried out in ponds in the Norfolk Broads. *Freshwater Biology*, **21**.

Bondi, H. (1985). A model of discontinuous change in a three-component community. *Proceedings of the Royal Society of London*, **B**, **224**, 1–6.

Broads Authority (1987). *Broads Strategy and Management Plan.* Norwich.

Broads Consortium (1971). *Broadland Study and Plan.* Norwich.

Brooks, J. L. & Dodson, S. I. (1965). Predation, body size and composition of plankton. *Science*, **150**, 28–35.

Clark, R. B. (1986). *Marine Pollution.* Clarendon, Oxford.

Collingwood, R. W. (1977). *A Survey of Eutrophication in Britain and its Effects on Water Supplies.* Technical Report TR 40. Water Research Centre, Medmenham.

Commission of the European Communities (1978). *Noxious Effects of Dangerous Substances in the Aquatic Environment.* Environment and Consumer Protection Service, EUR 5983 EN.

Countryside Commission (1977). *The Broads: Possible Courses of Action.* Countryside Commission, Cheltenham.

Croll, B. T. (1969). Organochlorine insecticides in water. Part 1. *Water Treatment and Examination,* **18**, 253–74.

Cushing, D. H. (1975). *Marine Ecology and the Fisheries.* Cambridge University Press.

Duffey, E. (1964). The Norfolk Broads. A regional study of wild life conservation in a wetland area with high tourist attraction. *Project Mar Conference. IUCN Publications, New Series,* **3**, 290–301.

Edmondson, W. T. (1970). Phosphorus, nitrogen and algae in Lake Washington after diversion of sewage. *Science,* **169**, 690–1.

Ellis, E. A. (1963). Some effects of selective feeding by the coypu (*Myocaster coypus*) on the vegetation of Broadland. *Transactions of the Norfolk and Norwich Naturalists Society,* **20**, 32–5.

Ellis, E. A. (1964). In the countryside. *Eastern Daily Press,* 15 September, p. 6.

Ellis, E. A. (1965). *The Broads.* Collins, London.

George, M. (1976). Land use and nature conservation in Broadland. *Geography,* **61**, 137–42.

Gliwicz, M. S. & Sieniawska, A. (1986). Filtering activity of *Daphnia* in low concentrations of a pesticide. *Limnology and Oceanography,* **31**, 1132–7.

Gosling, L. M. (1972). The coypu in East Anglia. *Transactions of the Norfolk and Norwich Naturalists Society,* **23**, 49–59.

Hall, D. J., Threlkeld, S. T., Burns, C. W. & Crawley, P. H. (1976). The size and efficiency hypothesis and the size structure of zooplankton communities. *Annual Review of Ecology and Systematics,* **7**, 177–208.

Hrbacek, J., Bvorakova, K., Korinek, V. & Prochazkova, L. (1961). Demonstration of the effect of the fish stock on the species composition of the zooplankton and the intensity of metabolism of the whole plankton association. *Verhandlungen internationale der Vereinigung theoretische und angewandte Limnologie,* **14**, 192–5.

Irvine, K., Moss, B. & Balls, H. R. (in press). The loss of aquatic plants on eutrophication II. Relationships between fish and zooplankton in a set of experimental ponds, and conclusions. *Freshwater Biology.*

Kerfoot, W. C. & Sih, A. (1987). *Predation, Direct and Indirect Impacts on Aquatic Communities.* University Press of New England, Hanover, New Hampshire.

Lack, T. J. (1981). Advances in the management of eutrophic reservoirs. *Notes on Water Research,* **27**, 1–4. Water Research Centre, Medmenham.

Lambert, J. M., Jennings, J. N., Smith, C. T., Green, C. & Hutchinson, J. N. (1960). *The Making of the Broads: a Reconstruction of their Origin in the light of New Evidence.* Royal Geographical Society, London.

Leah, R. T., Moss, B. & Forrest, D. E. (1980). The role of predation in causing major changes in the limnology of a hyper-eutrophic lake. *Internationale Revue der gesamten Hydrobiologie,* **65**, 223–47.

Likens, G. E. (Ed.) (1972). *Nutrients and Eutrophication.* Special Symposium 1 of the American Society of Limnology and Oceanography, Lawrence, Kansas.

Lovelock, J. E. (1979). *Gaia: a New Look at Life on Earth.* Oxford University Press.

Lowden, G. T., Saunders, C. L. & Edwards, R. W. (1969). Organochlorine insecticides in water. Part II. *Water Treatment and Examination,* **18**, 275–87.

Maberley, S. C. & Spence, D. H. N. (1983). Photosynthetic inorganic carbon use by freshwater plants. *Journal of Ecology*, **71**, 705–24.

McHugh, J. L. (1984). *Fishery Management*. Springer, Berlin.

Mason, C. F. (1977). Populations and production of benthic animals in two contrasting shallow lakes in Norfolk. *Journal of Animal Ecology*, **46**, 147–72.

Mason, C. F. (1981). *Biology of Freshwater Pollution*. Longman, London.

Mason, C. F. & Bryant, R. J. (1975). Changes in the ecology of the Norfolk Broads. *Freshwater Biology*, **5**, 257–70.

Medawar, P. B. (1982). *Pluto's Republic*. Oxford University Press.

Medawar, P. B. (1986). *Memoir of a Thinking Radish*. Oxford University Press.

Morgan, N. C. (1972). Problems of the conservation of freshwater ecosystems. *Symposium of the Zoological Society of London*, **29**, 135–54.

Moss, B. (1977). Conservation problems in the Norfolk Broads and rivers of East Anglia, England — phytoplankton, boats and the causes of turbidity. *Biological Conservation*, **12**, 95–114.

Moss, B. (1980). Further studies on the palaeolimnology and changes in the phosphorus budget of Barton Broad, Norfolk. *Freshwater Biology*, **10**, 261–79.

Moss, B. (1981). The composition and ecology of periphyton communities in freshwaters. II Inter-relationships between water chemistry, phytoplankton populations and periphyton populations in a shallow lake and associated experimental reservoirs ('Lund tubes') *British Phycological Journal*, **16**, 59–76.

Moss, B. (1983). The Norfolk Broadland: experiments in the restoration of a complex wetland. *Biological Reviews*, **58**, 521–61.

Moss, B. (1988). The palaeolimnology of Hoveton Great Broad, Norfolk: clues to the spoiling and restoration of Broadland. *Exploitation of Wetlands* (Ed. by P. Murphy & C. French), pp. 163–91. British Archaeological Reports, International Series.

Moss, B., Balls, H., Booker, I., Manson, K. & Timms, M. (1984). The River Bure, United Kingdom: patterns of change in chemistry and phytoplankton in a slow-flowing fertile river. *Verhandlungen internationale der Vereinigung theoretische und angewandte Limnologie*, **22**, 1959–64.

Moss, B., Balls, H., Irvine, K. & Stansfield, J. (1986). Restoration of two lowland lakes by isolation from nutrient-rich water sources with and without removal of sediment. *Journal of Applied Ecology*, **23**, 391–414.

Moss, B. Forrest, D. E. & Phillips, G. (1979). Eutrophication and palaeolimnology of two small mediaeval man-made lakes. *Archiv für Hydrobiologie*, **85**, 409–425.

Moss, B., Irvine, K. & Stansfield, J. (1988). Approaches to the restoration of shallow eutrophicated lakes in England. *Verhandlungen internationale der Vereinigung theoretische und angewandte Limnologie*, **23**, 414–18.

Moss, B. & Leah, R. T. (1982). Changes in the ecosystem of a guanotrophic and brackish shallow lake in Eastern England: potential problems in its restoration. *Internationale Revue der gesamten Hydrobiologie*, **67**, 625–59.

Moss, B., Leah, R. T. & Clough, B. (1979). Problems of the Norfolk Broads and their impact on freshwater fisheries. *Proceedings of the First British Freshwater Fisheries Conference (Liverpool)*. (Sponsored by the University of Liverpool and National Federation of Anglers), pp. 67–85.

Nature Conservancy (1965). *Report on Broadland*. London.

North, C. (1988). One day, everybody will be an ecologist. *The Independent*, 18 April 1988, p. 17.

Osborne, P. L. (1980). Prediction of phosphorus and nitrogen concentrations in lakes from both internal and external loading rates. *Hydrobiologia*, **69**, 229–33.

Osborne, P. L. (1981). Phosphorus and nitrogen budgets of Barton Broad and predicted effects of a reduction in nutrient loading on phytoplankton biomass in Barton, Sutton and Stalham Broads, Norfolk, United Kingdom. *Internationale Revue der gesamten Hydrobiologie,* **66**, 171–202.

Osborne, P. L. & Moss, B. (1977). Palaeolimnology and trends in the phosphorus and iron budgets of an old man-made lake, Barton Broad, Norfolk. *Freshwater Biology,* **7**, 213–33.

Osborne, P. L. & Phillips, G. L. (1978). Evidence for nutrient release from the sediments of two shallow and productive lakes. *Verhandlungen internationale der Vereinigung theoretische und angewandte Limnologie,* **20**, 654–8.

Phillips, G. L., Eminson, D. & Moss, B. (1978). A mechanism to account for macrophyte decline in progressively eutrophicated freshwaters. *Aquatic Botany,* **4**, 103–26.

Pitcher, T. J. & Hart, P. J. B. (1982). *Fisheries Ecology.* Croom Helm, London.

Russell, N. (1987). Breakthrough for the bloody-minded. *Times Higher Educational Supplement,* 25 September 1987, p. 14.

Sanders, H. O. & Cope, O. B. (1968). Toxicity of several pesticides to two species of cladocerans. *Transactions of the American Fisheries Society,* **95**, 165–9.

Sand-Jensen, K. & Borum, J. (1984). Epiphyte shading and its effect on photosynthesis and diel metabolism of *Lobelia dortmanna* L. during the spring bloom in a Danish lake. *Aquatic Botany,* **20**, 109–119.

Shapiro, J. (1980). The importance of trophic-level interactions to the abundance and species composition of algae in lakes. *Hypertrophic Ecosystems* (Ed. by J. Barica and L. R. Mur), pp. 362–88. Junk, The Hague.

Simmonds, A. (Ed.) **(1978).** *The Future of Broadland.* Centre of East Anglian Studies, University of East Anglia, Norwich.

Skinner, A. F. (1978). The work of the Anglian Water Authority. *The Future of Broadland* (Ed. by A. Simmonds), pp. 17–18. Centre of East Anglian Studies, University of East Anglia, Norwich.

Stansfield, J., Moss, B. & Irvine, K. (in press). The loss of aquatic plants on eutrophication. III. Potential role of organochlorine pesticides in the replacement of plant-dominated by phytoplankton-dominated communities—a palaeoecological study. *Freshwater Biology,* **21**.

Timms, R. M. & Moss, B. (1984). Prevention of growth of potentially dense phytoplankton populations by zooplankton grazing, in the presence of zooplanktivorous fish, in a shallow wetland ecosystem. *Limnology and Oceanography,* **29**, 472–86.

Tutin, T. G., Heywood, V. H., Burges, N. A., Valentine, D. H., Walters, S. M. & Webb, D. A. (1964–80). *Flora Europaea,* Vols. 1–5. Cambridge University Press.

Vallentyne, J. R. & Hamilton, A. L. (1987). Managing human uses and abuses of aquatic resources in the Canadian ecosystem. *Canadian Aquatic Resources* (Ed. by M. C. Healey and R. R. Wallace), pp. 513–33. *Canadian Bulletin of Fisheries and Aquatic Sciences,* **215**.

Van Vierssen, W. & Prins, T. C. (1985). On the relationship between the growth of algae and aquatic macrophytes in brackish water. *Aquatic Botany,* **21**, 165–79.

Vollenweider, R. A. (1975). Input-output models with special reference to the phosphorus loading concept in limnology. *Schweizerische Zhurnal der Limnologie,* **37**, 53–84.

Vollenweider, R. A. & Kerekes, J. J. (1981). Background and summary results of the OECD cooperative programme on eutrophication. Appendix 1 in *The OECD Cooperative Programme on Eutrophication. Canadian Contribution.* Environment Canada Scientific Series, 131.

Welch, E. B. (1980). *Ecological Effects of Waste Water.* Cambridge University Press.

Wilkinson, D. E. (1985). *Chemical and biological studies of a shallow freshwater interface.* Ph. D. thesis, University of East Anglia.

Wium-Andersen, S., Anthoni, U., Christophersen, C. & Honea, G. (1982). Allelopathic effects on phytoplankton by substances isolated from aquatic macrophytes (Charales). *Oikos,* **39**, 187–90.

18. TOWARD AN EXACT HUMAN ECOLOGY

M. SLESSER
Centre for Human Ecology, University of Edinburgh, Buccleuch Place, Edinburgh EH8 9LN, UK

INTRODUCTION

In order to be a 'human ecologist' one does not need to be an ecologist, even though it may help. Human ecology has to include the social as well as the natural sciences, since cultural, social, institutional and economic issues are a major factor in creating ecological impacts.

Although the distinction is necessarily blurred, ecology deals with the natural system, which is driven by solar energy, and man's perturbation and management of it. Human ecology is concerned more directly with man's interaction with this natural system and has a strong anthropocentric emphasis. There is no country in the world, and virtually no community left, which survives by solar energy alone. Our material standard of living, our economic system, even the machine by which this text is being captured, are made possible because we are exploiting the earth's inherited energy resources. This is not to downgrade the importance of solar energy fluxes, for it is through fossil and fissile energy that one may catalyse the capture of solar energy, for example, in using synthetic fertilizers to enhance photosynthesis.

Like much in ecology, these are rate processes, not merely flows to be measured in energy units per unit time, but fluxes as energy units per unit time per unit area.

Human ecology is an integrative approach to looking at the world's economic system in the context of the sustainability of the natural environment, upon which it depends. It holds that one cannot look at the one without assessing its impact on the other, and vice versa. To quantify these interactions inevitably requires aggregation if the resulting models are to be manageable. This will undoubtedly disturb the ecologist familiar with particular systems. A further problem in quantifying interactions has been to find some common numeraire so that activities in both may be computed simultaneously and so related quantitatively. Does this offer an approach 'toward an exact human ecology'? Let us see.

FINDING A NUMERAIRE

In everyday life our activities are quantified and interrelated in money terms, and have given rise to a set of conventions called 'accounting practice' and to a quasi-science called economics. I call it 'quasi' not out of disrespect nor because it is unscientific, but because it is the only intellectual activity I know which, while calling itself a science, uses a unit of measurement that changes in an unpredictable manner with time, namely money.

Economics is often self-described as the science of scarcity, for it seeks to deal with the appropriate allocation of scarce resources between competing demands. In order to quantify the choices implicit in any development process, economics uses the concept of money. The money numeraire provides a means of expressing relative human values and preferences, and certainly no other system has ever been invented which does this in such a sophisticated manner, or which offers such an instantaneous valuation. What do we do, however, when society has no valuation of an activity or situation, even when many individuals feel it should have? The standard response of the economics profession is to carry out a cost-benefit analysis. Yet, however brilliantly executed, such analyses imply value judgements, and are highly sensitive to the time horizon taken into account. This time horizon is set by the discount rate to be applied, which is undeterminable.

What we need, as one enlightened planner said to me recently, is transcendental cost–benefit analysis!

There would appear to be two key areas where the tools of the economists fail to offer a practical methodology. One is when looking beyond the immediate future — that is forecasting — and the other when seeking to embrace environmental factors. For example, a common natural good like air or species diversity, while underpinning the economy, has no marketable monetary value.

At the moment we deal with these issues by indulging in models of each, and then linking them by mental models — a most haphazard procedure, and one fraught with danger.

LOOKING TO THE FUTURE

Let us first look at economic forecasting. If what is forecast becomes the basis of policy, then it must have an ecological consequence. So the quality of the forecast is important to an ecologist. For example, an accurate forecast of the amount of fossil fuel to be burnt in the world must surely influence the policy towards the acid-rain problem. However, data are the essential problem of economic forecasting. Whereas economic study can determine the manner in which people's valuations have changed in the past, there is no certainty that such valuations will persist into the future, and experience has demonstrated that most do change with time and situation. Since development planning calls for estimation of how future choices will be made (and valued), development planning along these lines faces considerable uncertainty. Indeed, forecasts which are made on the basis of such assumptions have been notoriously inaccurate (Ascher 1978). Schrattenholzer, who has made himself the focus of a world survey of energy forecasting at the International Institute for Applied Systems Analysis, recently highlighted the issue with the words 'Long-term energy projections become obsolete quite rapidly' (Schrattenholzer 1984). One consequence has been that there is a growing distrust of economic models as a guide to development planning, and a trend away from their use. This is serious, for it implies the replacement of an explicit statement of one's assumptions by hunch, prejudice or, worse still, the expression of normative goals without any demonstration of their possibility.

While the use of the valuation concept has proved enormously useful in permitting economics to explain how people make and change their choices, and how development has taken place, its role as a forecasting tool is really limited to the immediate future, where current values can be expected to continue with little immediate change. The valuation concept has proved deceptive when it comes to considering matters ecological. First, there are many things to which valuation cannot be ascribed, or for which there is such uncertainty as to provide meaningless values — for example, a fine viewpoint. Secondly, nature is concerned with the long term. At today's discount rate, economic calculations, even if they dealt with acceptable values, are indifferent to any event 15 or more years hence.

If one attempts to introduce ecological considerations through the medium of money, then one has to make guesses about future values of aspects of ecological systems for which one cannot even compute present values.

This chapter proposes an alternative. It is to consider not what may happen, but what could happen — that is to say, to consider the set of feasible options determined by the physical aspects of the system, and how they interact with the environment. To do so a common numeraire to quantify physical determinants is required.

The research upon which I am embarked explores the use of energy as the common numeraire in place of money. Economists who look at this work allege, quite correctly, that it implies an energy theory of value. I find that in the majority of economic activities I have examined, such a theory explains what happens at least as well as any other, and often better. This approach has been made possible by the development over the last 14 years of the discipline of energy analysis, the procedure whereby one can compute the primary energy sequestered in order to bring a good or service to the market-place.

Though energy provides the numeraire it does not provide the method, for which a fresh paradigm was required. This turned out to be glaringly obvious. An economic activity cannot take place if it is not physically possible. Since all activities use energy, it cannot take place unless the necessary energy is there. Unlike money, one cannot borrow energy. It is therefore necessary to make a model of the economy in which the flows and conversions of raw materials to consumer goods and capital are expressed in terms of their embodied energy.

RESOURCE ACCOUNTING

The distinction between resource accounting and economic analysis is essentially one of numeraire. Money, reflecting value, subsumes all the inputs to production. Resource accounting, reflecting energy, subsumes only energy. This is often mistakenly assumed to imply that energy is held to be scarce. However, such implication is incorrect.

Consider the distinction between a model of economic growth expressed in money terms as opposed to energy. In a simple industrial growth cycle money flows round the system. It is a positive feedback system, so leading to exponential growth. If the same system is expressed in terms of resource accounting — that is, energy — though the flows and the feedback mechanisms remain unchanged, it is no longer possible to visualize a circular flow, for at each stage the energy used is dissipated. Only the embodied energy (the sum of all primary energies taken from the earth and dissipated in the process of delivering a good or service) in the output moves on. The idea of money not being used up as a result of a process involving money is one of its intriguing aspects. This is possible because money is not a reality, it is a convention. It has no physical property, either extensive or intensive, with which it may be measured. In order to put resource accounting to work, we now need to consider the underlying concepts and accepted conventions. Before doing so, however, it is necessary to make the connection between physical resources and energy.

ENERGY AND RESOURCES

Physical resources, with the possible exception of helium and mercury, are so abundant in the earth's crust that mankind cannot conceivably run out of them. Furthermore, though they may be used, unlike energy they are never used up. Many researchers have shown that, provided there is energy, know-how and capital, one can always access physical resources, though the energy requirement of so doing may rise to very high levels (Cook 1976; Goeller & Weinberg 1976; Chapman & Roberts 1983). As for labour, it is being progressively substituted by energy as a source of work. Labour is increasingly becoming a decision-making element in production rather than a source of effort.

The price that must be paid for this economic advance is not inconsiderable, for, as an economy develops, it becomes ever more dependent on a flow of fuel to maintain the effort required to drive the economy. If a given standard of living is to be maintained, then it is a *sine qua non* that a flow of energy be sustained. Nor can we take too much refuge in technical progress. Chapman & Roberts (1983) have shown that for many metals technical progress can no longer compensate for declining ore grades. My associates and I (unpublished) have shown that on thermodynamic grounds certain technologies have already reached the limits of their feasible technical advance. Thus thermodynamics can be used as a guide to threshold limitations. None the less, there are also changes possible in the way we do things that may produce more with less exhaustion of resources, for example by recycling of biological waste.

DERIVATION OF RESOURCE ACCOUNTING

Let us consider this through the medium of the production function. Traditionally it expresses output in terms of capital and labour. The OPEC oil price increase of 1973 induced a reappraisal in which energy was sometimes included as a separate factor. In a different approach Slesser (1978a) showed that, if one networked back all the inputs to a production process, every input could be traced to the twin actions of prior energy and labour use. Stated in another way, time (= labour) and energy (= resources) were the only inputs to production which were irretrievably dissipated.

Thus, for example, capital is represented by capital goods, themselves the result of manufacture, for which were needed prior capital goods, physical inputs and labour. Similar reasoning applies to all other inputs, including the life-support system for labour itself. It may be shown that of physical resources none is irretrievably dissipated except energy.

Thus the resource accounting (RA) form of a production function states:

$$P \text{ output/time} = \text{function of } (L, L', E, E'),$$

where L and E are operational labour and energy and L' and E' are prior labour and prior energy. It should be noted that, unlike economic production functions which use money as a numeraire, the labour terms are not couched in the same numeraire as the energy terms, and so cannot be added or multiplied. In fact, each term stands apart, each determining a rate process, so that it is possible for each to be a limiting factor — a constraint. An analagous example is the combustion of coal, a rate process, which at low temperatures is entirely dictated by the chemical reaction of oxygen with carbon, but at high temperatures is controlled by the mass transfer of oxygen to the coal surface. Only within a very circumscribed range of conditions is the rate of combustion controlled by both. In the case of an economy analysed by resource accounting it would be only where the fossil and fissile energy use per unit of result was of the same magnitude as the human energy effort, measured in energy terms; that is, where the national per capita human work effort was of the same order as the fossil plus solar energy use. This is of the order of 100 kg of coal energy per capita per year. There are virtually no economies at such a low level, and, if solar energy capture through plants is taken into account, there are none. In other words, beyond an energy use of about 100 kg coal equivalent per capita, labour, as a source of work, is not the principal means of enhancing output. The dissipation of energy in the production process is a measure of the non-renewable resource consumption of each and every economic activity. Though solar energy is renewable, when that energy source is harnessed it is nevertheless dissipated as it is used.

Labour

The foregoing does not imply that labour is irrelevant. It remains important in two senses. First, labour equals people, and people make decisions. Labour is used less and less to provide physical work. The workers on the assembly line using power tools are essentially making decisions. A developing economy needs an increasing proportion of educated and trained labour. Secondly, there are many criteria, often political, that enter into the decision-making process. Being a systems approach, it can only inform what are the long-term consequences of these decisions. In fact it identifies the feasible, not necessarily the probable.

Energy analysis

The procedure for evaluating economic activities in energy terms is well established (e.g. IFIAS. 1974; Slesser 1978a; Chapman & Roberts 1983). Each time energy is used, either wastefully or usefully, that energy is irrevocably dissipated. For this reason it is common jargon to talk of the 'embodied' energy in a good or service. It is this concept which has the potential of permitting a *rapprochement* between the ideas expressed in this article and the concerns of the field ecologist and conservationist (Wiegert 1988). Thus capital stock is, *inter alia*, the consequence of previously dissipated energy, the amount (and quality) of which is determinable. Industrial capital stock permits labour to achieve enhanced output. In doing so, physical resources are used up and energy, as an operational input, is dissipated. Every economic activity becomes a sink for energy. Thus, in an economy, energy does not circulate like money does; rather, as a rate of supply, it drives the economy.

Energy transformation system

But how is the energy provided in the first place? It arises through the ability of the industrial system to extract and refine energy resources, whether fossil, fissile or renewable, and so deliver the fuel where needed. In order to accommodate this fact, Gilliland (1978) has redrawn the conventional economic model of the economy; part of the system output (= embodied energy) is devoted to getting at energy resources. This prior energy expenditure yields current fuel supplies. The lower the ratio of primary energy extracted to fuel delivered, the greater the potential rate of economic expansion per unit investment. In energy analysis it is called the energy requirement for energy (ERE), values of which for many resources have been quoted in the literature, e.g. by Chapman, Leach & Slesser (1974) and Peckham & Klitz (1978).

THE ECOLOGICAL CONNECTION

So far little has been done to incorporate environmental factors into such RA models. The approach so far has been to seek to quantify, in energy terms, the 'effort' that would be needed to maintain a given environmental standard, or avoid some foreseeable environmental degradation. Here are two examples.

Soil erosion

The FAO in its international 'Agro-ecological Zone' studies in some hundred countries has sought to include a term for the effort to combat soil erosion. When this study was used in the development of an RA model of Kenya, those data were transferred into energy units, and readily incorporated.

Dam construction

A new dam is to be developed, let us say, to provide hydroelectricity. It will have at least two negative impacts. The first is the loss of land, usually fertile, and hence the displacement of human and animal populations and loss of crops. The second is that it slowly fills up with silt. With appropriate measurement beforehand it should be possible to estimate the lifetime energy production of the system, the lifetime losses of food output (whose substitution elsewhere has an energy component), and the biomass energy loss. The accumulating silt is valuable, and in due course can be excavated.

APPLICATION OF RESOURCE ACCOUNTING

Since no economic activity can be intensified or expanded without the dissipation of energy, one may choose to model the development of an economic system in energy terms. The objection often made to doing this is that energy cannot be used as an allocation mechanism since people, it is argued, do not make their choices on the basis of energy requirement per unit of output. It seems, however, that we may do so unawares. Roberts (1982), in a telling paper on energy and value, showed how close were the relationships between embodied energy and value for many goods. However, the merits of this argument are not crucial to the validity of resource accounting as a means of looking at the longer-term future.

The economic approach to development planning is to move iteratively through the cycle:

Wants → Technology → Resources → Feasible wants

till convergence is obtained. The reliability of the result rests very much on the assumptions made as to technological advance, resource costs and how people, collectively, may in future value aspects of life and goods.

Considerable use is made of the concept of income and price elasticity of demand. Yet such relationships, though valuable for short-term forecasting, are too fickle to carry one many years forward in time, as we see by the complete failure of energy forecasting in recent years. Thus, though economics seeks to analyse market forces and to look ahead, the time horizon it is able to handle with any confidence is very limited. Marchetti (1985) goes even further and suggests the underlying growth of the system is a systems rather than an economic phenomenon. If this be so, resource accounting is a means, perhaps rather a robust means, of assessing the potential for growth (or decline) of the system.

The implication of growth being a systems phenomenon in no way denies the undoubted influence of market forces on the short-term behaviour of the economy. Resource accounting can offer no helpful guidance in the short term. However, since one cannot spend joules one does not have, it does offer a way of quantifying the long-term consequences of present-day decisions, where the future is beyond the whim of short-term market forces. An example of this is the study by Phillips & Edwards (1976), who showed that, despite quite violent short-term fluctuations in the metal markets, the long-term price is set by the energy (more precisely, the Gibbs free energy) required to turn the ores into metals.

By its nature, resource accounting must analyse development in the reverse way to economic planning. Resources must be liberated in order to liberate further resources. That is:

Resources → Feasible output → Capital → Resources →
Feasible output technology →

Thus one cannot set up a set of 'wants', and work towards them. Rather, taking the system as it is now (a fact), RA can explore its tendency in systems terms. Some of the relationships that determine change within the system may be set by physical laws, others by the nature of the country in question, its resources, climate, environment, or socio-cultural attributes. Nevertheless one can impose certain basic normative aims, such as goal nutritional level per inhabitant, domestic energy provision, or minimal water supply.

The fact that resource accounting is a systems approach does not create an inevitable future; far from it. On the contrary it enables one to probe for an acceptable future. It can be used to incorporate ecological considerations. Examples are the nitrate ground-water pollution arising from desired degrees of cereal output and the need to look at alternative farming methods. Another would be soil erosion, taking into account capital and other requirements to offset this. Another would be to relate sulphur in fuels and energy demand to acid-rain production, and hence consequences. This list is endless. But in each case a prior step is to have a resource accounting model within which the more detailed issue can be embedded in order to establish the interactions with the economy as a whole. Being a systems approach it can only inform what are the long-term consequences of these decisions. In fact it identifies the feasible, not necessarily the probable. This aspect of resource accounting renders it very suitable for making carrying-capacity studies of a country.

CARRYING CAPACITY

In ecology carrying capacity is seen as a threshold. This would indeed be the case as well in the affairs of men were it not for our use of stored energy sources, like oil and coal, etc. Economic carrying capacity is seen as the number of people who, sharing a given territory, can be supported at any moment on a sustainable basis, taking into account its known resources, as well as factors of a socio-cultural nature. Using resource accounting one can assess whether investment matches supply needs, and output matches investment needs. If development is to occur at the best possible rate, yet remain sustainable, or move towards sustainability, then all sectors of the economy must grow in harmony. For example, investments in water and energy supply must match industrial, agricultural, domestic and service needs. There are many ways to feed a country's population, of which the extremes are self-sufficiency (e.g. USA) or total food imports (e.g. Hong Kong). Imports imply exports. Both have embodied energy. Energy needs may be met indigenously or by imports. But, just as many of the developing countries have come to realize that a high degree of food self-sufficiency must now replace the former priority for industrial development, the OPEC oil price increases have introduced a similar concern for energy. For example, self-sufficiency is a corner stone of EEC energy policy.

SUSTAINABILITY

The notion of sustainability is central to ecological thinking. As an objective it has been voiced increasingly, especially in UN circles, and was implied in the first recommendation of the world population conference in Mexico (Anon. 1983).

Energy analysis offers a means of making a precise statement of what is implied in the 'sustainability' of an economy. Economic growth depends on the growth of capital stock (= embodied energy) to enable labour to provide enhanced output (= dissipation of energy). If this is to be sustainable, then the rate of energy flow (more precisely, flux) must always be maintained (up to a point, it can be compensated for by technical progress). This in turn means that part of the energy flow must, through capital and operating energy, be recycled to drive the energy transformation system by which resources in the ground become economically useful fuel. As energy resources become less accessible, the energy that must be dissipated in the process of acquiring fuel rises, that is the energy requirement for energy rises. In money terms they become more expensive to extract. In a classic study, Peckham & Klitz (1978) have evaluated this for North Sea oilfields as a function of cumulative depletion.

Economic sustainability may thus be defined as a state where the rate of growth of energy flow into the economy is sufficient to liberate and harness the resources needed to maintain indefinitely or enhance the material standard of living of the people living within that economy.

In 1983 a carrying capacity approach to this question was developed as a result of an initiative by the Population Division of Unesco and with the support of the UN and UNDP. It was used to identify policies which might lead a country towards sustainable development. In the long run that means that a country can maintain its food and energy provision, perhaps with a measure of food and energy self-sufficiency (the degree chosen is a political option to be tested). Clearly population growth is a key parameter in the entire equation.

MODELLING CARRYING CAPACITY

Simulation rather than optimization is used in carrying-capacity assessment since in the real world many criteria, often political, enter into the decision-making process. Simulation also lends itself to modelling feedback, which is the basis of any system's response to external influence.

Resource accounting has been put to work as a dynamic simulation model called ECCO — enhancement of carrying-capacity options. The ECCO model is described by King & Slesser (1988).

ECCO is a feedback model — that is, any change (effect) within it takes place because change (cause) occurs elsewhere in the model. ECCO runs endogenously once the policy decisions have been inserted. For example in its initial form only two normative values (policies) were imposed upon it: a goal nutritional level for all inhabitants and a minimal per capita availability of domestic energy.

APPLICATION OF RESOURCE ACCOUNTING MODELS

The first test of the resource accounting concept, as encapsulated in ECCO, was for Kenya. The model was run from 1980 to 2020 for a number of policy options. The program had 420 statements expressed in systems dynamics language. The project was a joint one with FAO, which as a result of its existing agro-ecological zone and 'Agriculture towards 2000' studies (Anon. 1985) already had a large data base that could be adapted in energy analysis terms and incorporated in ECCO.

Subsequently studies have been launched in China, India, Mauritius and Thailand. A validation study of the UK economy using 1974 as a base year computed the state of the economy in 1984 and compared it with the actual 1984 situation. Using Theil's inequality coefficient (Theil 1965), the ECCO-based prediction was eight times better than that offered either by the UK Treasury model or that of the National Institute for Social and Economic Research. This work is not as yet published, but is available in report form by application to the author.

One of the uses of ECCO is to explore the consequences of a policy for increased food and energy self-sufficiency. In this mode it continually compares the actual population with the population that can be fed out of indigenous food production (the sustainable population) and, if noting an imbalance, drives investment towards agriculture. A full description of the model is available from M. Slesser.

LONG-TERM OUTCOMES OF POLICY OPTIONS

The policy options that may be tested by the model user are many and relevant to the responsibilities of a government. Consider the example of population growth. Curbing population growth is not an easy task, and no government wants to face the unpopularity of enforcing birth control to a greater degree than absolutely necessary. Thus, if it can assess the effects of various rates of population growth on the welfare of future populations, this can help it to decide the degree of priority it must give to controlling birth rate.

The model user specifies the population growth through time. It may be the continuation of current rates (3·8% in the case of Kenya) or a move towards reduced rates sometime in the future. Any known plans for development such as (in the case of Kenya) water catchment, hydroelectric facilities or biomass plantations are incorporated. Thereafter such investments are the topic of policy options to be explored.

ECCO is then run, and generates information on population, dependency ratio, sustainable population, cereals imports, food self-sufficiency, energy self-sufficiency, necessary and possible investment rates in the main economic sectors including energy and agriculture, material standard of living and many other outputs. It should be noted that, since the Kenya study is on pilot data, the outcomes must be regarded as no more than sensitivity-testing, and not yet representing real scenarios for Kenya. Moreover, many other policy options remain to be tested. The potential for incorporating environmental factors has yet to be fully exploited.

CONCLUSION

It has proved possible to model an economy using an energy numeraire. This procedure is called resource accounting. Such an approach provides, in principle, a means of relating those economic, environmental and ecological considerations that are resource-based, resource-constrained or dependent on additional physical capital to effect amelioration or change, through a single numeraire — energy. The procedure is hierarchical. Resource accounting provides the overview. Particular ecological aspects are dealt with through subsets.

The data problem has not proved insuperable, but traditional economic sources of data are rarely adequate in themselves. This can be improved as awareness increases of what constitute the relevant data to be collected. The potential for considering ecological aspects would appear to be considerable.

A basic generic version of a model for enhancement of carrying capacity options (ECCO) has been prepared with documentation, called RootEcco, which can be used to initiate an ECCO analysis of a fresh situation or country.

REFERENCES

Anon. (1983). Recommendation Number 1. *United Nations Conference on Population, Mexico.*

Anon. (1985). *Carrying Capacity Assessment: with a Pilot Study of Kenya.* Report W/R80 12 of FAO, Rome, and UNESCO, Paris.

Ascher, W. (1978). *Forecasting: an Appraisal for Policy-Makers and Planners.* Johns Hopkins University Press, Baltimore.

Chapman, P. F., Leach, G. & Slesser, M. (1974). The energy cost of fuels. *Energy Policy,* **2**, 231–243.

Chapman, P. F. & Roberts, F. (1983). *Metal Resources and Energy.* Butterworth, London.

Cook, E. (1976). Limits to the exploitation of non-renewable resources. *Science,* **191**, 677–83.

Gilliland, M. (1978). *Energy Analysis: a New Public Policy Tool.* American Association for the Advancement of Science Symposium Series, Washington, DC.

Goeller, H. E. & Weinberg, A. M. (1976). The age of substitutability. *Science,* **191**, 683–9.

IFIAS (1974). *Workshop on Energy Analysis, Methodology and Conventions. International Federation of Institutes for Advanced Study Report,* Vol. 6. Ulriksdals Slott, Solna, Sweden.

King, J. & Slesser, M. (1988). Resource accounting in development planning. *World Development,* **16**, 293–303.

Marchetti, C. (1985). On a fifty year pulsation in human affairs: analysis of some physical indicators...., pp. 83–85. International Institute for Applied Systems Analysis, Laxenburg, Austria.

Peckham, R. & Klitz, K. (1978). *Energy Requirement of North Sea Oil.* Euratom Joint Research Centre, ISPRA, Italy.

Phillips, D. P. & Edwards, W. G. (1976). Metal prices as a function of ore grade. *Resource Policy,* **1**, 167–78.

Roberts, P. O. (1982). Energy and value. *Energy Policy,* **10,** 171–80.

Schrattenholzer, L. (1984). International energy workshop poll response. *Energy Journal,* **5**, 45–64.

Slesser, M. (1978a). *Energy in the Economy,* p. 55. Macmillan, London.

Slesser, M. (1978b). Energy analysis: its utility and limits.... RM-78-46. International Institute for Applied Systems Analysis, Laxenburg, Austria.

Theil, H. (1965). *Techniques of Economic Forecasting.* OECD, Paris.

Wiegert, R. G. (1988). The past, present and future of ecological energetics. *Concepts of Ecosystem Ecology* (Ed. by L. R. Pomeroy & J. J. Alberts), pp. 29–55. Springer, New York.

AUTHOR INDEX

Page numbers shown in *italics* refer to the text, those in roman to the reference lists

SUBJECT INDEX

ORGANISM INDEX